J. H. Kruhl (Ed.)

Fractals and Dynamic Systems in Geoscience

Technical Revision and Layout
by Lars-Oliver Renftel

With 201 Figures and 16 Tables

Springer-Verlag Berlin Heidelberg GmbH

Jörn H. Kruhl
Geologisch-Paläontologisches Institut
Johann Wolfgang Goethe-Universität
Senckenberganlage 32–34
60325 Frankfurt/Main, Germany

ISBN 978-3-662-07306-3 ISBN 978-3-662-07304-9 (eBook)
DOI 10.1007/978-3-662-07304-9

Originally published by Springer-Verlag Berlin Heidelberg New York in 1994.

Softcover reprint of the hardcover 1st edition 1994

Production Editor: Renate Münzenmayer
Typesetting and Layout: Lars-Oliver Renftel, Maintal
32/3130-5 4 3 2 1 0 – Printed on acid-free paper

Preface

There is a continuing debate, both publicly and in the scientific community on whether or not fractal geometry is a new branch of scientific research – or even a new paradigm. In fact, despite these discussions, fractal geometry has penetrated into the daily work of scientists. Together with the theory of dynamic systems, it has proved its value in characterizing complex structures and in modelling processes in nature. Fractal geometry allows the description of natural patterns by simple numbers, to facilitate their comparison and to establish and test models of pattern formation. Therefore the anticipation arises – during optimistic moments, admittedly – that in this way we can catch a glimpse of how "nature works".

Fractal geometry has also turned out to be, in particular, a tool of the geosciences, the main topics of which are complex forms in nature and the dynamic processes leading to them. It is well known that geological structures can exhibit self-similarity over several, but not infinite, orders of magnitude. Therefore, during introductory courses we instruct the students that sketches or photographs of geological objects should always contain a scale. Moreover, it is taught that rocks are able to "flow" and that both small-scale structures in rock specimens and large-scale structures in the crust of the earth may resemble patterns in liquids or fluids – which are considered exemplary dynamic systems.

This book was initiated at an international conference on "Fractals and Dynamics Systems in Geosciences", held at the Johann Wolfgang Goethe University, Frankfurt am Main, April 1993. In addition to contributions presented at this conference, it contains papers which were submitted later or requested by the editor during the compilation of the book. It is the aim of this volume to give an overview of the applications of fractal geometry and the theory of dynamic systems in the geosciences which, for practical reasons but not for intrinsic reasons, is defined as the "science of the solid earth". Although the volume title also includes "Fractal Geometry" and "Dynamic Systems", most of the contributions deal with the application of fractal geometry in the geosciences. This correctly reflects that, now as before, in the geosciences the accurate description of structures and features is fundamental and forms the basis for studying the underlying dynamic processes. Certainly, 31 contributions cannot cover all the possibilities of application within the broad field of the geosciences. However, it is to be hoped that at least a view on the "state of the art" is provided.

The book is divided into four parts: (1) Deformation and tectonic structures; (2) physical features and behaviour of the earth; (3) formation, structure and distribution of minerals and matter; (4) methods. This division reflects the fields of study covered by the different contributions, but also indicates – other select o factors tacitly excluded – that, at present, in addition to mineralogical and geochemical applications, structural geology and physics of the solid earth are obviously those fields of geoscience with a particular relationship to fractal geometry.

It is one of the pleasant duties of the editor to thank the many people who made this book possible. Firstly, thanks are due to the authors who readily submitted their contributions, seriously tried to stay within the time and page limits and to respond quickly and benevolently to short-term inquiries. Above all, aspiring to produce a book of scientific integrity requires that the contributions be reviewed and accepted by knowledgeable and independent colleagues. Therefore, each contribution was reviewed at least twice. The 57 referees are listed in alphabetical order in the Appendix. They willingly accepted the rigorous time limits, tolerated my impatient demands, and some of them even reviewed up to five contributions. The timely completion of the volume is mainly their achievement. Thanks are due to all of them.

Undoubtedly, the book would have never appeared in its present form without Lars-Oliver Renftel. He converted the manuscripts into their final camera-ready forms, fought successfully without complaint with exotic text and graphic formats and carried out corrections and technical verification. He deserves our gratitude!

Acknowledgements are also due to all the public institutions which, through their financial support of the "Fractals and Dynamic Systems in Geosciences" conference, laid the foundations of the present book: The Deutsche Forschungsgemeinschaft (DFG), the Johann Wolfgang Goethe-Universität Frankfurt am Main, the Vereinigung von Freunden und Förderern der J.W. Goethe-Universität, the Stiftung zur Förderung der internationalen wissenschaftlichen Beziehungen der J.W. Goethe-Universität, the Ministry of Science and Art of the State of Hessen and the City of Gelnhausen. Moreover, thanks are extended to the Geologisch-Paläontologisches Institut of the J.W. Goethe-Universität, which allowed access to their facilities during the conference and the editing process of the book.

Lastly, thanks to the staff of Springer-Verlag who produced and distributed the book, in particular Wolfgang Engel who, through his interest in the project, initiated the volume in its present form.

Frankfurt am Main, March 1994 Jörn H. Kruhl

Contents

Part I
Deformation and Tectonic Structures

Crustal Deformation and Fractals, a Review

Donald L. Turcotte*
Institute for Theoretical Physics
University of California, Santa Barbara, California 93106, USA
*Permanent address: Department of Geological Sciences, Cornell University, Ithaca, NY 14853, USA

Abstract. The frequency-magnitude statistics of earthquakes have long been known to satisfy the Gutenberg-Richter relation; it is easy to show that this relation is fractal with D = 1.8 – 2. Since it is generally accepted that individual faults have characteristic earthquakes, it follows that the number-size statistics of faults are also fractal. Limited field studies indicate that the surface exposures have a two-dimensional fractal dimension of D ~ 1.6. (The seismic and fault fractal dimensions need not be equal since the repeat time of earthquakes can also have a power-law dependence on scale.) Fragments have a fractal number-size relation with D ~ 2.5 under a wide range of conditions. One model for tectonic deformation is comminution; such a scale-invariant model appears to be consistent with the fractal correlations discussed above. Seismicity has many of the characteristic features of „self-organized criticality". Energy (strain) is continuously added to the crust and is lost in discrete events (earthquakes) that have a fractal frequency-size distribution. It appears reasonable to hypothesize that the continental crust is everywhere in this critical state (similar to the critical stress associated with perfect plasticity). Evidence for this comes from the distribution of intraplate earthquake and from the occurrence of induced seismicity almost anywhere an artificial reservoir is filled.

Introduction

Seismicity is clearly the result of complex interactions between faults over a wide range of scales. Despite the complexity, there is order — in particular, the frequency-magnitude statistics of earthquakes are given by the Gutenberg-Richter relation.

$$\log \dot{N} = -bm + a \tag{1}$$

where $\dot{N}$ is the number of earthquakes per unit time in a specified region with magnitudes greater than m. The b-value is generally in the range $b = 0.9 \pm 0.1$. Equation (2) is equivalent to a fractal (power-law) relation between frequency and rupture area A (Aki 1981).

$$\dot{N} = CA^{-D/2} \tag{2}$$

This fractal relation is recognized to be applicable under a wide variety of tectonic settings from mid-ocean ridges to collision zones such as Tibet. In general, the fractal dimension is not diagnostic since it has such a small variation. It has also been argued (Pacheco et al. 1992) that the fractal dimension is substantially larger for very large earthquakes. The change is said to occur when the vertical limit on rupture propagation is reached; smaller earthquakes have equidimensional ruptures whereas larger earthquakes have horizontally elongated ruptures.

Despite limits and possible variations, the near universal applicability of the fractal relation (2) to regional seismicity is a remarkable feature. A simple explanation is that seismicity does not have a characteristic length scale and thus must be self-similar. The only statistical distribution that is self-similar is a power law. But this would not explain why the fractal dimension is nearly constant under wide variations in tectonic regime.

It is generally accepted as a working hypothesis that each fault is associated with a characteristic earthquake. Thus a fractal distribution of earthquakes implies a fractal distribution of faults. It does not follow, however, that the fractal dimension for the frequency-size distribution of faults is the same as that for earthquakes since this would imply that the interval of time between earthquakes is independent of scale. The intervals would also be expected to have a power-law dependence on scale.

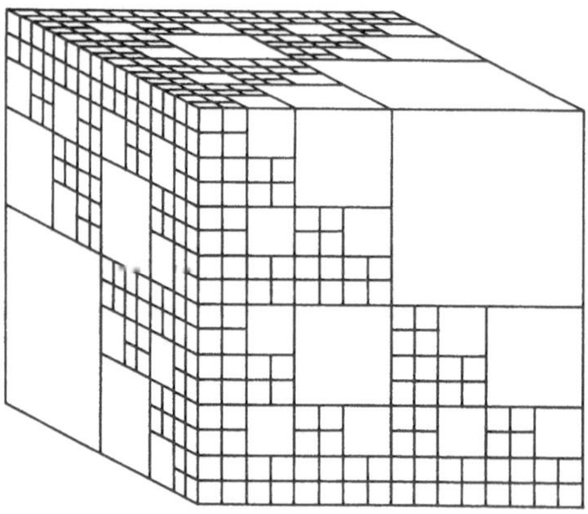

Fig. 1. Illustration of a discrete model for comminution. Diagonally opposite blocks are retained at each scale. The blocks satisfy a fractal relationship with D = ln 6/ln 2 = 2.585. The block surfaces satisfy a fractal relationship with D = ln 3 ln 2 = 1.585.

Sammis et al. (1986) have proposed that a comminution model for fragmentation is applicable to tectonic deformation. This hypothesis states that the direct contact between two blocks of near equal size during tectonic deformation will result in the breakup of one of the blocks. It is unlikely that small blocks will break large blocks or that large blocks will break small blocks. A deterministic model for comminution is given in Figure 1. Two diagonally opposed blocks are retained at each scale so that no two blocks of equal size are in contact with each other. We have two blocks with $r = 1/2$ and 12 blocks with $r = 1/4$ so that $D = \ln 6/\ln 2 = 2.585$. Turcotte (1986) has shown that many experimental studies of fragmentation yield fractal frequency-size distributions with $D \approx 2.5$. Considering each side of a block in Figure 8 to be a fault we have a fractal distribution of fault sizes with $D = \ln 3/\ln 2 = 1.585$. Barton and Hsieh (1989) considered the distribution of exposed joints and fractures near Yucca Mountain, Nevada, and found good agreement with a fractal distribution of block exposures taking $D = 1.6$.

Self-Organized Criticality

The concept of self-organized criticality was introduced by Bak et al. (1988) and is defined to be a natural system in a marginally stable state, when perturbed from that state it will evolve naturally back to the state of marginal stability. Energy input to the system is continuous but the energy loss is in a discrete set of events that satisfy fractal frequency-size statistics. Distributed seismicity is often taken as the classic example of a natural system that exhibits self-organized criticality. There is a continuous input of energy (strain) through the relative motion of tectonic plates. This energy is dissipated in a fractal distribution of earthquakes. Scholz (1991) has argued that the earth's entire crust is in a state of self-organized criticality. He makes the point that wherever a large dam is built, induced seismicity results from the filling of the reservoir. Thus the crust is everywhere on the brink of failure. Similar arguments have been given by Sornette et al. (1990).

In the original cellular-automata model for self-organized criticality a two-dimensional grid of boxes was considered. Particles were randomly added to the boxes until there were four particles in a box. The box was then considered unstable and the four particles were redistributed to the four adjacent boxes. If any of those boxes had four or more particles, a further redistribution of elements was carried out. Particles were lost only from the sides of the grid.

In the steady state, the number of particles in the grid of boxes fluctuates about a mean value that is considerably less than the maximum allowed value $3N^2$. The behavior of the model is characterized by the statistical frequency-size distribution of events. The size of a multiple redistribution event can be quantified in several ways. One measure is the number of boxes N_b that become

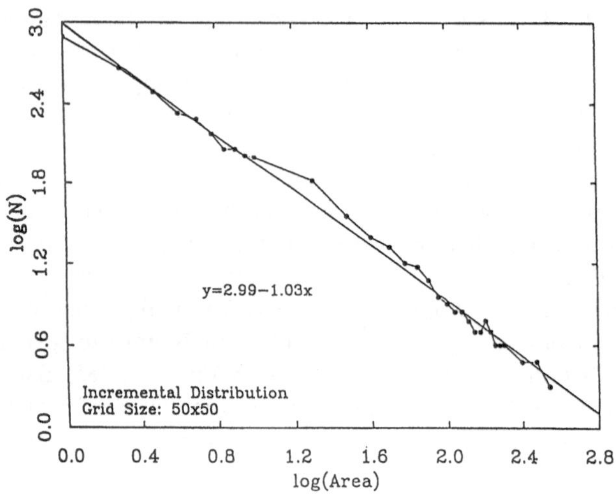

Fig. 2. Statistics for a cellular automata model on a 50 x 50 grid. The number N of events in which a specified area of boxes A become unstable is given as a function of A.

unstable in a multiple event. The results for a 50 x 50 grid of boxes are given in Figure 2. The number of events N in which a specified number of boxes A participated is given as a function of the number of boxes. A good correlation with a fractal power law is obtained with a slope of 1.03. Since the number of boxes A is equivalent to an area, the relevant fractal dimension is $D = 2.06$. Another measure is the number of particles N_p lost from the grid in a multiple event. Bak et al. (1988) and Kadanoff et al. (1989) have carried out extensive studies of the behavior of this model as a function of grid size. They found that the number of events N of a particular size, either N_b or N_p, satisfies the fractal relation (2) with $D \approx 2$.

Bak et al. (1988) made an analogy between their cellular-automata model and a sand pile on a table. Sand grains are dropped on the pile continuously and are lost from the table in discrete sand slides; some sand slides are large and some are small. A number of groups have studied actual sandpiles and in some cases fractal distributions of sandslides have been found (Nagel 1992, Rosendahl et al. 1993).

It has been argued that the events in the cellular-automata model strongly resemble the distribution of seismicity in a zone of crustal deformation (Bak & Tang 1989). In this analogy the addition of particles is analogous to an increase in the regional tectonic strain. The events are analogous to earthquakes, and the number of particles present is analogous to the mean regional stress. The applicability of the Gutenberg-Richter frequency-magnitude relation to earthquakes is taken as evidence that the earth's crust exhibits self-organized criticality.

Although there are important similarities between the cellular-automata model discussed above and distributed seismicity, there are also important differences. We note three of these:

1. The frequency-magnitude statistics for the cellular-automata models are not cumulative. When the number of events larger than a specified value is obtained the power law (fractal) relation is not applicable since the integral of A^{-1} becomes a logarithm.
2. The model does not generate "characteristic" earthquakes. A specified box in the grid may participate in a small event in one time step and a much larger event at a later time step. Large events are randomly distributed over the grid.
3. The model does not generate either foreshocks or aftershocks. A sequence of aftershocks is a universal feature of crustal earthquakes.

In order to address these differences Barriere & Turcotte (1991) introduced a cellular-automata model in which the boxes have a scale-invariant distribution of sizes. The objective was to model a scale-invariant distribution of fault sizes. When a redistribution from a box occurs it is equivalent to a characteristic earthquake on the fault. A redistribution from a small box (a foreshock) may trigger an instability in a large box (the main shock). A redistribution from a large box always triggers many instabilities in the smaller boxes (aftershocks).

A model of this type is illustrated in Figure 3. A square box is divided into four equal sized boxes at first order. At second order one box is retained and the other three are further divided into four boxes; the model has been extended to 5th order with a 32 x 32 grid. This model corresponds to the discrete model for comminution illustrated in Figure 1. For this example we have $N_1 = 108$ for $r_1 = 1$, $N_2 = 27$ for $r_2 = 2$, $N_3 = 9$ for $r_3 = 4$, $N_4 = 3$ for $r_4 = 8$, and $N_5 = 1$ for $r_5 = 16$. Except for N_1 the N_i ar related to the r_i by (2) with $D = \ln 3/\ln 2 = 1.5850$ in agreement with our previous discussion of Figure 1.

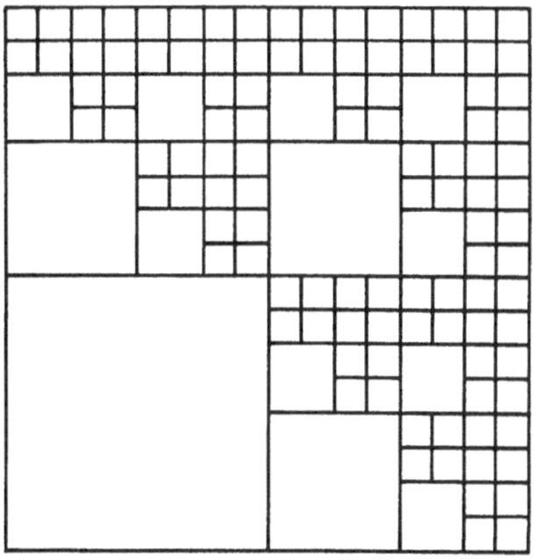

Fig. 3. Illustration of the fractal cellular-automata model corresponding to the discrete model for comminution illustrated in Figure 1 carried to fifth order $D = \ln 3/\ln 2 = 1.585$.

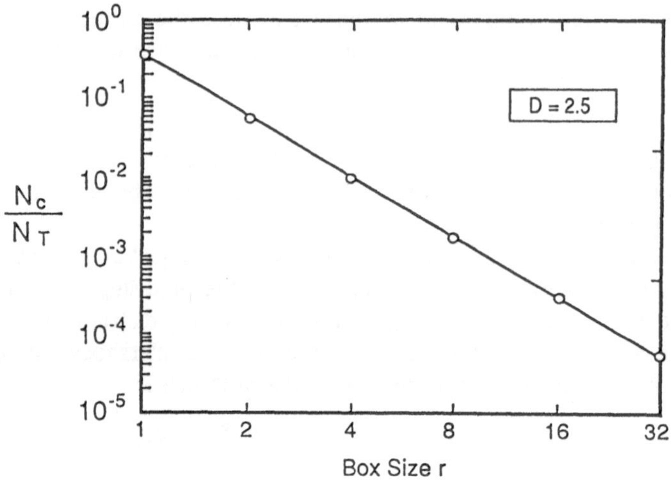

Fig. 4. Cumulative frequency-magnitude statistics for unstable events. The number of events N_c in boxes equal to or smaller than r is divided by the total number of events N_T and given as a function of r. The correlation is with (2).

We apply the standard cellular automata rules to this model:

1. Particles are added one at a time to randomly selected boxes. The probability that a particle is added to a box is proportional to the area A of the box.
2. A box becomes unstable when it contains 4A particles.
3. Particles are redistributed to immediately adjacent boxes or are lost from the grid. The number of particles redistributed to an adjacent box is proportional to the linear dimension of the box.
4. If after a redistribution of particles from a box any of the adjacent boxes are unstable, one or more further redistributions is carried out. In any redistribution the critical number of particles is redistributed. Redistributions are continued until all boxes are stable.

The cumulative frequency-magnitude statistics for main shocks of a seventh order (128 x 128) version of the model are given in Figure 4. We find an excellent correlation with the fractal relation (1) taking $D = 2.50$ ($b = 1.25$). This is significantly higher than observed values for distributed seismicity. Evernden (1970) has obtained b-values for regional seismicity and concludes that $b = 0.85 \pm 0.20$. It was also found that 31.5% of the largest events in the model had foreshocks. This is in reasonable agreement with studies of actual earthquakes; von Seggern et al. (1981) found that 21% of the earthquakes studied had foreshocks and Jones & Molnar (1979) found that 44% of larger shallow earthquakes that could be recorded teleseismically had foreshocks. The model aftershocks also correlate well with (1) taking $D = 2.02$ ($b = 1.01$).

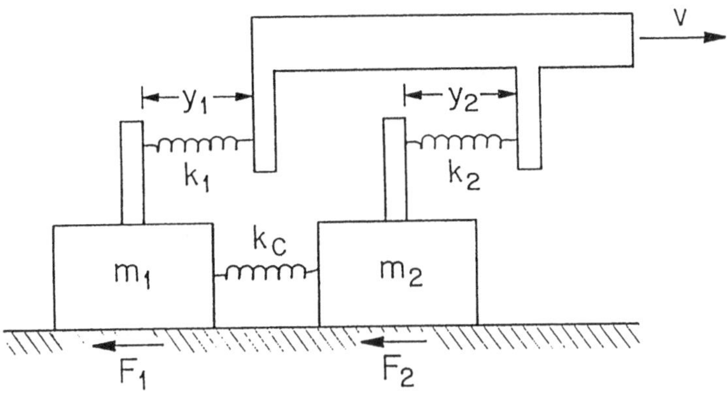

Fig. 5. Illustration of the two-block model. The constant velocity driver extends the springs until sliding of a block commences. In some cases sliding of one block induces the sliding of the second block.

Although the simple cellular-automata models have a number of quite remarkable similarities to distributed seismicity, they lack much of the basic physics such as stick-slip behavior and elastic rebound. A model that includes both was introduced by Burridge & Knopoff (1967). They considered a series of slider blocks which were pulled over a surface by driver springs and connected to each other by connector springs.

The simplest model for fault interactions is to consider the pair of interacting slider blocks illustrated in Figure 5. It has been shown that this system can exhibit chaotic behavior (Huang & Turcotte 1990a, 1992, Narkounskaia & Turcotte 1992). The two slider blocks are coupled to each other and to a constant velocity driver by linear elastic springs of stiffness k_c, k_1, and k_2, respectively. Other model parameters include the block masses m_1 and m_2, and the frictional forces F_1, and F_2. The position coordinates for each block, referred to the constant velocity driver, are y_1 and y_2.

The equations of motion for this two-block system are

$$m_1 \ddot{y}_1 + (k_1 + k_c)\, y_1 - k_c y_2 = F_1 \tag{3}$$

$$m_2 \ddot{y}_2 + (k_2 + k_c)\, y_2 - k_c y_1 = F_2 \tag{4}$$

Several additional assumptions are made to simplify the model. First, symmetry of the model is assumed except for the frictional forces. Accordingly, we take $m_1 = m_2 = m$, and $k_1 = k_2 = k$. In addition, the coupling spring constant is defined in terms of k: $k_c = \alpha k$. The parameter α is a measure of the stiffness of the system. A second assumption is that the loading velocity of the driver is sufficiently slow that we may consider it to be zero during the sliding of a block. The ratio of the frictional forces of the two blocks β is introduced such that

$F_2 = \beta F_1 = \beta F$. The parameter β is a measure of the asymmetry of the system. With these assumptions incorporated into the model, (3) and (4) can be written

$$m \ddot{y}_1 + (1+\alpha) k y_1 - \alpha k y_2 = F \tag{5}$$

$$m \ddot{y}_2 + (1+\alpha) k y_2 - \alpha k y_1 = \beta F \tag{6}$$

In order to obtain stick-slip behavior we choose a velocity-weakening friction law of the form

$$F = \frac{F_O}{1 + \left| \dot{y} / v_f \right|} \tag{7}$$

where F_O and v_f are constants. The frictional forces decrease monotonically as the slip velocity increases. In essence, this is a coupled oscillator system with damping from the dynamic friction.

It is convenient to introduce the nondimensional variables $Y_i = y_i k/F_O$ (i= 1,2) and $\tau = t\sqrt{k/m}$. The equations of motion during slip are

$$\ddot{Y}_1 + Y_1 + \alpha (Y_1 - Y_2) = \frac{1}{1+\gamma \left| \dot{Y}_1 \right|} \tag{8}$$

$$\ddot{Y}_2 + Y_2 + \alpha (Y_2 - Y_1) = \frac{\beta}{1+\gamma \left| \dot{Y}_2 \right|} \tag{9}$$

where $\gamma = F_O/V_f \sqrt{mk}$ is a measure of the amount of velocity weakening.

Numerical solutions for the behavior of the blocks have been obtained under a variety of conditions. The solutions for the behavior of the blocks can be represented in a four-dimensional phase plane consisting of Y_1, Y_2, $dY_1/d\tau$. For simplicity we consider the projection of the solution onto the Y_1 and Y_2 plane.

We first consider an example in which the system is symmetric taking $\beta = 1$, $\alpha = 1.2$, and $\gamma = 3$. The resulting phase diagram is given in Figure 6. The closed orbit indicates periodic motion and consists of two single-block failure events denoted by the vertical and horizontal segments of the orbit. Stress accumulation occurs along straight lines of unit slope, and failure occurs when the accumulation lines intersect the failure envelope, the intersecting straight lines.

When the system is asymmetric, i.e. the ratio of frictional forces β does not equal unity, the dynamical picture can change completely. An example taking $\beta = 2$, $\alpha = 1.2$, and $\gamma = 3$ is given in Figure 7. We note that there may be single-block failures, while at other times both blocks slide together. The trajectory does not settle into a closed orbit and apparently fills a section of the phase plane; this is typical for chaotic behavior.

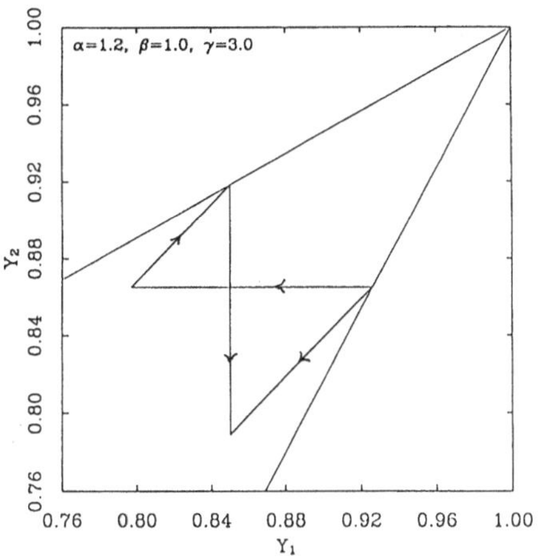

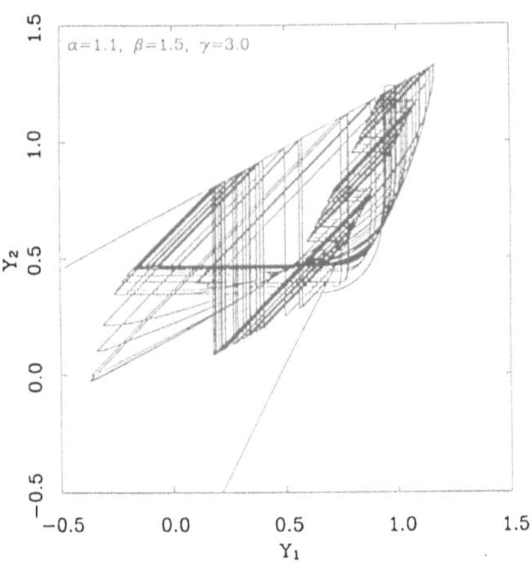

Fig. 6 (left). Trajectory of a symmetric system with parameters $\alpha = 1.2$, $\beta = 1.0$, $\gamma = 3.0$ in the Y_1, Y_2 phase plane. The diagonal converging lines are the failure envelope. Stress accumulation occurs on diagonal lines of unit slope.
Fig. 7 (right). Space-filling evolution of the system in the Y_1, Y_2 phase plane with system parameters $\alpha = 1.2$, $\beta = 2.0$, $\gamma = 3.0$. The plotted orbit consists of 100 events.

Huang & Turcotte (1990b) applied the chaotic behavior of the asymmetrical system to two examples of interacting fault segments. The descent of the Pacific plate beneath the Asian plate resulted in the formation of the Nankai trough along the coast of southwestern Japan. A sequence of great earthquakes have been documented in this region for the period 684-1946 AD. The sequence is marked by an irregular but somewhat repetitive pattern in which whole section failures occur following several alternate failures of single segments. In the two-block model the simultaneous slip of both blocks corresponds to an earthquake that ruptures the entire section and single block failures correspond to an earthquake on a single segment. Taking $\beta = 1.05$ and $\alpha = 0.81$ a chaotic model behavior was found that strongly resembled the observed sequence of earthquakes in the Nankai trough.

Another example is the interaction between the Parkfield segment and the rest of the south central locked segment of the San Andreas fault in California. A sequence of magnitude 6 earthquakes occurred on the Parkfield segment in 1881, 1901, 1922, 1934, and 1966. The last great earthquake on the locked segment to the south occurred in 1857 and is also associated with a rupture on the Parkfield segment. Taking $\beta = 2$ and $\alpha = 1.2$ a chaotic model behavior was found similar to that described above. A sequence of slip events on the weaker block often preceded the simultaneous slip of the weaker and stronger blocks. The model simulation suggested two alternative scenarios for a great southern California earthquake following a sequence of Parkfield earthquakes. In one case a Parkfield earthquake will transfer sufficient stress to trigger the great southern California earthquake. In the second case a small additional strain after a Parkfield earthquake will trigger an earthquake on the southern section and this results in an additional displacement on the Parkfield section. Because the system is chaotic its evolution is not predictable except in a statistical sense.

Spring-block models are a simple analogy to the behavior of faults in the earth's crust. However, the chaotic behavior of low-dimensional analog systems often indicates that natural systems will also behave chaotically. Thus it is reasonable to conclude that the interactions between faults that lead to the fractal frequency-magnitude statistics of earthquakes are examples of deterministic chaos. The prediction of earthquakes is not possible in the deterministic sense; only a probabilistic approach to the occurrence of earthquakes will be possible.

A model that combines the analog features of slider blocks and the high-order aspects of cellular automata models involves the use of many slider blocks. Carlson & Langer (1989) and Carlson et al. (1991) considered linear arrays of slider blocks with each block connected by springs to the two neighboring blocks and to a constant velocity driver. They used a velocity-weakening friction law and considered up to 400 blocks. Slip events involving large numbers of blocks were observed, the motion of all blocks involved in a slip event were coupled and the applicable equations of motion had to be solved simultaneously. Although the system is completely deterministic, the behavior was apparently chaotic. Frequency-size statistics were obtained for slip events and the events fell into two

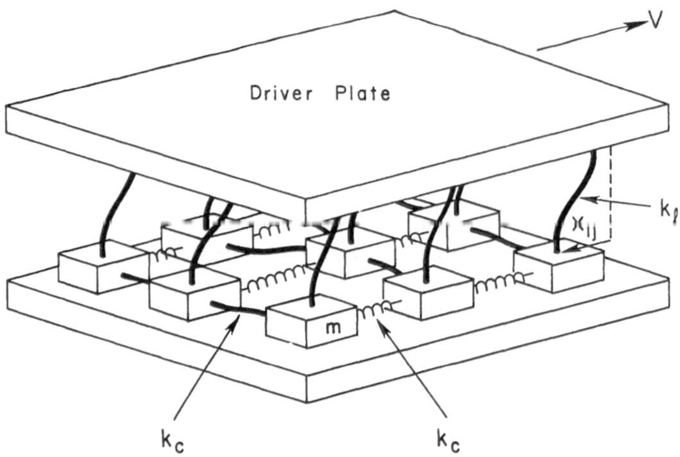

Fig. 8. Illustration of the two-dimensional slider-block model. An array of blocks each with mass m is pulled across a surface by a driver plate at a constant velocity V. Each block is coupled to the adjacent blocks with either leaf or coil springs of spring constant k_c, and to the driver plate with a leaf spring of spring constant k_l. The extension of the (i, j) pulling spring is x_{ij}.

groups, smaller events obeyed a power-law (fractal) relationship but there were an anomalously large number of large events that included all the slider blocks. This model was considered to be a model for the behavior of a single fault, not a model for distributed seismicity. The large events were associated with characteristic earthquakes on the fault and the smaller events with background seismicity on the fault between characteristic earthquakes.

Nakanishi (1990, 1991) proposed a model that combined features of the cellular-automata model and the slider-block model. A linear array of slider-blocks was considered but only one block was allowed to slide at a time. Thus interactions were only with nearest neighbors. The slip of one block could lead to the instability of either or both of the adjacent blocks, which would then be allowed to slip in a subsequent step, until all blocks were again stable. Brown et al. (1991) proposed a modification of this model involving a two-dimensional array of blocks. Other models of this type have been considered by by Takayasu & Matsuzaki (1988), Ito & Matsuzaki (1990), Sornette & Sornette (1989, 1990), Langer & Tang (1991), Carlson (1991a, b), Carlson et al. (1993), Matsuzaki & Takayasu (1991), Rundle & Brown (1991), Shaw et al. (1991), and Huang et al. (1992).

A typical high-order, slider-block model is illustrated in Figure 8. All blocks are connected to a constant-velocity driver via leaf springs, and the blocks slide on a frictional surface. All physical parameters of the two-dimensional system are homogeneous including the block mass m, coupling spring constant k_c, pulling spring constant k_l, and frictional force F. We assume that blocks are loaded in the x-direction only; there are no displacements in the y-direction. A second assumption is that the loading velocity of the driver is sufficiently slow that we may consider it to be zero during the sliding of a block.

The primary modification of the cellular-automaton, slider-block model from a two-dimensional, traditional slider-block system is that, during the sliding of an unstable block, all other blocks are assumed to be stationary. The requirement that only one block is allowed to slip at a given time limits the system to nearest neighbor interactions, which is a characteristic of cellular-automata systems. With this modification, it is no longer necessary to solve large numbers of equations simultaneously; using the static-dynamic friction law with f the ratio of the static to the dynamic friction, the amount of displacement for each failed blocks can be obtained analytically.

The evolution of the modified slider-block system was examined numerically using a square array of n x n blocks (maximum size 50 x 50, 2500 blocks). Frequency-size statistics for $\phi = 1.5$, $n = 50$, and several values of a are given in Figure 9. The ratio of the number of events N with size N_f to the total number of events N_O is given as a function of N_f, N_f is the number of blocks that participate in an event and is a measure of the area of an event. A good correlation is obtained with a fractal (power-law) dependence with a slope of -1.36. The frequency-size relation shows a roll-off from a power law near the large event end of the scaling region. This deviation is reduced as the stiffness parameter a

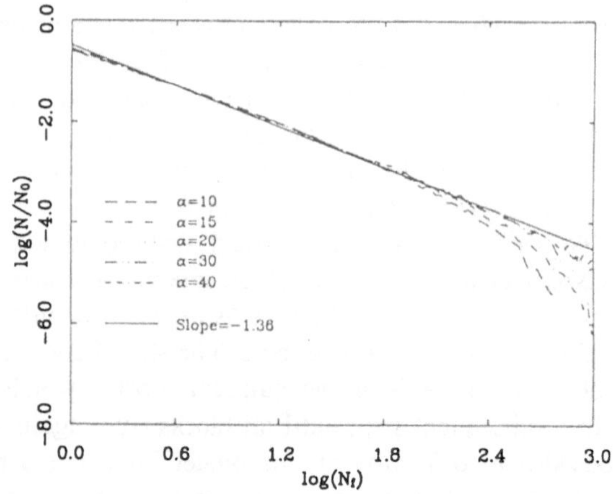

Fig. 9. The ratio of the number of events N with size Nf to the total number of events N_O is plotted against N_f. Results are given for $\phi = 1.5$ and $\alpha = 10, 15, 20, 30$, and 40. The solid line is the correlation with the power-law (fractal) relation.

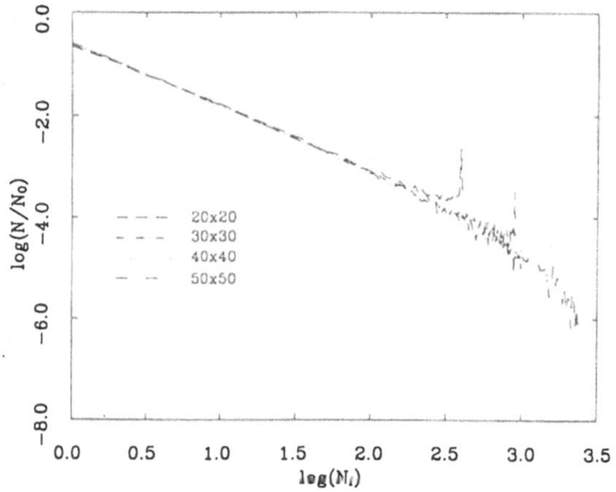

Fig. 10. The ratio of the number of events N size with N_f to the total number of events No is plotted against N_f. Results are given for systems of size 20 x 20, 30 x 30, 40 x 40, and 50 x 50 with parameters $\phi = 1.5$ and $\alpha = 50$. The peaks at log $N_f = 2.60$, 2.95, and 3.2 correspond to catastrophic events involving the entire system.

increases. It is also of interest to consider the behavior of the system when the parameter α/n is greater than one, i.e. when the system is stiff. Figure 10 shows the frequency-size statistics for $\alpha = 50$, $\phi = 1.5$, and several values of the linear size n. The peak at $N_f = 400$ (log $N_f = 2.60$) corresponds to catastrophic events with grid size $n = 20$, the peak at $N_f = 900$ (log $N_f = 2.95$) corresponds to catastrophic events with grid size $n = 30$, and the peak at $N_f = 1600$ ($\log N_f = 3.20$) corresponds to catastrophic events with grid size $n = 40$. Excess numbers of catastrophic events with $n = 50$ ($\alpha / n = 1$) are not observed. A catastrophic event is one in which all blocks participate. These results suggest that the number of catastrophic events exceeds the power-law statistics only when the parameter a/n is above a threshold value which is close to one; the system is sufficiently stiff that the range of correlation between blocks is larger than the linear size of the system.

Earthquake Hazard Assessment and Prediction

The fundamental aspects of self-organized criticality are also directly relevant to earthquake prediction and hazard assessment strategies. The thermodynamic and statistical mechanics aspects of critical phenomena appear to be directly relevant to distributed seismicity (Rundle 1988a, b, c). The trajectories of a spring-block model in phase space are abelian, i.e. all allowed points in phase space are occupied with equal probability. This is also the case for thermodynamic systems. This strongly suggests that the earth's crust resembles a thermodynamic system near a critical point. The fluctuations are earthquakes.

In order to better understand earthquakes, it is essential to determine whether concepts of self-organized criticality are applicable. If these concepts are applicable, then all earthquakes, including the largest, represent noise on a background level of stress. Self-organized criticality may result in a near-constant level of regional seismicity between major earthquakes. The temporal variation in seismicity prior to a major earthquake is a subject of considerable controversy. Some authors accept precursory quiescence as a widely applicable phenomena (Scholz 1988, Wyss & Haberman 1988, Schreider 1990), but others support a regional buildup in seismicity as a precursor to a major earthquake. The buildup of activity in the San Francisco Bay area prior to the 1906 earthquake has been discussed by Sykes & Jaume (1990).

Precursory seismicity is the basis for intermediate-term earthquake prediction (1-5 years) pattern recognition methods developed at the International Institute for the Theory of Earthquake Prediction and Theoretical Geophysics in Moscow under the direction of Academician V.I. Keilis-Borok (Keilis-Borok 1990, Keilis-Borok & Rotwain 1990, Keilis-Borok & Kossobokov 1990a, b). The pattern recognition includes increases in regional seismicity, increases in the clustering of earthquakes, and changes in aftershock statistics (Molchan et al. 1990). Premonitory seismicity patterns were found for strong earthquakes in California and Nevada (algorithm CN) and for earthquakes worldwide (algorithm M8). During the last 5 years, at least 3 strong earthquakes (Spitak, Loma-Prieta, Costa-Rica) were predicted in advance.

An important question is the association of large earthquakes with small earthquakes. Do large earthquakes only occur where small earthquakes occur? If so, the small earthquakes can be used to assess the regional seismic hazard. Johnson & Nava (1985) have used this approach to assess the hazard in the New Madrid area. Turcotte (1989, 1991) has discussed the approach in more general terms. A general association of large earthquakes with small earthquakes provides a rational basis for probabilistic earthquake hazard assessment. Such an approach would follow naturally from the applicability of self-organized criticality.

Table 1. Values of the Hausdorff measure H and flood intensity factor F for the ten benchmark stations considered by Benson (1968).

Station	Site	H	F
1-1805	Westfield River, Goss Heights, MA	0.513	3.31
2-2185	Oconee River, Greenboro, GA	0.540	3.47
5-3310	Mississippi River, St. Paul, MN	0.470	2.9
6-3440	Little Missouri River, Alzado, WY	0.520	3.31
6-8005	Elkhorn River, Waterloo, NE	0.540	3.4
7-2165	Mora River, Golondinas, NM	0.630	4.27
8-1500	Llano River, Junction, TX	0.719	5.24
10-3275	Humboldt River, Comus, NV	0.616	4.13
11-0980	Arroyo Seco, Pasadena, CA	0.875	7.4
12-1570	Wenatchee River, Plain, WA	0.310	2.04

Conclusions

Models that exhibit self-organized criticality have a number of strong similarities to distributed seismicity. The most universal is power-law (fractal) frequency-magnitude statistics. The slider-block models are completely deterministic but behave chaotically and exhibit a variety of statistical phenomena.

These models provide a test bed for prediction algorithms. The objective is to predict a large slip event on the model systems. If such predictions are successful, they may provide new approaches to using distributed seismicity to predict earthquakes. It is efficient and inexpensive to study the statistical behavior of these model systems.

Acknowledgements. This research was partially supported by NSF Grant PHY 89-04035 and by the US Geological Survey (USGS), Department of the Interior, under USGS award number 1434-92-G-2165. The views and conclusions contained in the document are those of the authors and should not be interpreted as necessarily representing the official policies, either expressed or implied, of the US Government.

References

Aki K (1981) A probabilistic synthesis of precursory phenomena, in Simpson DW and Richards PG, eds., Earthquake Prediction. American Geophysical Union, Washington, D.C. pp. 566-574.

Bak P, Tang C (1989) Earthquakes as a self-organized criticality. J Geophys Res 94: 15, 635-15,637.

Bak P, Tang C, Wiesenfeld K (1988) Self-organized criticality. Phys Rev A38: 364-374.

Barriere B, Turcotte DL (1991) A scale-invariant cellular-automata model for distributed seismicity. Geophys Res Let 18: 2011-2014.

Barton CC, Hsieh PA (1989) Physical and Hydrological-Flow Properties of Fractures, 28th International Geological Congress Field Trip Guidebook T385, American Geophysical Union, Washington, D.C.

Benson MA (1968) Uniform flood-frequency estimating methods for federal agencies. Water Resour Res 4: 891-908.

Brown SR, Scholz CH, Rundle JB (1991) A simplified spring-block model of earthquakes. Geophys Res Let 18: 215-218.

Burridge R, Knopoff L (1967) Model and theoretical seismicity. Seis Soc Am Bull 57: 341-371.

Carlson JM (1991a) Time intervals between characteristic earthquakes and correlations with smaller events: an analysis based on a mechanical model of a fault. J Geophys Res 96: 4255-4267.

Carlson JM (1991b) Two-dimensional model of a fault. Phys Rev A44: 6226-6232.

Carlson JM, Langer JS (1989) Mechanical model of earthquakes generated by fault dynamics. Phys Rev A40: 6470-6484.

Carlson J M, Langer JS, Shaw BE, Tang C (1991) Intrinsic properties of a Burridge-Knopoff model of an earthquake fault. Phys Rev A44: 884-897.

Carlson JM, Grannan ER, Swindle GH (1993) Self-organizing systems at finite driving rates. Phys Rev E47: 93-105.

Evernden JF (1970) Study of regional seismicity and associated problems. Seis Soc Am Bull 60: 393-446.

Huang J, Turcotte DL (1990a) Are earthquakes an example of deterministic chaos? Geophys Res Let 17: 223-226.

Huang J, Turcotte DL (1990b) Evidence for chaotic fault interactions in the seismicity of the San Andreas fault and Nankai trough. Nature 348: 234-236.

Huang J, Turcotte DL (1992) Chaotic seismic faulting with a mass-spring model and velocity-weakening friction. PAGEOPH 138: 569-589.

Huang J, Narkounskaia G, Turcotte DL (1992) A cellular automaton, slider-block model for earthquakes 2. Demonstration of self-organized criticality for a two dimensional system. Geophys J Int 111: 259-269.

Ito KM, Matsuzaki M (1990) Earthquakes as self-organized critical phenomena. J Geophys Res 95: 6853-6860.

Johnston AC, Nava SJ (1985) Recurrence rates and probability estimates for the new Madrid seismic zone. J Geophys Res 90: 6737-6753.

Jones LM, Molnar P (1979) Some characteristics of foreshocks and their possible relationship to earthquake prediction and premonitory slip of faults. J Geophys Res 84: 3596-3608.

Kadanoff LP, Nagel SR, Wu L, Zhou SM (1989) Scaling and universality in avalanches. Phys Rev A39: 6524-6533.

Keilis-Borok VI (1990) The lithosphere of the earth as a nonlinear system with implications for earthquake prediction. Rev Geophys 38: 19-34.

Keilis-Borok VI, Rotwain IM (1990) Diagnosis of time of increased probability of strong earthquakes in different regions of the world: Algorithm CN. Phys Earth Planet Int 61: 57-72.

Keilis-Borok VI, Kossobokov VG (1990a) Premonitory activation of earthquake flow: Algorithm M8. Phys Earth Planet Int 61: 73-83.

Keilis-Borok VI, Kossobokov VG (1990b). Times of increased probability of strong earthquakes (M>7.5) diagnosed by algorithm M8 in Japan and adjacent territories. J Geophys Res 95: 12, 413-12, 422.

Langer JS, Tang C (1991) Rupture propagation in a model of an earthquake fault. Phys Rev Let 67: 1043-1046.

Matsuzaki M, Takayasu H (1991) Fractal features of the earthquake phenomenon and a simple mechanical model. J Geophys Res 96: 19,925-19,931.

Molchan GM, Dmitrieve OE, Rotwain IM, Dewey J (1990) Statistical analysis of the results of earthquake prediction, based on bursts of aftershocks. Phys Earth Planet Int 61: 128-139.

Nagel SR (1992) Instabilities in a sandpile. Rev Mod Phys 64: 321-325.

Nakanishi H (1990) Cellular automation model of earthquakes with deterministic dynamics. Phys Rev A41: 7086-7089.

Nakanishi H (1991) Statistical properties of the cellular automata model for earthquakes. Phys Rev A43: 6613-6621.

Narkounskaia G, Turcotte DL (1992) A cellular-automata, slider-block model for earthquakes 1. Demonstration of chaotic behavior for a low order system. Geophys J Int 111: 250-258.

Pacheco JF, Scholz CH, Sykes LR (1992) Changes in frequency-size relationship from small to large earthquakes. Nature 355: 71-73.

Rosendahl J, Vekie M, Kelley J (1993) Persistent self-organization of sandpiles. Phys Rev E47: 1401-1404.

Rundle JB (1988a) A physical model for earthquakes. 1. Fluctuation and interaction. J Geophys Res 93: 6237-6254.

Rundle JB (1988b) A physical model for earthquakes. 2. Applications to Southern California. J Geophys Res 93: 6255-6274.

Rundle JB (1988c) A physical model for earthquakes. 3. Thermo-dynamical approach and its relation to nonclassical theories of nucleation. J Geophys Res 94: 2839-2855.

Rundle JB, Brown SR (1991) Origin of rate dependence in frictional sliding. J Stat Phys 65: 403-412.

Sammis CG, Osborne RH, Anderson JL, Banerdt M, White P (1986) Self-similar cataclasis in the formation of fault gauge. PAGEOPH 124: 53-78.

Scholz CH (1988) Mechanisms of seismic quiescences. PAGEOPH 126: 701-718.
Scholz CH (1991) Earthquakes and faulting: Self-organized critcal phenomena with a characteristic dimension, In: Riste T and Sherrington D (eds), Spontaneous Formation of Space-Time Structures and Criticality, Kluwer Dordrecht, pp. 41-56.
Schreider SYU (1990) Formal definition of premonitory seismic quiescence. Phys Earth Planet Int 61: 113-127.
Shaw BE, Carlson JM, Langer JS (1992) Patterns of seismic activity preceding large earthquakes. J Geophys Res 97: 479-488.
Sornette A, Sornette D (1989) Self-organized criticality and earthquakes. Europhys Let 9: 197-202.
Sornette A, Sornette D (1990) Earthquake rupture as a critical point: Consequences for telluric precursors. Tectonophysics 179: 327-334.
Sornette D, Davy P, Sornette A (1990) Structuration of the lithosphere in plate tectonics as a self-organized critical phenomenon. J Geophys Res 95: 17,353-17,361.
Sykes LR, Jaume SC (1990) Seismic activity on neighboring faults as a long-term precursor to large earthquakes in the San Francisco Bay area. Nature 348: 595-599.
Takayasu H, Matsuzaki M (1988) Dynamical transition in threshold elements. Phys Let A131: 224-247.
Turcotte DL (1986) Fractals and fragmentation. J Geophys Res 91: 1921-1926.
Turcotte DL (1989) A fractal approach to probabilistic seismic hazard assessment. Tectonophysics 167: 171-177.
Turcotte DL (1991) Earthquake prediction. An Rev Earth Planet Sci 19: 263-281.
von Seggern D, Alexander SS, Baag CE (1981) Seismicity parameters preceding moderate to major earthquakes. J Geophys Res 86: 9325-9351.
Wyss M, Haberman RE (1988) Precursory seismic quiescence. PAGEOPH 126: 319-332.

Scaling Laws of Fragmentation

Hiroyuki Nagahama[1] and Kyoko Yoshii[2]
[1] Institute of Geosciences, School of Science, Shizuoka University,
836 Oya, Shizuoka 422, Japan
[2] Japan Overseas Cooperation Volunteers, Hiroo, Shibuya,
Tokyo 150, Japan

Abstract. Fragmentation over a wide range of size scales has been discussed in previous papers. A variety of statistical power-law relations have often been used successfully to describe scaling laws of the size distribution and shape (surface roughness) of fragments, indicating that fragmentation is a scale invariant process. But it has been assumed that fracture shapes do not have any effect on the fractal dimension of fragment size distributions. New scaling laws of the fracturing energy for rock fragmentation can be derived from concepts of fractal geometry and the Griffith energy balance. These laws show that the two fractal dimensions describing size distributions and roughness in the shape of rock fragments increase as the fracturing energy increases. A relation between the two kinds of fractal dimensions indicates that the fractal dimension of roughness is the mean value of the fractal dimension of fracture-size distribution and the Euclidean space dimension of specimen volume. This relation indicates a constraint of fractal geometry for fragmentation and is expected to be a powerful tool for the fractography analysis in the fields of tectonics and seismicity.

Introduction

Several empirical studies on fragments have demonstrated a power-law dependence of the cumulative number $N(r)$ of fragments whose sizes are larger than size r,

$$N(R) \propto r^{-D_S} \tag{1}$$

where D_S is a constant (Takeuchi & Mizutani 1968, Hartmann 1969). This equation says that the size distribution of the fragments obeys a power-law characteristic of fractals, and D_S represents the fractal dimension of the size distributions of the fragments (Mandelbrot 1982, Matsushita 1985, Turcotte 1986a, b, 1992, Poulton et al. 1990, Shimamoto & Nagahama 1992, Nagahama 1993). Size-distribution represented by eq.(1) can be obtained from an equation known as a master equation (Haken 1978), and a scaling assumption (McGrady & Ziff 1982, Cheng & Redner 1988, Campi 1989, Redner 1990). It is assumed that the fragment size distribution is statistically scale invariant (Mandelbrot

1977, 1982, Matsushita 1985, Turcotte 1986a, b, 1992, Sammis et al. 1986, Zhao et al. 1990, Nagahama 1991).

The cumulative mass $M(r)$ of fragments with a radius less than r has been described by an empirical power-law equation

$$M(r) \propto r^{h} \tag{2}$$

where h is a positive constant (Gaudin 1926, Schuhmann 1940, Fujiwara et al. 1977, Turcotte 1986a, b, 1992, Nagahama 1991). When the 3-D fracture surface is modelled well by a fractal surface, i.e., it is composed of self-similar shapes, the fractal dimension D_R of the roughness of the fracture surface (the roughness of the fractal surface) (Mandelbrot 1977, 1982, Feder 1988) can range from 2, when the surface is smooth, up to 3 (Mandelbrot 1977, 1982, Avnir et al. 1983). If all fragments in some collection have a common fractal dimension D_R, the cumulative mass becomes

$$M(r) \propto \int_0^r r^{D_R - D_S - 1} dr \propto r^{D_R - D_S} \tag{3}$$

Comparison of eq.(2) with eq.(3) yields

$$h = D_R - D_S \tag{4}$$

Therefore, h is a parameter which depends on two fractal dimensions, the fracture shape (roughness) and the size distributions of fragments (Matsushita 1985). Using eq.(4), we can estimate the fractal dimension of the roughness from the fragmentation in any collection which exhibits power-law dependences like eqs.(1) and (2). But it has been assumed that the fracture shape does not have any effect on the fractal dimension of size distributions (Zhao et al. 1990).

In this chapter, we would like first to review a part of the results reported in Nagahama (1991, 1992, 1993) and Nagahama & Yoshii (1993). Next, we will introduce new scaling laws of fracturing energy for the rock fragmentation that are obtained from (1) the fractal concept; (2) the Griffith energy balance concept; and (3) the relationship between fractal dimensions of the fracture-size distribution and the surface roughness of fragments. Moreover, we propose a relation between two kinds of fractal dimensions of crustal and rock fragmentation and suggest that it is expected to be a powerful tool for the application of fractography to tectonics and seismicity.

Energy-Size Reduction Relationship in Fracturing

Since Rittinger (1867) proposed his theory on the size reduction of rocks, a number of theories on size reduction have been developed empirically (Kick 1885, Bond 1952, Rittinger 1867). Among them, Rittinger's, Kick's and Bond's theories have been widely referred to. However, many discrepancies between these theories and empirical measurements of size reduction have arisen. For example, Bergstrom et al. (1963) proposed the corresponding energy-size relation between Kick's and Bond's relationships. In general, the empirical energy-size relation is expressed by the Walker-Lewis' relation:

$$E = Cr^{-n+1} \tag{5}$$

where E is the energy per unit mass required for size reduction of feed having particle size r, and C and n are constants (Walker et al. 1937, Charles 1957, Schuhmann 1960, Tartaron 1963). In this relation, E depends on r.

Charles (1957) had discovered the relation between h and n to be

$$h - n + 1 = 0 \tag{6}$$

Eqs.(5) and (6) can be derived directly from eq.(2) (Schuhmann 1940). The derivations were made without assuming any of specific relationships between energy and particle size in previous literature (Kick 1885, Bond 1952, Rittinger 1867). But eq.(5) is unreliable and the basic underlying concept remains poorly known. The present status of these theories of size reduction of rocks is extremely unsatisfactory. By what kind of mechanism is the energy-size reduction relationship determined? To answer these questions, we will (1) derive the relation between two fractal dimension for the fragmentation and (2) derive energy density for fracturing and the theoretical relationship of size reduction of rocks from the fractal and Griffith energy balance concepts (Griffith 1920, 1924). Moreover, we would like to discuss the relation from the view point of strength.

Fractal Theory on Energy-Size Reduction

Fractals are a concern of the New Geometry, whose primary objective is to describe many natural structures with irregularities and roughness of various scales (Mandelbrot 1982, Cahn 1989). Crashed fragments have irregular and rough fracture surfaces (Mandelbrot et al. 1984, Brown & Scholz 1985, Scholz & Aviles 1986, Pande et al. 1987a, b, Mu & Lung 1988, Mecholsky et al. 1989, Chelidze & Guenguen 1990).

Nii et al. (1985) showed that a fracture surface becomes rougher as the energy density for the fracturing of the glass increases. Based on their fracture experiments on glass, the fractal dimension of the fracture surface becomes steeper as the energy density increases (Fig. 1). Moreover, the fracture experiments on basalt by Matsui et al. (1982) also shows that the fractal dimension of the size distribution of fragments increases as the energy density increases (Fig. 2). Why do the two fractal dimensions of the fracture surface with increased energy density in these two cases? We examine this problem from the Griffith energy balance and fractal concepts as follow.

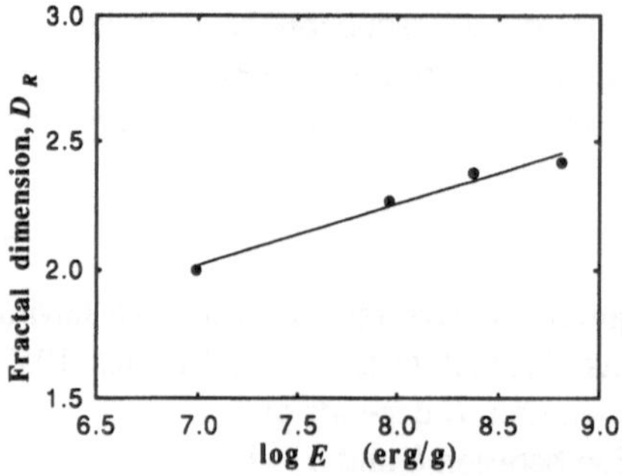

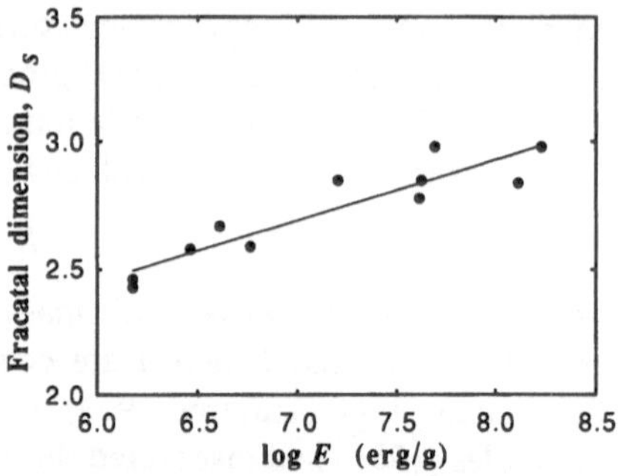

Fig. 1 (left). Fractal dimensions of roughness as a function of the energy density for fracturing glass. The solid line is fitted to the data by least squares regression (partly modified from Nagahama & Yoshii 1993 Fig. 1, data from Nii et al. 1985).
Fig. 2 (right). Fractal dimensions of the size distributions of fragments as a function of the energy density for the fracturing of tuff. The solid line is fitted to the data by least squares regression. (partly modified from Nagahama 1993 Fig. 1, data from Matsui et al. 1982).

Let us consider the case in which fracturing obeys the Griffith energy balance concept. The total potential energy of the system U, for a static fracture, is

$$U = (-W + U_B) + U_S \tag{7}$$

where W is the amount of work done by the external forces, U_B is the internal strain potential energy reserved in the elastic body and U_S is the expenditure energy in creating a new fracture surface (the surface energy per unit area). Mechanical energy must decrease with fracturing. However, the surface energy will increase then, because work must be done against the force of cohesion. Therefore, there are these competing influences; for fracturing there must be a reduction of the total energy of the system, and hence at equilibrium there is a balance between competing influences. The condition for equilibrium is

$$J = \frac{\partial U}{\partial A_T} = 0 \tag{8}$$

where J is identical to the potential energy release rate and A_T is the total area of the fracture surface (Griffith 1920, 1924).

If U_B is almost constant during fracturing and the surface energy per unit area $\partial U_S / \partial A_T$ is constant for minerals or rocks, the Griffith energy balance concept indicates that W is proportional to the total area of the fracture surface produced by fracturing. If the 3-D fracture surface is modelled well by a fractal surface, the total area of the fracture surface A_T is proportional to r^{D_r} where D_R is the fractal dimension of the fracture surface ($2 < D_R < 3$) and r is a characteristic linear dimension of the fracture surface (Avnir et al. 1983). Therefore, W can be expressed by

$$W \propto A_T \propto r^{D_R} \tag{9}$$

From eq.(9), we can get the energy per unit mass for fragmentation:

$$E \propto r^{\omega}, \quad \omega = D_R - 3 \tag{10}$$

Eq.(10) indicates that the fractal dimension of the fracture surface increases as the energy density increases. In brittle materials with cleavage, as in a variety of minerals, new fractures are created as the energy increases. In this case, Rittinger's theory is empirically suitable. In contrast, when the elastic energy stored in very hard rocks reaches a threshold, the rocks broken into fragments are similar in shape. In this case, Kick's theory is empirically suitable. But in general both of these mechanisms operate during fracturing. In this case, the input energy per unit mass for size reduction can be expressed by eq.(10). Using eq.(10), one can at least have a handle on the fracture surface topography. Moreover, eqs.(4), (5), (6) and (10) can be combined to give the relationship

$$E \propto r^{\varphi}, \quad \varphi = \frac{1}{2}(D_S - 3) \tag{11}$$

Eqs.(10) and (11) indicate that the two fractal dimensions, the surface roughness and the fragment size distribution, respectively, increase as the energy density for fracturing increases and are consistent with three previous size reduction theories: Kick's theory (D_R = 3.0; D_S = 3.0; n = 1.0), Bond's theory (D_R= 2.5; D_S = 2.0; n = 1.5) and Rittinger's theory (D_R= 2.0; D_S= 1.0; n = 2.0). But from only the fractal concept and the Griffith energy balance concept, the metallurgical basis for eq.(5) cannot be understood theoretically. Thus, next we would like to demonstrate the fractal dimensions of fragmentation based upon the rock strength.

Fractal Dimensions and Rock Strength

If the distribution of crack strength obeys a Weibull distribution, the size effect of the material's strength can be represented by

$$\sigma = \sigma_0 \, (r/r_0)^{-3/w} \tag{12}$$

where σ is the mean breaking strength of material of size r, σ_0 is the standard strength of material size of r_0, and is Weibull's coefficient of uniformity of materials (Weibull 1939a, b). Eq.(12) indicates that the mean breaking strength of materials decreases as material size increases. Moreover, the standard deviation s^2 of the distribution can be given by

$$s^2 \propto r^{-6/w} \tag{13}$$

so that σ^2/s^2 should be independent of the specimen volume size (Yokobori 1965, Jaeger & Cook 1969). This equation indicates that the standard deviation of strength increases as the w-value increases when the specimen volume is constant. Eq.(13) is proposed for the strength of rocks or minerals (Mogi 1962, Kanda et al. 1969, Paterson 1978).

On the assumption that fracture is due mainly to the tensile stress generated by the action of external forces, Majima & Oka (1969) and Oka & Majima (1969) proposed that the energy for size reduction of particle size r is proportional to $r^{3(1-2/w)}$. Then, the energy per unit mass E can be given by

$$E = C_M r^{-6/w} \tag{14}$$

where C_M is a constant relating to the tensile strength and Young's modulus of rocks. Kanda and coworkers obtained the relation between the energy per unit mass and Weibull's coefficient:

$$E = C_Y r^{-k/w} \tag{15}$$

where C_Y is a constant and k is the shape factor of rock specimens (Kanda et al. 1969, 1970). In eq.(15), the k-value was determined to be 5 based upon a disk splitting experiment and 6 upon a sphere splitting experiment.

From eqs.(10), (11) and (15), we can define two fractal dimensions:

$$D_R = 3 - \frac{k}{w}, \quad D_S = 3 - \frac{2k}{w} \tag{16}$$

Equation (16) indicates that these two fractal dimensions of fragments are related to the shape factor of rock specimens and the uniformity coefficient of rocks when the size effect of the tensile strength of rocks is taken into consideration and two fractal dimensions increase as the standard deviation of the rock strength increases. Therefore, if the fracture surface of rocks is smooth in the sphere splitting experiment, we can get the Hall-Petch's relation from eq.(12). Thus eq.(16) permits an accurate calculation of size reduction and these two fractal dimensions of fragments, using simple laboratory test information such as tensile tests yielding Young's modulus and fracture stress of rock specimens.

Relationship Between Two Fractal Dimensions and its Application to the Geosciences

Next we would like to obtain a relationship between the two fractal dimensions for fragmentation and apply it to the fields of tectonics and seismicity.

From eqs.(10) and (11), we can find a relationship between the two fractal dimensions of fragmentation:

$$2D_R = D_S + 3 \tag{17}$$

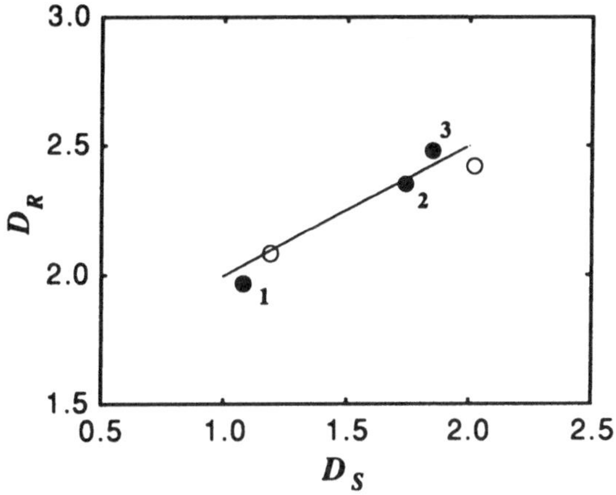

Fig. 3. Correlation between fractal dimensions of the roughness and of the fracture-size distribution of Abukuma granite specimens (open circles) and active fault systems in Japan (closed circle). *1*: Tohoku regions, *2*: Kinki regions, *3*: Chubu regions.

By eq.(17), we can estimate the fractal dimension D_R on the surface roughness from fragments in any collection that exhibit power-law dependences such as eq.(1). Eq.(17) expresses that the fractal dimension of the surface roughness in three dimensions is equal to the mean value of the fractal dimension of the size distribution and the Euclidean space dimension of the specimen volume, and indicates a constraint on the fractal geometry of fragmentation. Thus, we can derive the relationship (eq.17) between two fractal dimensions from eqs.(10) and (11). Eq.(17) is expected to be a powerful tool for the analysis of the surface roughness of fragments. One should note that this relation is reminiscent of (but different from) the hyperscaling relation in percolation, which relates the size-distribution exponent with the fractal dimension (Kapitulnik et al. 1984).

From geometrical considerations, the geological relation equivalent to eq.(17) has been suggested empirically for the relation between the fractal dimension of the geometry of fault systems and the power law distribution of the fault lengths (Main et al. 1990). If the entire crust is fragmented and faults are the edges of these fragments (one current hypothesis being entertained by earth scientists [King 1983, Turcotte 1986c, Haung & Turcotte 1988, Dubois & Nouaili 1989]), we can derive theoretically this empirically-founded geological relation from eq.(17). Fig. 3 shows the relationship between the fractal dimensions of the roughness and of the size distribution of rock fractures and active fault systems. This result is consistent with eq.(17).

Under many circumstances, the number of earthquakes $N(M)$ with a magnitude greater than M satisfies the empirical relation (Gutenberg & Richter 1954)

$$\log N(M) = -bM + a \tag{18}$$

where a and b are constants. The b-value is widely used as a measure of seismicity. Several authors (Caputo 1976, Aki 1981, King 1983, Turcotte 1986a, b, c, Main & Burton 1984, Huang & Turcotte 1988, Main et al. 1989, 1990, Dubois & Nouaili 1989, Nagahama 1991) have noted that the b-values of the log-linear frequency-magnitude distribution of earthquakes are consistent with the negative exponent of a power law distribution of fault lengths:

$$D_S = \frac{3b}{\delta} \tag{19}$$

where δ is a constant which depends on the relative durations of the seismic source and the time constant of the recording system. From the experimental studies on tensile crack propagation for a variety of crystalline rocks, $\delta = 3.0$ is appropriate (Main et al. 1989), so that $D_S = b$ (Main et al. 1990). Moreover, for most earthquake studies $\delta = 1.5$ has been appropriate (Kanamori & Anderson 1975), so that $D_S = 2b$ (Main et al. 1989, 1990). However, Dubois & Novaili (1989) obtained $\delta = 2.4$ from the seismicity of subduction zones between 100 and 700 km depth, so that $D_S = 1.25b$. Eq.(19) indicates that the b-value is related to the fractal dimension of crustal fragmentation. Moreover, this shows that the frequency-magnitude relation for earthquakes is also equivalent to eq.(1). The existing studies have introduced eq.(19) without distinguishing between the fractal dimension of size distribution and that of fragments' shape. Combining eqs.(17) and (19) supports theoretically this seismic relation, given by

$$D_R = \frac{3}{2}\left(\frac{b}{\delta} + 1\right) \tag{20}$$

which is expected to be valid for the roughness analysis of seismic rupture zones. By providing the basis for a possible framework to unify the interpretation of temporal variations in seismic b-values (eq.19), a relation similar to eq.(20) has already been suggested empirically for rough fault surfaces with many subsidiary faults (Main et al. 1990). This technique of roughness analysis will lead to advances in our understanding of earthquakes within the fields of tectonics.

Conclusions

Our primary conclusion is that fragmentation is a scale invariant process. The existence of two fractal dimensions describing the size distribution and roughness of fragments is direct evidence for scale invariance. Crushed fragments have irregular and rough surfaces at first sight but they are sieved into several fractions under the constraints of fractal geometry for fragmentation that are provided from fractal and Griffith energy balance concepts.

We can also obtain the new energy size reduction relationship by the fracture experiments on different brittle materials (Figs.1 and 2). This result solves the major drawback of eq.(5) and is consistent with three theories on size reduction and Hall-Petch's relation. The use of this relationship enables the estimation of the energy in the fracture system. Moreover, it is shown from the viewpoint of material strength that the two fractal dimensions for fragmentation are functions of a shape factor and the uniformity coefficient of rocks. This relation reflects the fact that the fractal dimension of the roughness is the mean value of the fractal dimension of the fracture-size distribution and the Euclidean space dimension of the specimen volume, and indicates a constraint of fractal geometry for fragmentation. This relation is expected to be a powerful tool for the fractal analysis in tectonics and seismicity.

Acknowledgments. We thank J. H. Kruhl and two anonymous reviewers for insightful comments which improved the manuscript and Robert M. Ross for improving the English from an earlier version of this manuscript. The authors would also like to thank J. A. Hudson for helpful reviews of early versions of the manuscript and M. Toriumi & Y. Nakajima for helpful discussions.

References

Aki KA (1981) Probabilistic synthesis of precursory phenomena. In: Simpson DW, Richards PG (eds) Earthquake Prediction: An International Review. AGU, Washington D.C., pp 566-574 (Maurice Ewing Series 4).

Avnir D, Farin D, Pfeifer P (1983) Chemistry in noninteger dimensions between two and three. II. Fractal surfaces of adsorbents. J Chem Phys 79: 3566-3571.

Bergstrom BH, Crabtree DD, Sollenberger CL (1963) Feed size effects in single particle crushing. Trans Amer Inst Mining Metall Petrol Engrs 226: 433-441.

Bond FC (1952) The third theory of comminution. Trans Amer Inst Mining Metall Petrol Engrs 193: 484-494.

Brown SR, Scholz CH (1985) Broad bandwidth study of the topography of natural rock surfaces. J Geophys Res 90: 12575-12582.

Cahn RW (1989) Fractal dimension and fracture. Nature 338: 201-201.

Campi X (1989) Geometric and kinetic models of fragmentation. J Physique 50: 183-189.

Caputo M (1976) Model and observed seismicity represented in a two-dimensional space. Ann Geophys (Rome) 4: 277-288.

Charles RJ (1957) Energy-size reduction relationships in comminution. Trans Amer Inst Mining Metall Petrol Engrs 208: 80-88.

Chelidze T, Guenguen Y (1990) Evidence of fractal fracture. Int J Rock Mech Min & Geomech Abstr 27: 223-225.

Cheng Z, Redner S (1988) Scaling theory of fragmentation. Phys Rev Lett 60: 2450-2453.

Dubois J, Novaili L (1989) Quantification of the fracturing of the slab using a fractal approach. Earth Planet Sci Lett 94: 97-108.

Feder J (1988) Fractals. Plenum, New York.

Fujiwara A, Kamimoto G, Tsukamoto A (1977) Destruction of basaltic bodies by high-velocity impact. Icarus 31: 277-288.
Gaudin AM (1926) An investigation of crushing phenomena. Trans Amer Inst Mining Metall Petrol Engrs 73: 253-316.
Griffith AA (1920) The theory of rupture and flow in solids. Phil Trans Roy Soc Ser A 221: 163-198.
Griffith AA (1924) The theory of rupture. In: Biezeno CB, Burgers JM (eds) Proc 1st Int Congr Appl Mech Boekhandel en Drukkerij J Walter Jr, pp54-63.
Gutenberg B, Richter CF (1954) Seismicity of the earth and associated phenomena, 2nd ed. Princeton Univ. Press, Princeton.
Haken H (1978) Synergetics: An Introduction, 2nd ed. Springer, Berlin Heidelberg.
Hartmann WK (1969) Terrestrial, lunar, and interplanetary rock fragmentation. Icarus 10: 201-213.
Huang J, Turcotte DL (1988) Fractal distributions of stress and strength and variations of b-value. Earth Planet Sci Lett 91: 223-230.
Jaeger JC, Cook NGW (1969) Fundamentals of Rock Mechanics. John Wiley & Sons, New York.
Kanamori H, Anderson DL (1975) Theoretical basis of some empirical relations in seismology. Bull Seis Soc Am 65: 1073-1095.
Kanda Y, Yashima S, Shimoiizaka J (1969) Size effects and energy laws of single sphere crushing. J Mining Metall Inst Japan 85: 987-992 (In Japanese with English abstract).
Kanda Y, Yashima S, Shimoiizaka J (1970) Size effects and energy laws of single disk crushing. J Mining Metall Inst Japan 86: 435-440. (In Japanese with English abstract).
Kapitulnik A, Gefen Y, Aharony A (1984) On the fractal dimension and correlations in percolation theory. J Statis Phys 36: 807-814.
Kick F (1885) Das Gesetz der proportionalem Widerstand und seine Anwendung. Arthus Felix, Leipzig.
King G (1983) The accommodation of large strains in the upper lithosphere of the Earth and other solids by self-similar fault systems: the geometric origin of b-value. PAGEOPH 121: 761-815.
Main IG, Burton PW (1984) Information theory and the earthquake frequency-magnitude distribution. Bull Seis Soc Am 74: 1409-1426.
Main IG, Meredith PG, Jones C (1989) A reinterpretation of the precursory seismic b-value anomaly from fracture mechanics. Geophys J 96: 131-138.
Main IG, Peacock S, Meredith PG (1990) Scattering attenuation and the fractal geometry of fracture systems. Pageoph 133: 283-304.
Majima H, Oka Y (1969) A new theory of size reduction. Flotation 38: 5-10 (In Japanese with English abstract).
Mandelbrot BB (1977) Fractals: Form, Chance and Dimension. Freeman, San Francisco.
Mandelbrot BB (1982) The fractal geometry of nature. Freeman, New York.
Mandelbrot BB, Passoja DE, Paullay AJ (1984) Fractal character of fractal surfaces of metals. Nature 308: 721-722.
Matsui T, Waza T, Kani K, Suzuki S (1982) Laboratory simulation of planetesimal collision. J Geophys Res 87: 10968-10982.
Matsushita M (1985) Fractal viewpoint of fracture and accretion. J Phys Soc Japan 54: 857-860.
McGrady ED, Ziff RM (1987) "Shattering" transition in fragmentation. Phys Rev Lett 58: 892-895.

Mecholsky JJ, Passoja DE, Feiberg-Ringel KS (1989) Quantitative analysis of brittle fracture surface using fractal geometry. J Am Ceram Soc 72: 60-65.
Mogi K (1962) The influence of the dimensions of specimens on the fracture strength of rocks: comparison between the strength of rock specimens and that of the Earth's crust. Bull Earthq Res Inst 40: 175-185.
Mu ZQ, Lung CW (1988) Studies on the fractal dimension and fracture toughness of steel. J Phys D: Appl Phys 21: 848-850.
Nagahama H (1991) Fracturing in the solid earth. Sci Repts Tohoku Univ, 2nd ser (Geol) 61: 103-126.
Nagahama H (1992) Scaling laws of rock fragmentations. Earth Monthly 14: 611-615 (In Japanese).
Nagahama H (1993) Fractal fragment size distribution for brittle rocks. Int J Rock Mech Min Sci & Geomech Abstr 30: 469-471.
Nagahama H, Yoshii K (1993) Fractal dimension and fracture of brittle rocks. Int J Rock Mech Min Sci & Geomech Abstr 30: 173-175.
Nii Y, Nakamura K, Ito K, Fujii N, Matsuda J (1985) The fractal dimension of fracture fragments in relation with fracturing energy. Proc 18th ISAS Lunar and Planetary Symp. 58-59.
Oka Y, Majima H (1969) Energy requirements in size reduction. Trans Soc Mining Engrs AIME 244: 249-251.
Pande CS, Richards LE, Louat N, Dempsey BD, Schwoeble AJ (1987a) Fractal characterization of fracture surfaces. Acta metall 35: 1633-1637.
Pande CS, Richards LE, Smith S (1987b) Fractal characteristic of fractured surfaces. J Mater Sci Lett. 6: 295-297.
Paterson MS (1978) Experimental rock deformation: the brittle field. Springer, Berlin Heidelberg New York.
Poulton MM, Mojtaba N, Fabmer IW (1990) Scale invariant behavior of massive and fragmented rock. Int J Rock Mech Min & Geomech Abstr 27: 219-221.
Redner S (1990) Fragmentation. In: Herrmann HJ, Roux S (eds) Statistical model for the fracture of disordered media. Elsevier, Amsterdam, pp 321-348.
Rittinger PR (1867) Lehbuch der Aufbereitungskunde. Ernst und Korn, Berlin.
Sammis CG, Osborne RH, Anderson JL, Banerdt M, White, P (1986) Self-similar cataclasis in the formation of fault gouge. PAGEOPH 124: 53-78.
Scholz CH, Aviles CA (1986) The fractal geometry of faults and faulting. In: Das S, Boatwright J, Scholz CH (eds) Earthquake Source Mechanics. AGU, Washington, D.C. pp147-155 (Maurice Ewing Volume 6).
Schuhmann RJr (1940) Principles of comminution, I - Size distribution and surface calculations. Mining Engrs AIME, New York (AIME Tech Pub no1189).
Schuhmann RJr (1960) Energy input and size distribution in comminution. Trans Amer Inst Mining Metall Petrol Engrs 217: 22-25.
Shimamoto T, Nagahama H(1992) An argument against the crush origin of pseudotachylytes based on the analysis of clast-size distribution. J Struct Geol 14: 999-1006.
Takeuchi H, Mizutani H. (1968) Relation between earthquake occurrence and brittle fracture. Kagaku (Science), Iwanami Shotten, Tokyo 38: 622-624 (In Japanese).
Tartaron FX (1963) A general theory of comminution. Trans Amer Inst Mining Metall Petrol Engrs 226: 183-190.
Turcotte DL (1986a) Fractals and fragmentation. J Geophys Res 91: 1921-1926.

Turcotte DL (1986b) Fractals in geology and geophysics. PAGEOPH 131: 171-196.
Turcotte DL (1986c) A fractal model for crustal deformation. Tectonophysics 132: 261-269.
Turcotte DL (1992) Fractals and chaos in geological and geophysics. Cambrige Univ Press, New York.
Walker WH, Lewis WK, McAdams WH, Gilliland ER (1937) Principles of chemical engineering. Mc Graw-Hill, London.
Weibull W (1939a) A statical theory of the strength of materials. Ing Vetenskaps Akad, Stockholm Handlingar, no 150.
Weibull W (1939b) The phenomenon of rupture in solids. Ing Vetenskaps Akad, Stockholm Handlingar, no 153.
Yokobori T (1965) Strength, Fracture and Fatigue of Materials. Wolters-Noordhoff, Groningen, The Netherlands.
Zhao ZY, Wang Y, Liu XH (1990) Fractal analysis applied to cataclastic rocks. Tectonophysics 178: 373-377.

Fractal Structure and Deformation of Fractured Rock Masses

Xing Zhang & David J. Sanderson
Department of Geology, University of Southampton,
Southampton, SO9 5NH, United Kingdom

Abstract. A numerical simulation method is used to predict connectivity of fractured rock masses. There is a threshold of fracture density, below which fractures are poorly connected. Where fracture density is at or above the threshold, there is a continuous fracture cluster (i.e. the largest cluster) throughout the fractured rock mass. Fractal dimension, D_f, is used to describe quantitatively the connectivity and compactness of the largest fracture cluster in the fractured rock mass and increases with fracture density. The critical fractal dimension, D_{fc}, describes the geometry of the largest fracture cluster at the threshold of fracture density, and has a rather constant value of 1.22 to 1.38 for wide variations in the distribution of size and orientations of the fractures.

Simulation of biaxial compressive tests of fractured rock masses has been carried out using a numerical method, UDEC (Universal Distinct Element Code). The deformation of fractured rocks increases greatly with fractal dimension and is mainly created by the shear displacements and openings along fractures. A link between fracture density and deformability of a fractured rock mass is established through the fractal dimension.

Introduction

Fractures of all scales have a dominant effect on many physical properties of the upper crust, such as the mechanical, hydraulic, thermal, and seismic properties. For example, the permeability of fractured rock masses may increase super-linearly with increase in fracture density (Englman et al. 1983, Shimo & Long 1987, Rasmussen 1987). It is desirable for many projects, including hazardous waste disposal, efficient oil recovery from fractured reservoirs, thermal energy extraction and ore deposition study, to have a thorough understanding of fracture features. The connectivity of fractures plays a key role in the deformability and permeability of a fractured rock mass. For example, Zhang et al. (1992) proposed a parameter, the connectivity ratio, to describe fracture connectivity and found that this connectivity ratio had a major effect on the volume flow rate of fluid into an excavation. Zhang (1993) related the connectivity ratio of fractures to the shear resistance of fractured rock masses and found that connectivity of fractures plays a key role in slope stability. Zhang et al. (1993) also found that the deformation of a fractured rock mass increases with increasing fracture density and that at a critical fracture density the rock structure may become unstable.

This study emphasizes the link between characteristics of fractured rock masses, such as connectivity, fractal dimension and fracture density, and deformation. Fracture density, the total length of fractures per unit area of rock, is easily obtained from mapping, core data or remote geophysical techniques. In this study, a 2-dimensional numerical model is used to investigate the connectivity and fractal dimension of a fractured rock mass and to determine the critical stage or percolation threshold at which rock properties show a sudden phase change. Distinct element methods are used to investigate the deformation characteristics at this phase change.

Modelling of 2-Dimensional Fractured Rock

Fractures are commonly observed as traces intersecting an outcrop, and the size of a fracture is described by its trace length in 2-dimensional problems. In this paper, a numerical model of fracture systems, based on self-avoiding random generation of fractures, is used to simulate fracture evolution in rock masses. This model involves the following procedures:

i) The simulated region of the rock mass is a square of unit length. The largest fracture length is fixed at 0.25, hence the fractures are small in relation to the simulated region. The lower length limit of fracture-traces (S_l) varies from 0.0005 to 0.15 in different simulations.
ii) The trace lengths of some natural fractures have been shown to follow a power-law distribution (Segall & Pollard 1983, Barton & Hsieh 1989, Heffer & Bevan 1990). In this model, the trace-lengths are sampled from a power-law distribution where the number of fractures (N) of length (L) has form $N \propto a L^E$. The exponent (E) is varied from 1.2 to 1.8 and this parameter, together with upper and lower limits of fracture size, controls the length distribution.
iii) Fracture orientation is defined as the angle of the fracture-trace to a reference axis. Angles may be selected randomly in the range 0-180° or in sets with a normal distribution in orientation, defined by the mean and standard deviation, here termed dispersion angle (V_s) and varying from 0° to 50°. Where V_s is zero, all fractures are parallel within a set of fractures. Several sets of fractures with different mean orientations can be simulated.
iv) The coordinates of the centre of a fracture are randomly selected from a uniform distribution within the square. A procedure of self-avoiding generation (Rives et al. 1992) is used, such that new fractures are selected only if they are located at a minimum distance to previously generated fractures. The minimum distance may be varied but is usually set at a tenth of the smallest fracture-length, i.e. much less than the smallest fracture-length.

Fractures are generated sequentially according to the above rules. As more fractures are added, density of a fractured rock mass increases. Fracture density (d) is defined as the total length of fracture-traces per unit area and is simply the sum of the length of all fractures generated since the square has unit area. Although these simulation procedures allow a wide variety of fracture patterns to be produced, it should be pointed out that by simply adding new fractures to increase fracture density no attempt is made to simulate growth of pre-existing fractures.

Connectivity and Percolation Threshold

Within a fractured rock mass, some fractures are isolated (unconnected), whilst others intersect. In practice, it is very difficult to identify if fractures are connected or unconnected in 3-dimensional samples of naturally fractured rocks. However, in the 2-dimensional simulations, it is relatively straightforward to determine if a fracture is connected to its neighbours.

At low fracture densities, the fractures are generally isolated, although some may be connected locally (Fig. 1a). As the fracture density increases, more fractures become connected and relatively large clusters form (Fig. 1b). At this stage, there is not a continuous cluster throughout the fractured rock mass, since the largest cluster within the simulated square area does not intersect all the boundaries (Fig. 1b).

With further increase in fracture density, more fractures and fracture clusters become connected, until a critical stage is reached where the largest cluster intersects all the boundaries of the simulated square area (Fig. 1c). At such a point, it is possible to define a critical fracture density and corresponding connectivity. The attainment of a critical fracture cluster which interconnects the boundaries of the rock mass at any scale corresponds to the percolation threshold, familiar from studies of lattice models (Stauffer 1985). If more fractures are added to the fractured rock mass in Fig. 1c, the largest cluster will grow to cover the whole area and the rock mass is fragmented, (Fig. 1d).

At this critical stage, the fracture-related properties would be expected to suddenly change, i.e. there would be a form of phase transition. Gueguen et al. (1990) have already pointed out that, as the number of transmissive fractures increases, more and more infinite fracture paths will develop and the rock mass as a whole may suddenly become permeable. In a similar manner, deformation in a fractured rock structure may become unstable where the fracture density is above some critical value (Zhang et al. 1993). Thus, fracture density may be expected to play an important role in the characterization of both the geometry and associated physical properties of fractured rock masses.

The key to understanding such behaviour involves the determination of the connectivity of fractures. During modelling of the evolution of a fracture network,

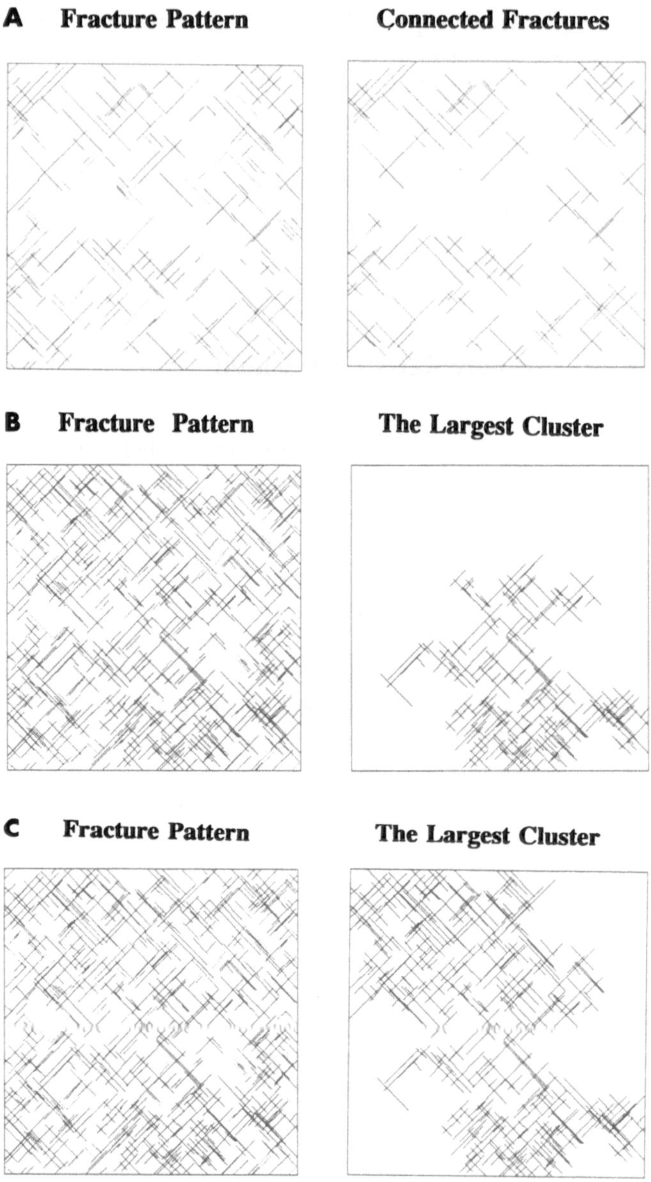

Fig. 1 a-d. Fracture patterns with different fracture densities; all fractures are shown on the left, whereas only connected fractures or the largest cluster are shown on the right. **a** Low density fractures with isolated or locally connected fractures; **b** Just below the critical density, where more fractures connect together, but the largest cluster does not intersect all the boundaries of the simulated square area; **c** At the critical density, where the largest cluster intersects all the boundaries; **d** Above the critical density, where more fractures connect to form the largest cluster and the rock becomes increasingly fragmented.

D Fracture Pattern **The largest cluster**

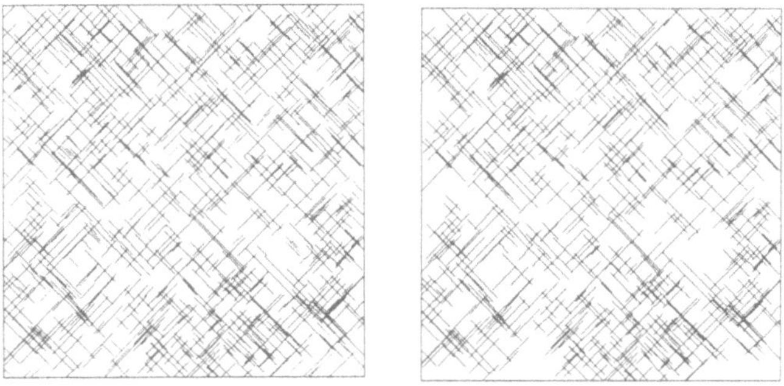

Fig. 1 (continued)

it is possible to determine which fractures are connected in two-dimensions. At any given stage, all unconnected (isolated) fractures can be identified and all the connected fractures can be assigned to a cluster. Normally, both isolated fractures and clusters would be present in any fractured rock. The percolation threshold can be recognised when the largest cluster just intersects all the boundaries of the simulated square area. The associated density is recorded and termed the critical density.

Figure 2 shows the results of a series of simulation consists of 2 orthogonal sets of parallel fractures. The fracture lengths were sampled from a power-law distribution with different exponents (E). The upper limit is fixed at 0.25 and the lower limit (S_l) varies from 0.0005 to 0.15. The number of fractures in each simulation was increased until the percolation threshold was reached, which represents a change in the connectivity status of the rock masses in which the S_l-N space is divided into two domains of connected and unconnected fracture clusters (Fig. 2). The relationship between fracture number, N, and the lower limit of fracture size, S_l, at the percolation threshold follows an approximate power-law. The slight curvature in Fig.2 can be attributed to the censoring of large fractures in the sample, i.e. those > 0.25 (Jackson & Sanderson 1992). Thus, for a fractured rock-mass, the connectivity can be predicted from the number of fractures, the fracture distribution and the range of fracture lengths.

Fractal Dimension of the Largest Fracture Cluster

Essentially, the largest fracture-cluster in the 2-dimensional simulation of the rock-mass represents a complex network which partially covers the area and is a kind of fractal structure. Obviously, the fractal dimension of the largest cluster is greater than 1, the dimension of an individual fracture trace, since it partially

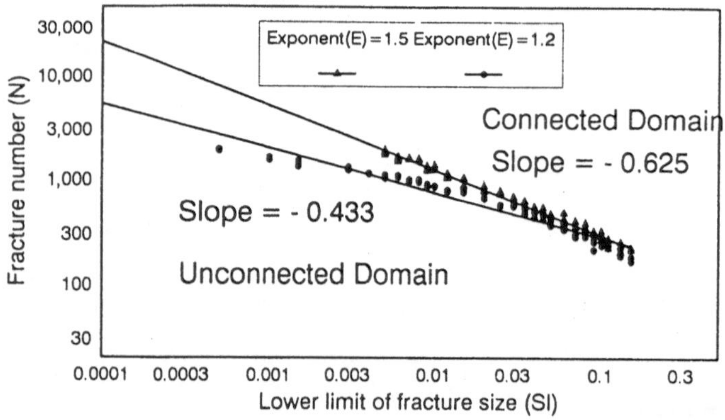

Fig. 2. Connectivity status of fractured rock masses. Above the critical density, connectivity of fractured rocks is above the percolation threshold. The lower limit (S_l) of size varies from 0.0005 to 0.15 for an exponent E = 1.2, and from 0.005 to 0.15 for E = 1.5.

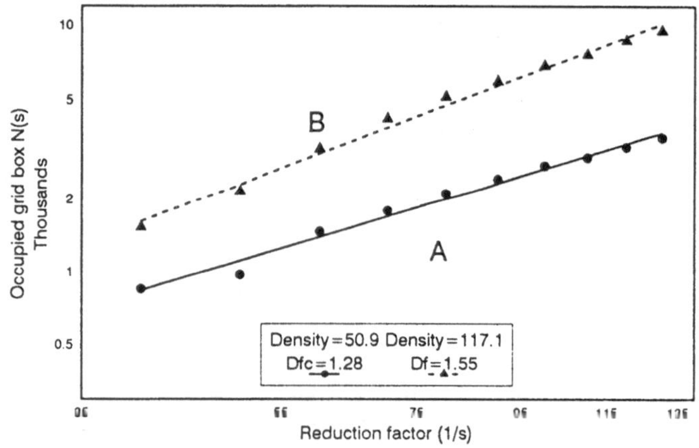

Fig. 3. Fractal dimension of two fractured rocks using box-counting principle. *A* represents the largest cluster in Fig.1c, and has a fractal dimension, D_{fc} = 1.28; *B* represents the largest cluster in Fig.1d and has a fractal dimension, D_f = 1.55. The sub-square size varies from 1/40 to 1/140.

covers the sample area. On the other hand, the dimension must be less than 2, the dimension of the square area itself. A 2-dimensional cluster can be described by its fractal dimension, establishing a measure of the connectivity and compactness (local density) of fractures. A largest cluster of fractures with a higher fractal dimension would be expected to have more interconnected fractures. For those fractured rock masses which have a fracture density below the critical value, there is no continuous fracture cluster.

There are a variety of methods of determining the fractal dimension (e.g. Peitgen et al. 1992), but the box-counting dimension (denoted by D_f here) is used in this study, since it is particularly useful for the description of random structures. D_f is the slope of a best fit straight line obtained by plotting Log N(s) against Log 1/s, where N(s) is the number of boxes of size (s) which contain at least one fracture belong to the largest cluster (Fig. 3).

By using this method the fractal dimension of the largest cluster can be determined, either at the critical density (Fig. 1c) or at higher fracture densities (Fig. 1d). The relationship between N(s) and (s) of two fracture clusters is expressed by the best fit lines in the log-log plot (Fig.3) with a correlation coefficient, $\gamma = 0.998$ and $\gamma = 0.999$. The critical cluster has a fractal dimension, $D_{fc} = 1.28$, whereas the cluster with a higher density has a significantly higher fractal dimension, $D_f = 1.55$. Where the relative density of fractures is above the critical density, (d_c), the largest cluster of fractures becomes more space-filling, i.e. the cluster consists of more fractures and covers more area.

In the previous example, 2 orthogonal fracture sets were used. An alternative group of models based on random fracture orientation was also developed. Fig.4 shows the relationship between fracture density, (d), and fractal dimension, (D_f), for the two types of models. For each group, two approaches are used to increase fracture density, 1) increasing fracture number within a fixed range of fracture size and 2) increasing the lower limit of fracture size with fixed fracture number. D_f was found to increase with the increase in fracture density, independent of the fracture pattern (two orthogonal sets or random) and fracture size range. Hence, similar fractal dimension (i.e. connectivity) of fractured rock masses is predicted for a given fracture density from a variety of common fracture patterns.

The relationship between the fractal dimension of critical clusters (D_{fc}) and different fracture size ranges (S_l) for a given exponent of the power-law distribution ($E = 1.5$) is shown in Fig.5. The critical fractal dimension (D_{fc}) for different power-law exponents ($E = 1.2$ to 1.8) with a fixed size range from 0.04 to 0.25 is shown in Fig.6. For the same size range and size distribution, the critical fractal dimension against different standard deviation of fracture orientation (V_s) from 0° to 50° is shown in Fig.7. These results indicate that S_l, E and V_s have no systematic effect on the critical fractal dimension (D_{fc}) of the largest clusters. This means that the fractal dimension of the critical fracture clusters is dominated by fracture density in rock masses and is largely independent of other geometric parameters, although there are some random variations. For those critical clusters with variations in size range from 0.0005 to 0.15, in standard deviation of fracture orientation from 0° to 50° and in exponents from 1.2 to 1.8, the critical fractal dimensions lie in a narrow range from 1.22 to 1.38 (average 1.30).

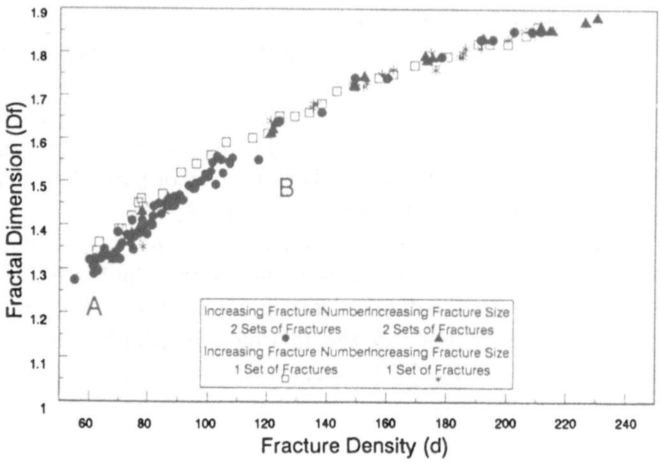

Fig. 4. Fractal dimension (D_f) against fracture density (d). The fractal dimension increases with increasing fracture density, independent of fracture direction set and fracture size range. The values of two fractal dimensions denoted by *A* and *B* correspond to the fitting lines in Fig.3.

Deformation of Fractured Rock Masses

Numerical experiments, using the distinct element method, have been carried out to study the deformation of the simulated fractured rock-masses. The UDEC code (The Universal Distinct Element Code, Version 1.6) was used for the simulations; it was written by Cundall and is marketed by ITASCA Consulting Group, USA. The source code has been made available to the authors by BP research and has been modified so that appropriate simulations can be specified in the study. The rock mass was modelled as a series of elastic blocks of intact rock, separated by fractures along and across which displacement can occur when the model is loaded (Cundall et al. 1978, Last & Harper, 1990). Table 1 lists the material properties used for the blocks and fractures. The amount of normal and tangential displacements between two adjacent blocks was determined directly from block geometry and block centroid translation and rotation. The force-displacement law relates incremental normal and shear forces (∂F_n, ∂F_s) to the amount of incremental relative displacement (∂u_n, ∂u_s):

$$\partial F_n = K_n \partial u_n \tag{1}$$

$$\partial F_s = K_s \partial u_s \tag{2}$$

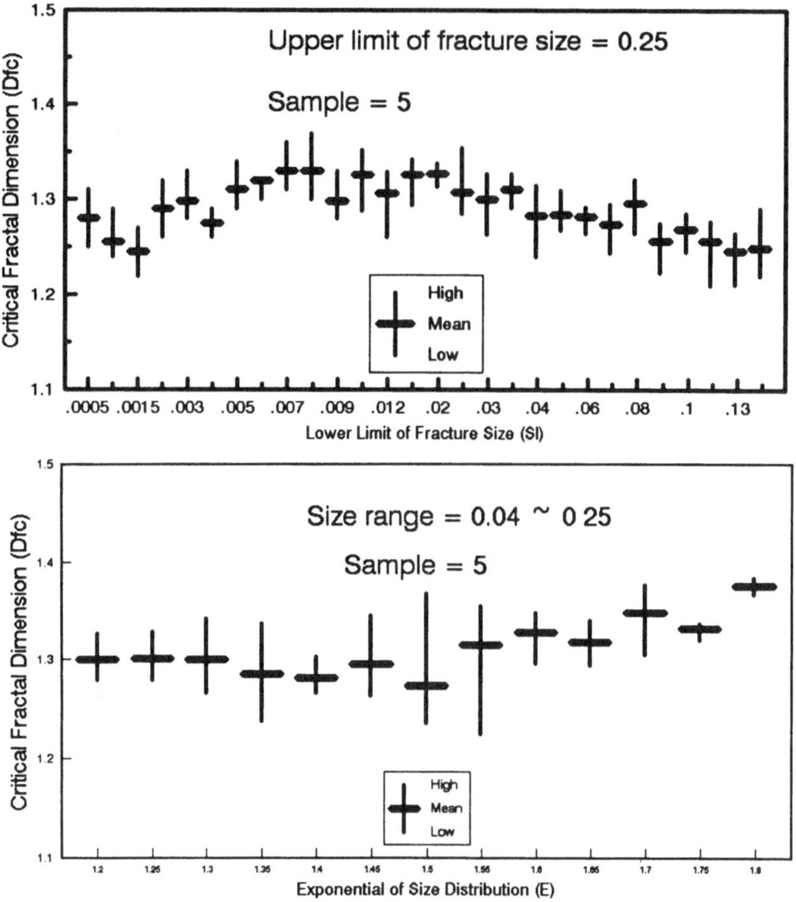

Fig. 5-6. upper: Relationship between critical fractal dimension (D_{fc}) and size range of fractures (S_l) with a given exponent $E = 1.5$; the upper limit of size is fixed at 0.25 and the lower limit (S_l) varies from 0.0005 to 0.15. The horizontal bar indicates the mean value of 5 samples at given parameters, and the vertical line the range. **lower:** Relationship between critical fractal dimension (D_{fc}) and the exponent of size distribution (E) for given size range of 0.04 to 0.15. The horizontal bar indicates the mean value of 5 samples at given parameters, and the vertical line the range.

where Kn and Ks are the contact normal and shear stiffness respectively. Stress increments can be expressed in terms of the fracture stiffness k_n and k_s, as:

$$\partial \sigma_n = k_n \partial u_n \tag{3}$$

$$\partial \sigma_s = k_s \partial u_s \tag{4}$$

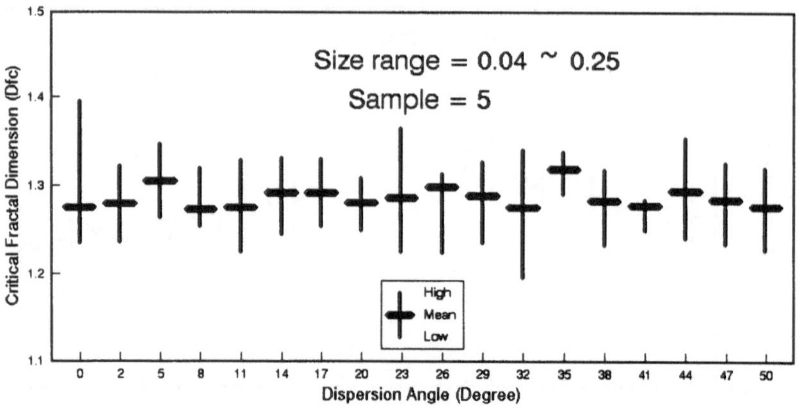

Fig. 7. Relationship between critical fractal dimension (D_{fc}) and dispersion angle of fractures (V_s) for a size distribution (E = 1.5) and size range of 0.04 to 0.15. The horizontal bar indicates the mean value of 5 samples at given parameters, and the vertical line the range.

Such force-displacement relationships allow the evaluation of shear and normal forces between the intact blocks in a deformed region of the model.

Table 1. Material parameters used in this modelling

Block Property	Value	Units
Density	2500	kg/m^3
Shear modulus	15.4	GPa
Bulk modulus	33.3	GPa
Young's modulus	40	Gpa
Poisson's ratio	0.3	
Fracture Property	**Value**	**Units**
Tensile strength	0	MPa
Cohesion	0	MPa
Friction angle	25	Degree
Dilation angle	0	Degree
Joint normal stiffness	33.3	GPa/m
Joint shear stiffness	15.4	GPa/m

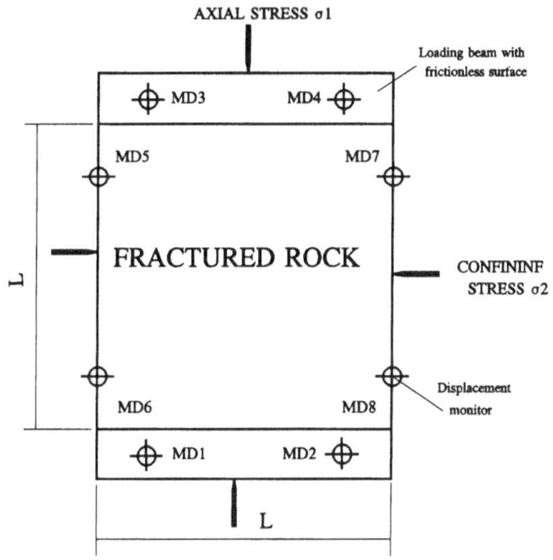

Fig. 8. Schematic illustration for loading and monitoring system of numerical experiments. Two deformation components in X- and Y-direction at eight points are measured.

Simulations of biaxial compressive tests have been run, with fixed confining stress (σ2) of 0 or 0.3 MPa and different deviatoric stresses (σ1 - σ2). This enables the study of the deformability of fractured rocks with different geometric and mechanical parameters. The rock and fracture parameters were selected to represent a tight sandstone (e.g. Barton et al. 1985, Yoshinaka & Yamabe 1986, Last & Harper 1990). Fig.8 shows the schematic diagram of the loading and monitoring system of the numerical experiments. The displacements in two directions are measured at eight points so that average deformation in axial and lateral directions could be obtained.

For rocks with a fracture density below the critical density, i.e. no continuous fracture cluster, the rock mass showed only a small amount of deformation, mainly contributed by elastic deformation of intact blocks. Where fractures just form a completely connected cluster throughout a fractured rock mass (Fig.9a), which corresponds to a connectivity at the percolation threshold with critical fractal dimension of 1.30, larger deformation occurs due to shear displacements and openings along a few fractures. At still higher fracture density, the connected fracture cluster, with a fractal dimension of 1.39, covers more of the simulated area and the resulting rock mass is more fragmented (Fig.9b). Deformation of the fractured rock increases abruptly and is dominantly contributed by shear displacements and openings rather than elastic deformation of the blocks.

Stress-strain behaviour of the three models with different fracture densities is shown in Fig.10. Below the critical density (Model 1), the stress-strain curve shows an approximate elastic behaviour, since the deformation is mainly contributed by the elastic deformation of the intact blocks. At the critical fracture density (Model 2), the stress-strain behaviour presents softening, i.e. the ratio of stress to strain drops, at deviatoric stress of 2.7 MPa and flow at deviatoric stress

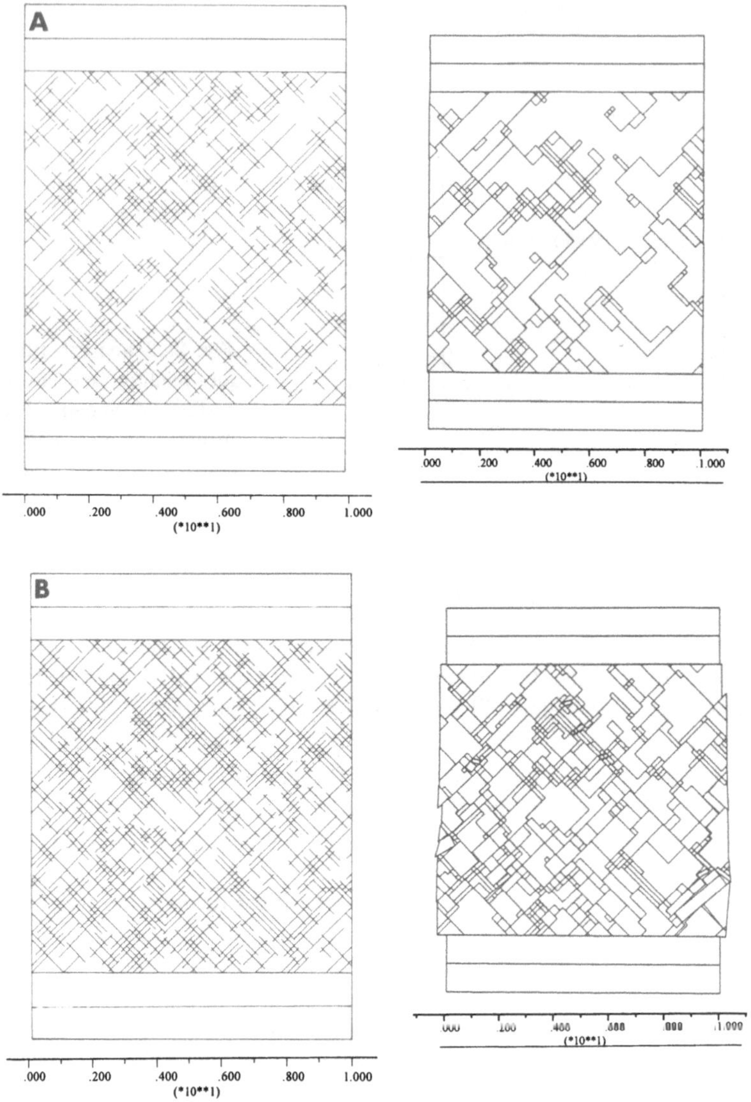

Fig. 9 a, b. Geometry before loading (**left**) and deformation after loading (**right**) of fractured rock masses. **a** Fractured rock mass with the critical fractal dimension of 1.30 and fracture density of 55.5. After loading, some displacements and openings occur along a few fractures. **b** Fractured rock mass with a higher fractal dimension of 1.39 and fracture density of 75.5. After loading, more displacements and openings occur along many fractures.

of 3.6 MPa, with the development of the shear and normal displacements along fractures. At a higher fracture density (Model 3), strain-softening occurs at a lower deviatoric stress level (1.8 MPa) and the flow occurs at a relatively low deviatoric stress of 2.4 MPa since more fractures contributed displacements to the deformation of the fractured rock. In these models, stress concentration occurs at more local areas and develops a higher value, which results in larger deformation, including the elastic deformation of blocks at stress concentration areas as well as the shear displacements and openings along fractures.

At different levels of confining stress (0 and 0.3 MPa), the relationship between deformation and fractal dimension is summarized in Fig.11. Although deformation increases with fractal dimension, a rapid increase occurs when the fractal dimension exceeds a value of 1.35 (the associated fracture density is about 65), independent of confining stress level.

Conclusions and Discussions

A numerical simulation approach is used to predict connectivity and to measure fractal dimension of 2-dimensional fractured rock masses as fracture density increases. Measurement of the fractal dimension of fracture clusters provides a powerful descriptive parameter and predictor of rock-mass properties. The concept of a critical fractal dimension could constitute a new failure criterion, although this would need more testing. The results obtained from the simulations enable the following conclusions to be made:

1. The critical density of fractures at the percolation threshold has been determined for a variety of fracture patterns using a numerical simulation method. It can be used to indicate connectivity status of fractured rock masses.
2. Fractal dimension, D_f, is used to measure the connectivity and compactness of the largest fracture clusters. D_f increases with the increasing fracture density, and is largely independent of the size distribution of fractures and fracture pattern.
3. The fractal dimension, D_{fc}, of the critical fracture clusters has a rather constant value. D_{fc} lies from 1.22 to 1.38 for the variations in the size distribution (E = 1.2 to 1.8), the standard deviation of fracture orientation, ($V_s = 0^o$ to 50°) and the lower limit of fracture size (Sl = 0.0005 to 0.15).
4. Distinct Element modelling of the simulated fracture patterns shows that there are sudden changes in the deformability of the rock mass as the critical fracture cluster forms. For elastic behaviour of the rock matrix, there is an elastic response of the rock-mass below the critical stage, but where a continuous fracture cluster forms, the rock-mass exhibits Mohr-Coulomb behaviour and flows at relatively low deviatoric stress. The deformability of the rock-mass increases where the fractal dimension is above some value of 1.35 (associated fracture density about 65), which corresponds to fragmentation of fractured rock masses.

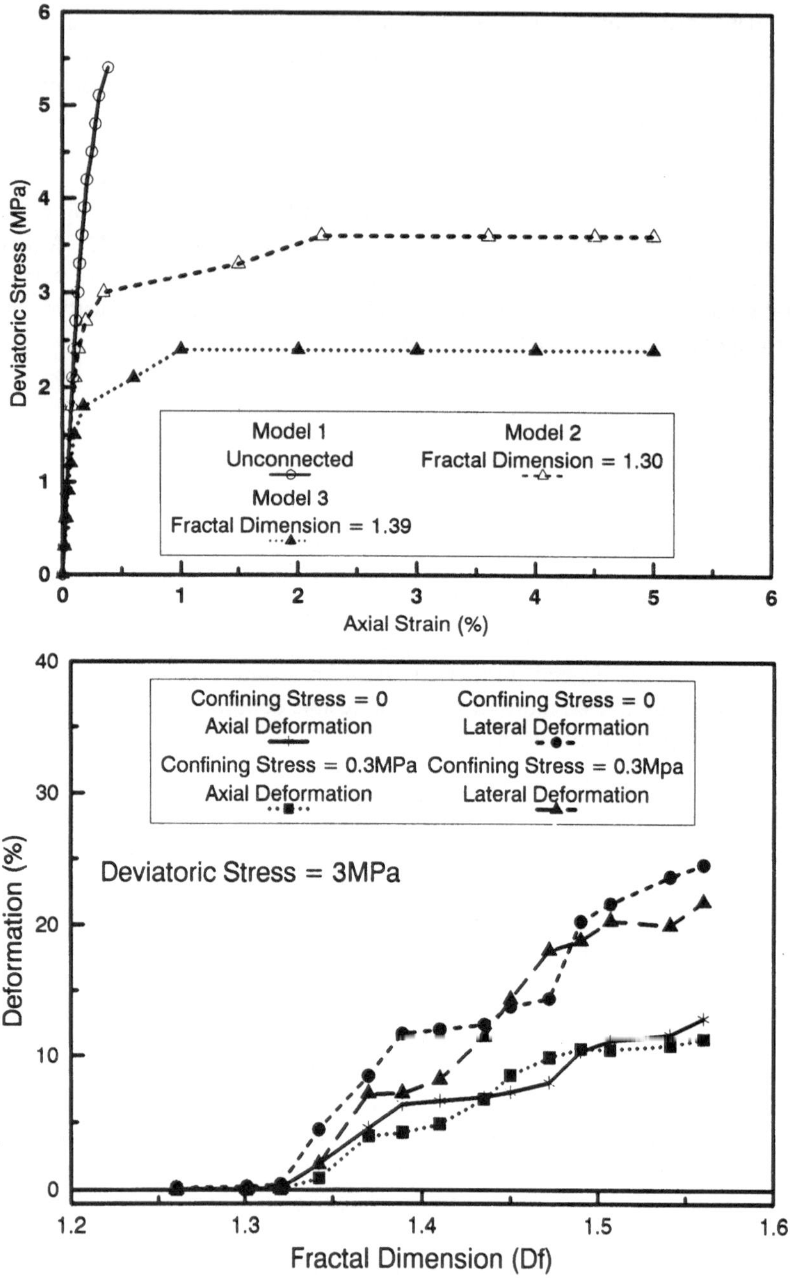

Fig. 10-11. upper: Stress-strain behaviours of 3 simulated fractured rock-masses with different fractal dimensions (i.e. different fracture densities). Model 1 (open circles) has a fracture density below the critical density, the stress-strain presents an approximate elastic behaviour. Model 2 (open triangles) is at the critical fracture density and

corresponds to Fig.9 with a fractal dimension of 1.30 and a fracture density of 55.5. Model 3 (filled triangles) is above the critical fracture density and corresponds to Fig.10 with a fractal dimension of 1.39 and a fracture density of 75.5. lower: Deformation of simulated fractured rock-masses at different confining and deviatoric stress. The deformation increases rapidly above the critical fractal dimension (approximately 1.35)

The models, although simple, have great relevance to a range of rock deformation and permeability studies, including particle size reduction processes (e.g. Sammis et al. 1987, Blenkinsop 1992). However, the lack of realistic simulation of fracture growth is a significant limitation of the model. In addition, a wider range of length scales could help reduce error in the determination D_f. Clearly models could be developed to overcome these problems and to extend the investigation to a three-dimensional model.

Acknowledgements. The authors wish to express their thanks to Richard M. Harkness and Nigel C. Last for their help with UDEC.

References

Barton CC & Hsieh PA (1989) Physical and hydraulic-flow properties of fractures. 28th Int. Geological Congress, Field Trip Guidebook T-385: 1-31.

Barton NR, Bandis S & Bakhtar K (1985) Strength, deformation and conductivity coupling of rock joints. Int J Rock Mech Min Sci Geomech Abstr 22: 121-140.

Blenkinsop TG (1992) Cataclasis and processes of particle size reduction. PAGEOPH 136: 59-86.

Cundall PA, Marti J. Beresford P, Last NC & Asgian M (1978) Computer modelling of jointed rock masses. US Army Eng WEST Tech Rep, N-74-8.

Englman R, Gur Y & Jaeger Z (1983) Fluid flow through a crack network in rocks. J App Mech 50: 707-11.

Gueguen Y, Reuschle Y & Darot M (1990) Single-crack behaviour and crack statistic. In:Barber PG et al. (ed) Deformation Processes in Minerals, Ceramics and Rock. Unwin Hyman, London, pp 51-71.

Heffer KJ & Bevan TG (1990) Scaling relationship in natural fractures: data, theory and application. Soc Pet Eng SPE 20981: 367-376.

Jackson P & Sanderson DJ (1992) Scaling of fault displacements from the Badajoz-Córdoba shear zone, SW Spain. Tectonophysics 210: 179-190.

Last NC & Harper TR (1990) Response of fractured rock subject to fluid injection. Part I: Development of a numerical model. Tectonophysics 172: 1-31.

Peitgen HO, Jürgens H & Saupe D (1992) Fractals for Classroom, Springer, New York.

Rasmussen TC (1987) Computer simulation model of steady fluid flow and solute transport through three-dimensional networks of variably saturated, discrete fractures. In: Evans, DD & Nicholson TJ (ed) Flow and transport through unsaturated rock. American Geophysical Union Monograph 42:

Rives T, Razak M, Petit JP & Rawnsley KD (1992) Joint spacing: analogue and numerical simulations. J Struct Geol 14: 925-937.

Sammis CG, King G & Biegel R (1987) The kinematics of gouge deformation. PAGEOPH 125: 777-812.

Segall SW & Pollard DD (1983) Joint formation in granitic rock of the Sierta Nevada. GSA Bull 94: 563-575.

Shimo M & Long J (1987) A numerical study of transport parameters in fracture networks. In: Evans DD & Nicholson TJ (ed) Flow and transport through unsaturated rock. Am Geophys Union Monogr 42.

Stauffer D (1985) Introduction to percolation theory. Taylor & Francis, London.

Yoshinaka R & Yamabe T (1986) Joint stiffness and the deformation behaviour of discontinuous rock. Int J Rock Mech Min Sci Geomech Abstr 23: 19-28.

Zhang X (1993) Shear resistance of jointed rock masses and stability calculation of rock slopes. Geotech Geol Eng 11: 107-124.

Zhang X, Harkness RM & Last NC (1992) Evaluation of connectivity characteristics of naturally jointed rock masses. Eng Geol 33: 11-30.

Zhang X, Sanderson DJ Harkness RM & Last NC (1993) Connectivity and instability of fractured rock masses. Proc 2nd Int Conf Non-Linear Mechanics, Beijing, pp 929-932.

Multi-Scale Model of Damage Evolution in Stochastic Rocks: Fractal Approach

Vadim V. Silberschmidt[1,2] & Vladimir G. Silberschmidt[2]
[1] Institut für Strukturmechanik DLR
Postfach 3267, D-38022 Braunschweig, Germany
[2] Mining Institute, Ural Dept. of the Russian Academy of Sciences
78A Karl Marx St., 614007 Perm, Russia

Abstract. The spatiotemporal evolution of damage in stochastic rocks is studied in terms of the multi-scale approach with the fractal tree used as an analogue of a discretization scheme for the region under study. Such an approach, based on the fractal theory, allows quantitative information on spatial (morphology of fracture surface) and temporal scaling of the failure process to be obtained. The numerical simulation for different types of stochasticity has shown that the fractal dimension of the fractured zone is the invariant characteristic of the failure process and depends totally on the stochastic properties of material. The analysis of the load vs time-to-fracture relation for a vast interval of loads proved the scale-invariance of the fracture process. The multifractal properties of the load distribution near fractures are studied.

Introduction

The development of failure in rocks is the result of the spatiotemporal evolution of defects at various scale levels under the action of gravity forces and load redistribution linked with natural and/or artificial cavities. The full description of all types of defects within the uniform approach is a sufficiently complicated (if not 'overcomplicated') task. Still there is a necessity for description of the macroscopic failure process which is greatly influenced by the generation, accumulation and coalescence of microdefects. One of the traditional ways to solve such a problem is to describe the rock with microdefects in terms of its effective properties, thus using a modified continuum instead of the system 'rock + defects'. It allows the common models (linear and non-linear elasticity, plasticity, viscoplastictity, etc.) to be exploited for simulation, but the usual 'price' is the necessity to introduce (and, consequently, to obtain from the experiments) the complicated dependence of the effective rock parameters on the set of factors: level of load, rate of loading, temperature, etc., and/or to formulate various criteria (usually as yield surfaces) for a transition to failure.

The alternative is to use continuum damage mechanics (CDM) (Kachanov 1986, Lemaitre & Chaboche 1990, Krajcinovic & Fonseka 1981, Chaboche 1988) which exploits an additional external variable - damage level - and a kinetic equation for it, thus, naturally introducing the time axis in the model (note, that traditional mechanical models of slow fracture are formulated only in terms of the stress-strain state). However, CDM also uses the approximation of the deterministic continuum which should be reformulated for analysis of brittle rocks which exhibit the random distribution of properties linked with several factors (Silberschmidt V.G. et al. 1992): relatively large grain size (if compared to metals and alloys); marked heterogeneity; random distribution of components and/or defects, etc. These factors result in a complicated morphology of the fracture surface and in a high level of fragmentation. The theory of fractals (Mandelbrot 1977, 1982, Takayasu 1989) together with the knowledge of the fractal nature of fracture (Mandelbrot et al. 1984, Louis & Guinea 1987) provides new techniques for analysis of such complicated cases of fracture. This chapter describes multi-scale fractal model of stochastic rock fracture regard to for the damage accumulation and the study of spatial and temporal characteristics of the failure process.

Multi-Scale Model

Fractal Tree and Discretization of Continuum

The description of stochastic rocks presupposes the transition from the continuum (with deterministically changing parameters) to its discrete analogue in order to distribute random properties over the set of cells. This procedure can be performed in different ways (Hansen 1990). We shall use the fractal Cayley tree as an analogue of the region under study (Silberschmidt 1990, 1993a, Silberschmidt & Silberschmidt 1990). It is obvious that the region can be subdivided into elements, which, in their turn, can also be divided into subelements, etc. Let the structure of the fractal tree correspond to this division of a region into more and more small subelements. Then the largest branch (the level $n = 0$) of the Cayley tree corresponds to a whole region under study. Transfer from the k-th level of the tree to the lower $(k+1)$-th level corresponds to the division of each element of the higher level into L parts, L being a tree's coordination number. Thus, each branch of the n-th level represents the structural element of the region making up $1/L^n$ part of the entire volume (considering that at each level of the tree we have elements of equal dimensions).

Such a division is performed until the structural element's dimensions of the level $n = N^*$ correspond to the requirement of the representative volume (unit cell): the smallest element is large enough to contain a great number of microscopic objects but at the same time its dimensions are negligible in comparison to the scale of change of macroscopic parameters (stress field, etc.).

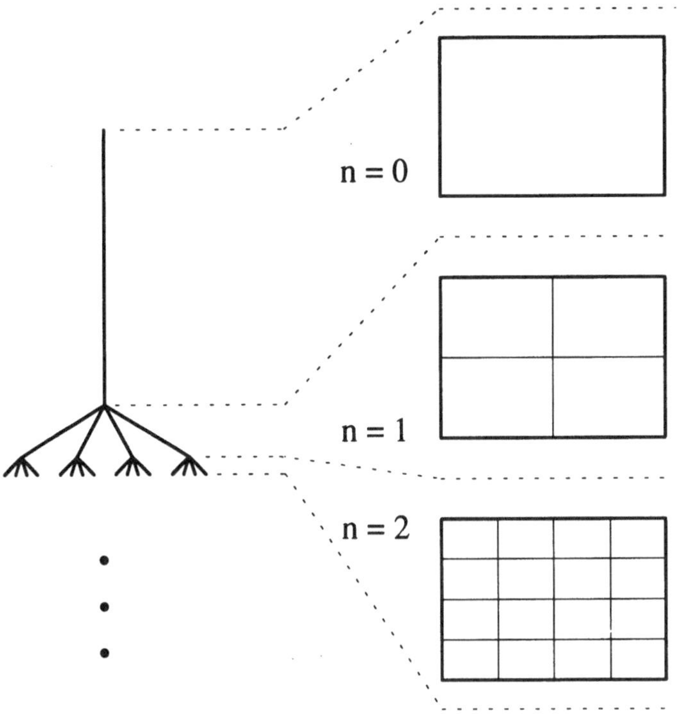

Fig. 1. Fractal tree as analogue of the discretization scheme. First stages of discretization.

Such a construction of the fractal tree allows the multi-scale behaviour to be described. It is especially important for the rock description that the number of scales under consideration is larger than in models for traditional construction materials. Besides, the exact structure of the rock massif can influence the type of Cayley tree used. For instance, the layered massif can be modelled by a tree where branches of the upper levels represent the layers (and, if necessary, sublayers) of rocks with different properties (rheology, strength, etc.). The difference of fracture mechanisms acting at various structural levels and/or in various types of rocks building a massif can be naturally accounted for in terms of various constitutive equations for different scales and conditions for failure transition from one level to another (up the fractal tree). In general, the processes at interfaces can also be described by the respective relations for a certain level of the fractal tree. The natural and/or artificial cavities in rocks can be modelled by the absence of the respective tree's branches, representing the regions occupied by such objects.

For the sake of simplicity let us study the spatiotemporal evolution of damage for the 2D-problem - a rectangular region with dimensions $a \times b$. Then, for a coordination number $L = 4$ with n increasing from 0 to k (transfer to the k-th level of the fractal tree), one has 4^k rectangular elements with dimensions

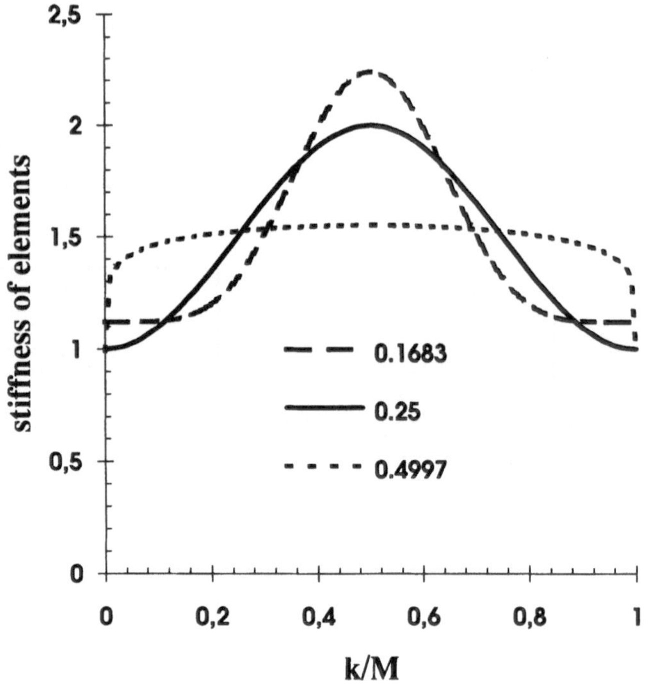

Fig. 2. Distribution functions for elements' stiffness for various types of rock's stochasticity (coefficient of the distribution's half-width $K_{0.5}$). M is the total number of elements at the N^*-th level of the fractal tree, k is the random number of the element.

$a/2^k \times b/2^k$ (Fig. 1); correlating nodes of the tree to the centres of the elements the direct linkage of the tree with the coordinates of the region thus being introduced.

In contrast to the renormalization group approach of Turcotte et al. (1985) where the collapse of the fractal tree itself was studied, the proposed model uses the fractal tree only as analogue of the multi-scale spatial discretization process and implement to the load redistribution process in the region under study.

Damage Parameter and Constitutive Equations

The specificity of the stress state of rocks results in the development of a multi-mode fracture. An adequate description of such a process in terms of damage evolution is possible with the introduction of damage parameters responsible for respective fracture mode (Silberschmidt & Chaboche 1993). These parameters are the second-order tensors. In order not to overcomplicate the analysis here, let us consider the case of uniaxial tension caused by the constant load S perpendicular to the rectangular region under study. This simplification is still physically meaningful: because of the large difference in yield stresses under tension and compression for rocks (Silberschmidt V.G. et al. 1992) even a low level of tensile load could result in the initiation of macroscopic fracture. The dangerous zones with horizontal tension stress often form in the roofs of underground mines. In this case instead of the full second-order form (Naimark & Silberschmidt 1991) the scalar presentation for the damage parameter can be used (Silberschmidt 1990, 1992, Silberschmidt & Silberschmidt 1990).

All the rock's stochasticity we shall link in the model approach with the random distribution of stiffness over the elements. This is a characteristic feature, for instance, of salt rocks. Three various cases (with different values of the coefficients of effective half-width $K_{0.5}$) of stiffness distribution functions, which will be analysed in our chapter, are shown in Fig. 2. Note, that the total stiffness of the cross-section G - the space under curves - is equal for all three cases.

For the case under study the system of constitutive equations can be reduced to the load distribution law for structural elements according to their stiffness and to the kinetic equation of damage accumulation. The first relation in approximation of uniform axial strain is

$$S_i^n = S\frac{G_i^n}{G} \tag{1}$$

the load is being distributed according to the random stiffness of elements. The damage accumulation law has the quasi-linear form (Silberschmidt 1992) and is formulated only for the level N^* , thus the lowest level of the fractal Cayley tree describing the macroscopic result of the microscopic processes:

$$\dot{p}_i^{N^*} = AS_i^{N^*} + Bp_i^{N^*} \tag{2}$$

In eqs. (1),(2) S_i^n, G_i^n are load and stiffness, respectively, of the i-th element of the n-th level ($0 \leq n \leq N^*$); $p_i^{N^*}$ is the level of damage in the i-th element of the level N^*; A, B are the rock parameters which in a general case can also be random magnitudes. The second equation introduces to the model description the time axis, thus providing a possibility for the temporal characteristics, omitted by the traditional mechanical approaches for non-dynamic fracture, to be studied. For the description of spatiotemporal fracture development the constitutive equations must be accompanied by the local fracture criterion. As known from the experimental observations, the local pre-fracture volume concentration of defects is practically independent of the level of acting load (Betekhtin et al. 1989). Thus, one can use the concentration failure criterion which for the damage parameter p is formulated as follows:

$$p_i^{N^*} = p_{cr} \tag{3}$$

The development of the fracture process in the region under study will correspond to the motion up the fractal tree; that is the condition of the transition process to the k-th level can be fracture of all branches (structural elements) of the $(k+1)$-th level coming from the common node of the k-th level.

Results and Discussion

Multiplicative Cascades and Multifractality of Load Distribution

The constitutive relation in form (1) together with the concept of the fractal Cayley tree as an analogue of the region's discretization process results in a procedure of load distribution over elements of various levels which is analogous to a multiplicative cascade used in the theory of multifractals (Feder 1988, Mandelbrot 1989, Mandelbrot & Evertsz 1991, Evertsz & Mandelbrot 1992). This process is implemented down the fractal tree: from the top level $n = 0$ up to the lowest level $n = N^*$. The load carried by the branch of the k-th level is divided into L parts and distributed to the branches of the level $(k+1)$ which enters the common node with the portion of loading distributed to the element being proportional to its stiffness.

It is obvious that the stiffness of the element of the k-th level is equal to the sum of the stiffnesses of the elements of the $(k+1)$-th level building this element. Thus such a distribution of load means the transition from the self-similar tree to a stochastic one: one could consider the thickness of a tree's branches to be linked with their stiffness.

For analysis of the load distribution character resulting from this procedure the method of moments (Evertsz & Mandelbrot 1992) was used. The partition function in this case can be written as (Silberschmidt 1993c)

$$\chi_q(\varepsilon) = \sum_{i=1}^{N(\varepsilon)} S_i^q \,, \qquad q \in R \tag{4}$$

Here ε and $N(\varepsilon)$ are the box size and the number of boxes necessary to cover the set, S_i is the part of the load of the i-th box (its measure). Then, the multifractal spectrum $f(\alpha)$ is obtained by the Legendre transform (Evertsz & Mandelbrot 1992)

$$f[\alpha(q)] = q\,\alpha(q) - \tau(q) \tag{5}$$

where

$$\alpha(q) = \frac{\log S_i}{\log \varepsilon_i} \tag{6}$$

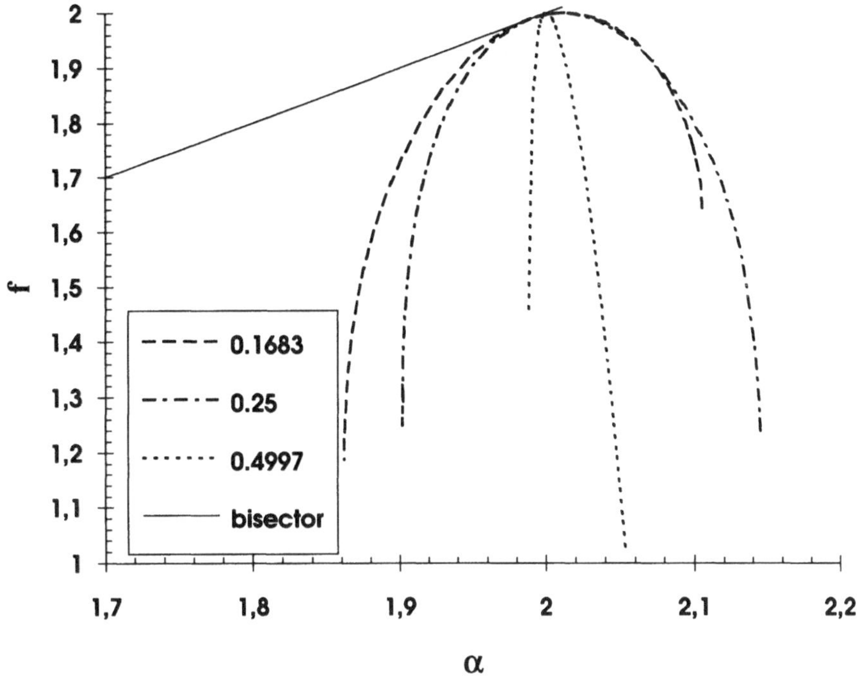

Fig. 3. Multifractal spectrum of the initial distribution of the load for various types of rocks' stochasticity (coefficient of the distribution's half-width $K_{0.5}$).

is a coarse Hölder exponent, $\tau(q)$ is the slope of the dependence $\chi_q(\varepsilon)$ versus ε in double logarithmic coordinates. For calculations based on the method of moments the structure of the fractal tree was used: the boxes coincide with the elements of its respective level. The obtained $f(\alpha)$ dependence (Fig. 3) for the initial load distribution is a multifractal one. $f(\alpha)$ has all the properties characteristic of multifractal measures: $f(\alpha) \leq \alpha$ and has only one common point with the bisector $f = \alpha$ (straight line in Fig. 3). The type of media's stochasticity influences the multifractal properties of the initial load distributio: increase in uniformity (growth of $K_{0.5}$) results in the decrease of the width of the interval $[\alpha_{min}, \alpha_{max}]$. Within the a limit of a uniform rock, the multifractal spectrum tends to the point asymptote $f(d) = d$, d being the dimension of geometric support of the measure.

The non-uniform distribution of the initial load over elements results in a difference in damage accumulation rate. Consequently, the process of overcoming the threshold value of damage [fulfilment of the concentration failure criteria (3)] is by no means simultaneous for all elements. With the stiffness of the locally failed element tending to zero and with the unchanged external load, the part of it which was carried by this element is redistributed to its neighbours. The redistributed parts of the load are also proportional to the elements' stiffness.

Thus failure of each element (breakage of the corresponding branch) leads to a load redistribution on the other elements within the limits of a joint node of the fractal tree. In this way the macrodefect disturbs a stress state in the region, with the radius correlated with the size of the defect itself (Silberschmidt 1990, Silberschmidt & Silberschmidt 1990). Each load redistribution process (linked with the local failure of the element) down the fractal tree starts from the highest level k of the Cayley tree which has at least one failed element (branch) and is carried out for all lower levels. The fracture of the elements at various levels of the fractal tree thus results in initiation of a set of "local" random multiplicative processes down the corresponding branches of the tree.

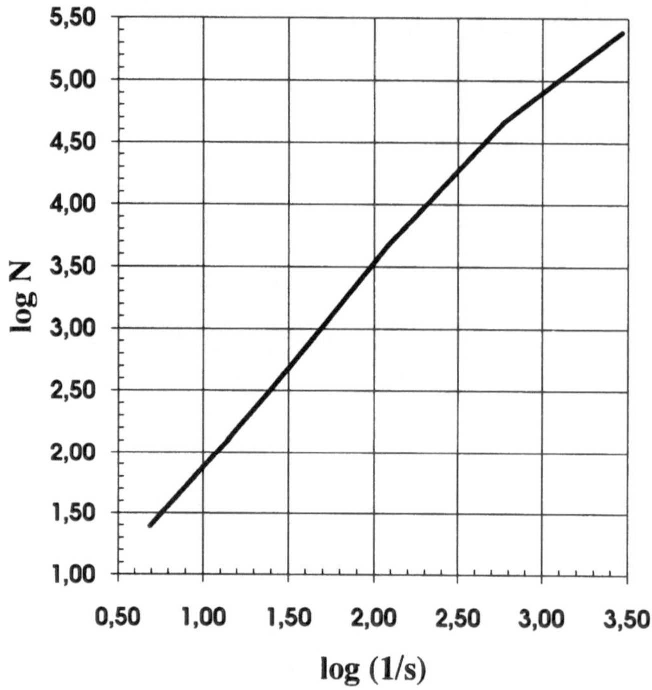

Fig. 4. Determination of the box-counting dimension for the fractal cluster of locally failed elements.

Fractal Dimension of Fractures

Additional loading of the elements caused by the load redistribution from its neighbours results in an activation of the damage accumulation in them [this process is accounted for by Eq.(2)]. As a result a cluster is formed from the locally failed elements, leading to fragmentation or complete fracture (breaking off) of the cross-section. The fractal dimension of the cluster was obtained for various values of the relative half-width of the distribution $K_{0.5}$ using the box-counting method according to which the fractal dimension D is the slope of the curve $N(s) \propto 1/s$ in double logarithmic coordinates [$N(s)$ being the number of elements of the mesh - with mesh size s - occupied by the fractal cluster]. Graph of $N(s) \propto 1/s$ dependence for one of statistical realizations for $K_{0.5} = 0.25$ is shown in Fig. 4. With nearly constant slope, even for those data non-averaged over the set of statistical realization, the fractal character of the fracture morphology is proved. The values of box-counting dimension are

$$
\begin{aligned}
D &= 1.52 \pm 0.04 \quad \text{for} \quad K_{0.5} = 0.1683, \\
D &= 1.60 \pm 0.03 \quad \text{for} \quad K_{0.5} = 0.25 \quad , \\
D &= 1.71 \pm 0.03 \quad \text{for} \quad K_{0.5} = 0.4997.
\end{aligned}
\tag{7}
$$

The increase in the fractal dimension correlates with the growth of the material's homogeneity with a natural limit $D \to 2$ for uniform stiffness distribution. For such a case (and uniform external loading conditions) all the elements of the region break simultaneously.

Load Distribution near Fractures

The final – pre-fracture – state of the system is the result of multiple random multiplicative cascades. This process is linked with the initial stochasticity of material properties distribution and thus the load distribution near the fracture surface – the fractal cluster of locally failed elements – is rather complicated. Still, the theory of fractals provides the possibility to estimate the fractal (multifractal) properties even for a stochastic situation. The obtained multifractal spectra for load distribution (based on the above procedure) near the fractal cluster (Fig. 5) proved its multifractality. The tendency to decrease the interval of α values with the growth of the rock's uniformity remains, but it is not as strict as for the initial state (Fig. 3). It is linked with the additional randomizing mechanism (even for relatively uniform case) – load redistribution under the multiple failure of elements at various levels of multi-scale description.

Temporal Scaling of Fracture

The previous analysis was performed for the same level of the external load S. Still it is obvious that the change of load must sufficiently influence the damage accumulation process and, respectively, the fracture of rocks. Traditionally the temporal scaling is studied in terms of time-to-fracture – external load dependence. In our case the moment of the formation of the fractal cluster corresponds to the time-to-fracture parameter. Thus the numerical simulation was

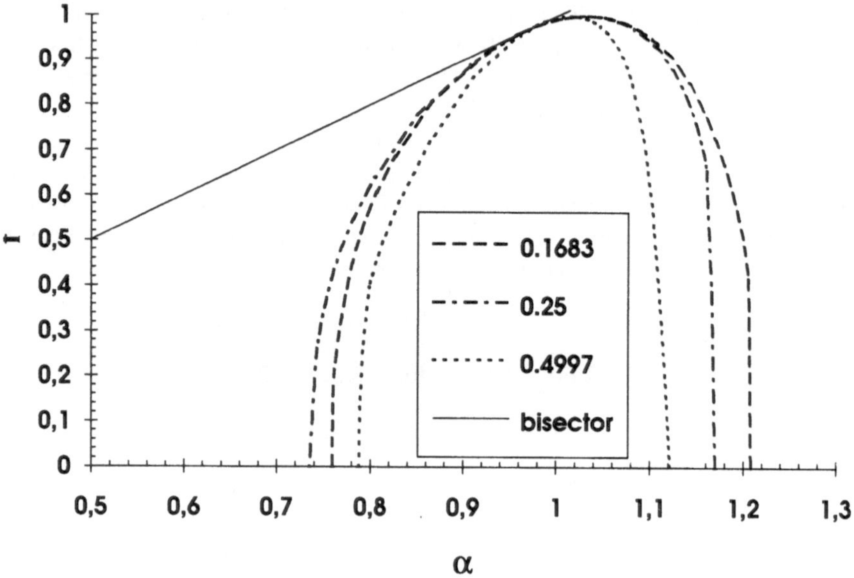

Fig. 5. Multifractal spectrum of the load distribution near fractures (clusters of locally failed elements for various types of rock stochasticity (coefficient of the distribution's half-width $K_{0.5}$).

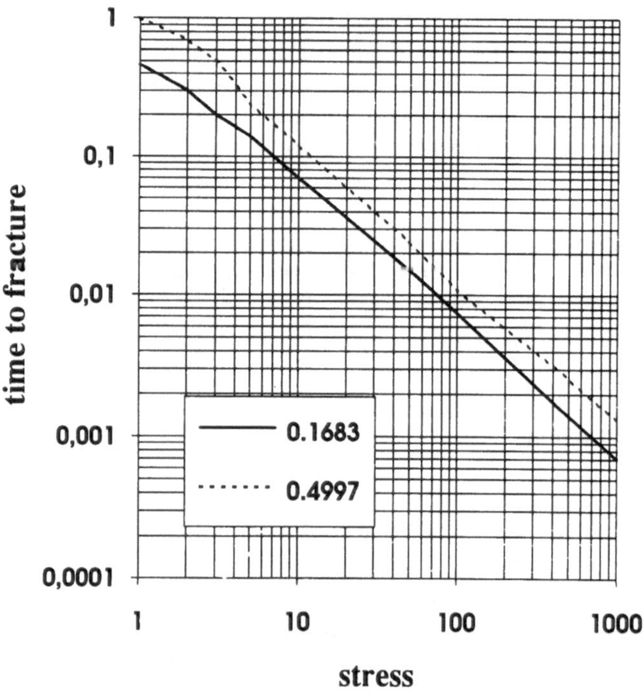

Fig. 6. Temporal scaling of fracture for two types of rock stochasticity (coefficient of the distribution's half-width $K_{0.5}$).

fulfilled for the level of the external load (stress) changing in an interval of three orders. The results are shown in Fig. 6 for two marginal cases of stochasticity. Thus, the time-to-fracture vs. external load relation can be approximated by the power law $t_{fr} \propto \sigma^{-\gamma}$ with the scaling exponent $\gamma = 0.94 \pm 0.02$, which is very close to one. This is in a good accordance with experimental observations. It is necessary to note that the temporal scaling exponent is the same for various cases of rock stochasticity. Even a macroscopic crack, initiated by the V-shape notch, does not influence this value (Silberschmidt 1993b).

Acknowledgement. One of the authors (V.V.S.) is grateful to the Alexander von Humboldt Foundation for the possibility to carry out the research project in Germany.

References

Betekhtin VI, Naimark OB, Silberschmidt VV (1989) On fracture of solids with microcracks: experiments, statistical thermodynamics and constitutive equations. In: Salama K, Chandar KP (eds) Advances in fracture research. Proc 7th Int Conf Fracture, Houston. Pergamon Press, Oxford, Vol.6: 38-45.

Chaboche J-L (1988) Continuum damage mechanics. Trans ASME J Appl Mech 55: 59-72.

Evertsz CJG, Mandelbrot BB (1992) Multifractal measures. In: Peitigen HO, Jürgens H, Saupe D. Chaos and fractals. New frontiers of science. Springer, New York , Berlin, Heidelberg, pp 921-953.

Feder J (1988) Fractals. Plenum Press, New York.

Hansen A (1990) Disorder. In: Herrmann HJ, Roux S (eds) Statistical models for the fracture of disordered media. Elsevier, Amsterdam, pp 115-158.

Kachanov LM (1986) Introduction to continuum damage mechanics. Martinus Nijhof, Boston.

Krajcinovic D, Fonseka GV (1981) The continuous damage theory of brittle materials. Trans ASME J Appl Mech 48: 809-824.

Lemaitre J, Chaboche J-L (1990) Mechanics of solid materials. Cambridge University Press, Cambridge.

Louis E, Guinea F (1987) The fractal nature of fracture. Europhys Lett 3: 871-877.

Mandelbrot BB (1977) Fractals, form, chance and dimension. Freeman, San Francisco.

Mandelbrot BB (1982) The fractal geometry of nature. Freeman, San Francisco.

Mandelbrot BB (1989) Multifractal measures, especially for the geophysicist. PAGEOPH 131: 5-42.

Mandelbrot BB, Evertsz CJG (1991) Multifractality of the harmonic measure on fractal aggregates, and extended self-similarity. Phys A 177: 386-393.

Mandelbrot BB, Passoja DE, Paullay AJ (1984) The fractal character of fracture surfaces of metals. Nature 308: 721-722.

Naimark OB, Silberschmidt VV (1991) On fracture of solids with microcracks. Europ J Mech A/Solids 10: 607-619.

Roux S (1990) Continuum and discrete description of elasticity and other rheological behaviour. In: Herrmann HJ, Roux S (eds) Statistical models for the fracture of disordered media. Elsevier, Amsterdam, pp 87-114.

Silberschmidt VG, Silberschmidt VV, Naimark OB (1992) Fracture of salt rocks. Nauka Publishing House, Moscow (in Russian).

Silberschmidt VV (1990) Fractal models in fracture analysis. Sverdlovsk (in Russian).

Silberschmidt VV (1992) Mathematical modelling and fractal analysis of stochastic fracture process. In: Samarsky AA, Sapagovas MP (eds) Mathematical modelling and applied mathematics. Elsevier, Amsterdam, pp 399-406.

Silberschmidt VV (1993a) Micro - macro transition for stochastic fracture process: fractal approach. In: MECAMAT 93. International seminar on micromechanics of materials. Moret-sur-Loing, 6-8 July, 1993. Editions Eyrolles, Paris, pp 72-82.

Silberschmidt VV (1993b) Scale invariance in stochastic fracture of rocks. In: A Pinto da Cunha (ed) Scale effects in rock masses 93. Balkema, Rotterdam, 1993, pp 49-54.

Silberschmidt VV (1993c) On the multifractal character of load distribution near the cracks. Europhys Lett 23: 593-598.

Silberschmidt VV, Chaboche J-L (1993) Effect of stochasticity on the damage accumulation in solids. Int J Continuum Damage Mech 3: 57-70.

Silberschmidt VV, Silberschmidt VG (1990) Fractal models in rock fracture analysis. Terra Nova 2: 483-487.

Silberschmidt VV, Silberschmidt VG (1993) Fractal method in analysis of fracture development in rocks. In: Pasamehmetoglu AG, Kawamoto T, Whittaker BN, Aydan Ö (eds) Assessment and prevention of failure phenomena in rock engineering. Balkema, Rotterdam Brookfield, pp 733-738.

Silberschmidt VV, Yakubovich, YuM (1993) Effect of stochasticity on scaling properties of crack propagation. Int J Fracture 61: R35-40.

Takayasu H (1989) Fractals in physical sciences. Manchester University Press, Manchester New York.

Turcotte DL, Smalley RF Jr, Solla SA (1985) Collapse of loaded fractal trees. Nature 313: 691-692.

Fractal Characteristics of Joint Development in Stochastic Rocks

Vadim V. Silberschmidt
Institut für Strukturmechanik DLR
Postfach 3267, D-38022 Braunschweig, Germany
Permanent address:
Mining Institute, Ural Department of the Russian Academy of Sciences,
78A Karl Marx Street, 614007 Perm, Russia

Abstract. Generation and propagation of macroscopic joints in brittle rocks depend on the evolution of defects and their interaction with cracks. The proposed approach accounts for damage accumulation and joint-damage interaction in terms of local stress intensity factors (SIF). The numerical simulation of the joint's development in the 2D region is carried out for the various kinds of the rock stochasticity. It is shown that the fractal dimension of the crack front is the invariant of the joints' propagation process: the fractal dimension increases with the uniformity of the material properties distribution.

Introduction

The description of joint generation and propagation in brittle rocks is a complicated problem. In contrast to the traditional schemes of linear fracture mechanics (Liebowitz 1968, Tada et al. 1973, Sih 1973, Knott 1973, Atluri 1986, Atkinson 1987) which are widely used for analysis of the failure of metals and alloys, a set of specific factors of rock fracture demands new approaches for its study. These factors are:

1) marked heterogeneity of real rocks;
2) presence of natural and technogeneric defects of various scales;
3) several scales of rock non-uniformity;
4) specific stress-strain state of rocks around underground cavities, etc.

All these factors result in the complicated scenario of joint development and in specific crack morphology. Traditional methods of fracture mechanics deal, as a rule, with relatively simple geometry of joints: the analysis of branching cracks is usually excluded from consideration. Besides, additional parameters are necessary to account for the interaction of joints with the arrays of defects of smaller size. The last problem has been intensively studied during the past years (Chudnovsky & Kachanov 1983, Rubinstein 1986, Rose 1986, Hutchinson 1987, Chudnovsky & Wu 1992, Gorelik & Chudnovsky 1992). Following Chudnovsky & Wu (1992) the approaches of crack-microcrack interactions can be divided into three main groups:

1) models of microcrack array as an inclusion of an effective elastic media;
2) a detailed description of a microcrack array (by postulation of position and geometry for each microcrack);
3) description of the array in terms of statistical distribution of microcrack densities, sizes and orientations.

Despite the great progress achieved in the analysis of crack-microcrack interaction, there are still open questions, and among them the description of spatiotemporal evolution of joints in real stochastic materials, brittle rocks being one of the vivid examples. The development of the theory of fractals (Mandelbrot 1977, 1982, Takayasu 1989, Aharony & Feder 1990, Peitgen et al. 1992) gave new impetus to such investigations.

This chapter is dedicated to the description of the evolution of joint in brittle rocks (on the basis of 2D-simulation) with an account in the interaction of and the stochasticity of material properties distribution in crack-microdefects.

Fractal Analysis of Propagating Joints

Model and Constitutive Equations

Though the stress state of rocks results in a multi-mode type of fracture (linked with the presence of the compressive components of stress), the great difference in the yield stresses in tension and compression (up to thousands per cent) causes in several cases the dominance of the I-mode fracture (in terms of fracture mechanics). For instance, in salt mines the generation of joints is usually observed in the roof where there is a horizontal tension stress component (Silberschmidt V.G. et al. 1992). The propagation of such a joint to the upper water layers can result in the flooding of the mine (Silberschmidt & Silberschmidt 1992). This presupposes the study of the I-mode joints, which are traditionally initiated in fracture mechanics by the V-shape notch.

Let us study the common case (Fig.1) of an I-mode crack propagation: the rectangular beam with a V-shaped notch loaded at its ends with a constant force S. Stochasticity of spatial distribution of the rock's properties transforms the traditional 1D problem to a 2D one: we shall examine the joint evolution in the cross-section ABCD, which contains the symmetry plane (and an apex) of the V-shaped notch (Fig.1). This cross-section is divided into elements whose dimensions correspond to the requirements of the elementary volume (all macroscopic parameters can be considered to be constant within such an element). All the non-uniformity of the mechanical properties and a stochastic character of the defects could be taken into account by setting the distribution of these parameters along the elementary volumes in a random way. The character of such a distribution should reflect the results of experimental data treatment.

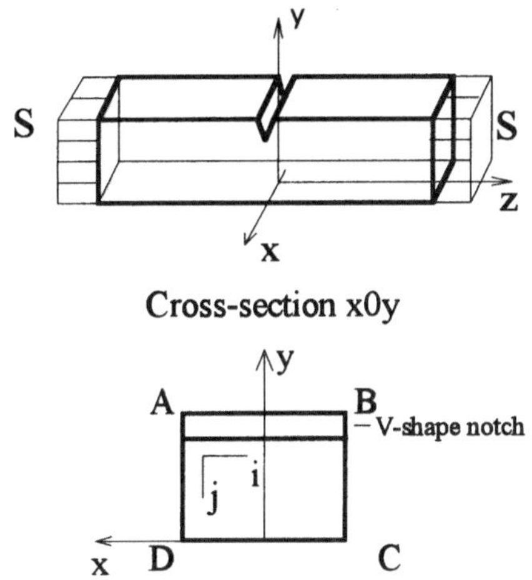

Fig. 1. The scheme of the problem under study.

The study of the fracture development in 2D regions was at the centre of interest of various scientific groups. The detailed analysis of lattice models was made by Hermann & Roux (1990), Duxbury (1990), de Arcangelis (1990) (see also corresponding references in these papers). The traditional lattice models represent the medium as the net of the elements with relatively simple behaviour – elastic springs or rods, thus excluding the account for the structural processes which govern macroscopic fracture development: evolution of ensembles of defects. The fracture in such models is usually generated by the cutting of some links in the net.

The evolution of microstructure can be described in terms of continuum damage mechanics (Kachanov 1986, Lemaitre & Chaboche 1990), the damage parameter – additional internal variable – characterizing the macroscopic result of the evolution of ensembles of defects. In a common case this parameter is the second order tensor (Naimark & Silberschmidt 1991) but for the examined case of uni-axial loading it can be reduced to the scalar parameter (Silberschmidt 1991, 1993a).

For our problem the system of constitutive equations can be reduced to the load redistribution law over structural elements linked with the crack propagation and to the kinetic quasi-linear equation of damage accumulation. The latter can be written in the form (Silberschmidt & Silberschmidt 1990, Silberschmidt 1991, 1992a, b, c)

$$\frac{dp^{ij}}{dt} = A\sigma^{ij} + Bp^{ij} \tag{1}$$

where a pair of indices (i,j) corresponds to the element of the i-th row and j-th column (Fig. 1), rows are perpendicular to the crack front; p^{ij} is the level of damage in element (i,j); $\sigma^{ij} = \sigma^{ij}_{zz}$ is a component of a macroscopic stress tensor in this element, z being the load direction; A and B are parameters of material (in a common case, the random values).

The load redistribution law must reflect the stress field disturbance caused by a joint propagation and its interaction with microdefects. The spatial non-uniformity of this process, linked with the random character of mechanical properties, results in the difference of the crack length along its front. The latter is accounted in this chapter in terms of local SIF. In general form the constitutive equation for macroscopic stress can be written as follows:

$$\sigma^{ij} = K^{ij} \sigma_{in}^{ij} \tag{2}$$

where s_{in}^{ij} is the initial stress value, K^{ij} are the reloading coefficients.

Description of Rock Stochasticity

In a model all the spatial stochasticity of material properties is considered to be linked with random stiffness distribution over elements. This is a characteristic feature of salt rocks, the variance of stiffness being more than 100 per cent (Silberschmidt V.G. et al. 1992). For the model case the stiffness distribution can be introduced as follows (Silberschmidt 1993a, b)

$$G^{ij} = G_q \left[1 + \left(\sin \frac{\pi k}{M} \right)^q \right] K_{ren}(q) \qquad (\mathrm{i,j} = \overline{1,\mathrm{N}}) \tag{3}$$

where G^{ij} is the stiffness of the element (i,j), G_q is a parameter of approximation which is the minimum value of element stiffness in the main case of $q = 2$, k is a random integer, $1<k<M$; $M = N^2$ is a total number of elements for a square lattice of N rows by N columns; the ratio of maximum to minimum stiffness equals 2. Parameter q is linked with the effective coefficient of distribution's half-width $K_{0.5}$ by the relation (Silberschmidt 1991, 1993b):

$$K_{0.5} - \frac{1}{2} - \frac{1}{\pi} \arcsin \frac{1}{\sqrt[q]{2}} \tag{4}$$

and thus characterises the extent of the rock's non-uniformity: $q = 0$ corresponds to the uniform distribution; its increase diminishes the distribution's half-width. Three cases of stiffness distribution which are analyzed in this chapter are shown in Fig. 2. To exclude the influence of macroscopic parameters on the character of the stochastic damage evolution, the total stiffness of the cross-section of the beam was considered to be the same for all three cases of distribution. In order to fulfil this, the renormalizing multiplier K_{ren} is introduced in the form (Silberschmidt 1993b)

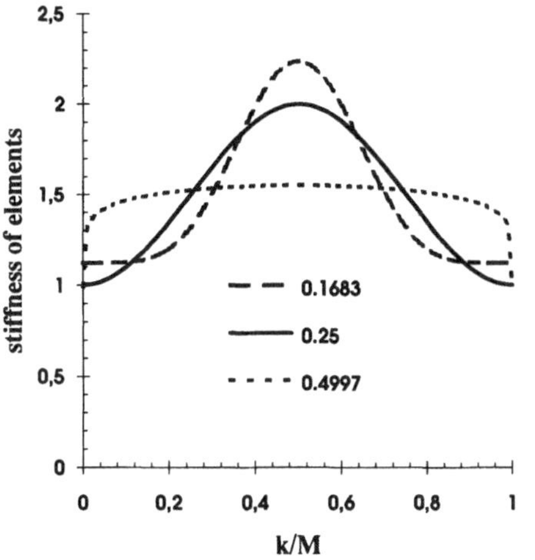

Fig. 2. Three distributions of the elements' stiffness for various cases of

$$K_{ren}(q) = \frac{3G_2}{2G_q}\left[1+\frac{1}{M}\sum_{k=1}^{M}\left(\sin\frac{\pi k}{M}\right)^q\right]^{-1} \tag{5}$$

where G_2 is the value of G_q for the main case of $q = 2$. Thus, the joints' propagation in three macroscopically identical beams (with equal total stiffness) but with different structure is examined.

Stress Redistribution Process

The random distribution of the rock's stiffness results in the non-uniformity of the initial stress distribution over elements. With a part of the external load carried by the element being proportional to its stiffness, the relation for the initial stress σ_{in}^{ij} has the following form for the case of the uniform axial strain

$$s_{in}^{ij} = S\frac{G^{ij}}{Gl_x l_y} \tag{6}$$

Here S is the external force, $G = \sum_{i,j=1}^{N} G^{ij}$ is the total stiffness of the cross-section ABCD; l_x and l_y are the dimensions of the element along x and y axes, respectively.

The reloading coefficients K^{ij} in Eq.(2) should account for two interacting fracture mechanisms: the crack propagation under a constant load and the local failure of the elements in the crack front caused by the damage accumulation. The system of constitutive equations (1) and (2) should be accompanied by the local fracture criterion for the element. It can be formulated in the form analogous to the concentration criterion, widely used in fracture physics: $p^{ij} = p_{cr}$ (Silberschmidt & Silberschmidt 1994 in this book). We shall assume also that for the failed element its stiffness tends to zero.

In a common case the relation for the reloading coefficients can be written as (Silberschmidt 1992b, 1993a)

$$K^{ij} = \frac{\overline{G}^i}{\widetilde{G}^i} K_I^i \sqrt{pl_y} (\sqrt{j+1-n_i} - \sqrt{j-n_i}) \tag{7}$$

Eq.(7) is obtained by integrating the well known relation of fracture mechanics (Liebowitz 1968, Sih 1973, Atluri 1986) $\sigma_{zz} = K_I / \sqrt{2\pi y}$ along the y axis in the rows. In (7) n_i is the number of columns in the i-th row occupied by the crack. The first multiplier in Eq.(7) accounts for the decrease of the total stiffness of the i-th row under the crack propagation and local failure of elements: $\overline{G}^i$ and $\widetilde{G}^i$ are the initial and current stiffness of the i-th row, respectively. The value of the local SIF K_I^i can be approximated by relations from the handbooks of SIF. In numerical simulation, the following presentation was used (obtained by reformulating an approximation by Cherepanov 1979):

$$K_I^i = S\overline{G}^i \frac{1.11 - 5(n_i / N)^4}{G(1 - n_i / N)} \tag{8}$$

The joint-damage interaction is accounted for in terms of load redistribution mechanism: the stress concentration near the initial notch results in the activation of the damage accumulation in adjoining elements. The stochasticity of stiffness distribution and consequent randomness of initial stress values in

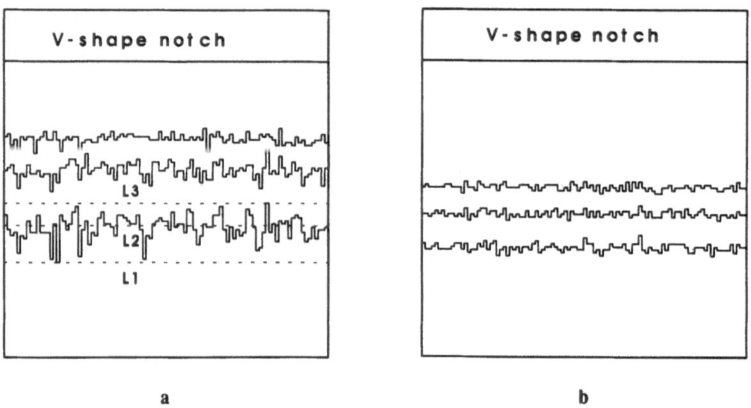

Fig. 3. a Position of joints' fronts at different moments of time for $K_{0.5} = 0.1683$ and **b** $K_{0.5} = 0.4997$.

elements cause non-uniformity in damage accumulation and failure of elements in accordance with the local fracture criterion. The coalescence of the failed elements in a crack front results in its shift. Such a shift leads to the new stress redistribution process.

Numerical Simulation

The numerical simulation was fulfilled for the case of $N = 100$ (*i.e.* the square lattice of 10000 elements) and the V-shaped notch occupying 15 per cent of specimen's width. The character of joints' propagation along the cross-section ABCD is shown for two types of stochasticity (different values of $K_{0.5}$) in Fig. 3. It is clearly seen that the process is sufficiently non-uniform and varies for transition from one type of rock stochasticity to another. In order to describe more precisely the stochastic joint three crack fronts are introduced. The first (shown as L1 in Fig.3a) is the column of elements, containing at least one locally failed element, which is situated at the maximum distance from the notch apex. The second (L2 in Fig.3a) and the third (L3 in Fig.3a) are the columns, which contain 50% and 100% of failed elements, respectively, again, the farthest from the notch apex).

Note that the increase in the rock's uniformity results in sufficient decrease of difference in the position of three fronts. It could be naturally explained: in the limit $K_{0.5} \to 0.5$ (the case of uniform material) this difference tends to zero, all three fronts coincide and propagate as the straight line parallel to the initial notch line. This situation is a traditional one for fracture mechanics analysis.

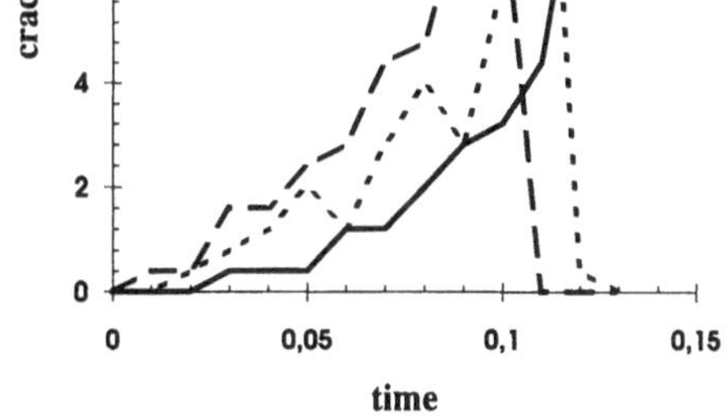

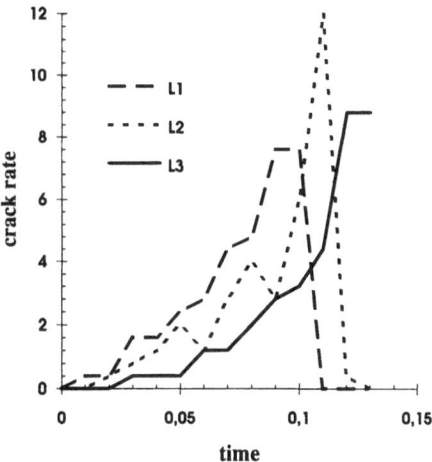

Fig. 4. Change of rates of three fronts with time ($K_{0.5} = 0.1683$).

Figure 4, presenting the change of fronts' rate with time, approves the high level of non-uniformity of joint propagation. It also reflects the step-like character of crack evolution. In cases of more heterogeneous distributions of rock properties (small values of $K_{0.5}$) this character of crack propagation disappears. It is linked with the greater influence of local fluctuations, the random character of which makes the macroscopic process temporally more "smooth" (Silberschmidt 1992a, 1992c).

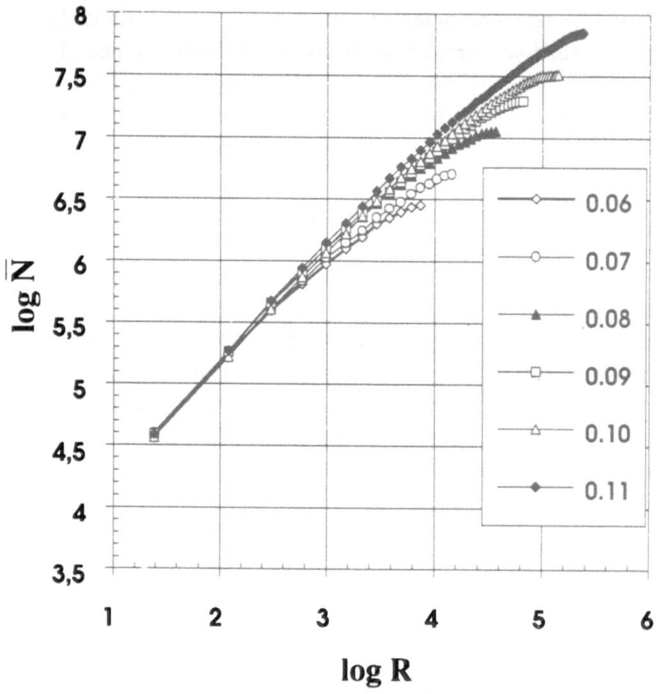

Fig. 5. Graphs of log $\overline{\mathrm{N}}$ *vs* log R for various moments of time ($K_{0.5} = 0.25$).

Fractal Characteristics of Joints

The fractal dimension of the joint's front was calculated as the slope of the curve $\overline{N} \propto R$ in double logarithmic coordinates, R being the distance from the L3 front and $\overline{\mathrm{N}}$ the number of failed elements in joint's tip within R. The results of such calculations are shown in Fig.5 for the case of $q = 2$ ($K_{0.5} = 0.25$). Calculations carried out for various statistical realizations give the value of the fractal dimension for a propagating joint

$$D = 0.59 \pm 0.02 \quad \text{for} \quad K_{0.5} = 0.1683,$$

$$D = 0.68 \pm 0.03 \quad \text{for} \quad K_{0.5} = 0.25\,, \qquad (9)$$

$$D = 0.78 \pm 0.04 \quad \text{for} \quad K_{0.5} = 0.4997.$$

It is obvious that the fractal dimension changes insufficiently with the joint's propagation (its deviation from the straight line in double logarithmic coordinates for the large R values is linked with the finiteness of the region under study) and depends upon the type of rock stochasticity. The increase of the $K_{0.5}$ results in the growth of the fractal dimension with a natural limit $D \xrightarrow[K_{0.5} \to 0.5]{}$

Another characteristic of the fractal joint's front - its radius of gyration R_g:

$$R_g = \frac{l_y}{\overline{N}} \left(\sum_{i=1}^{N} \sum_{j=1}^{n_i - L3} j^2 \right)^{1/2} \qquad (10)$$

where the summing is carried out only for failed elements. It changes non-monotonously (Fig.6) increasing at the first stages of joint propagation and diminishing before the total rupture. The latter is linked with the reaching of the opposite side of the specimen by the joint. The difference in the type of stochasticity results in the change of value of the gyration radius: the level of R_g for $K_{0.5} = 0.1683$ is approximately 4-5 times higher than for the $K_{0.5} = 0.4997$. The figure also reflects the step-like character of the joint's propagation in a case of more uniform distribution of rock properties: the gyration radius in such a case has a set of maxima and minima.

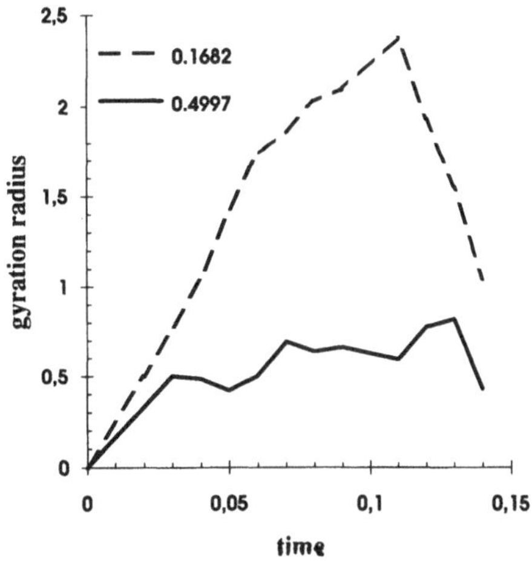

Fig. 6. $K_{0.5}$ change of gyration radius with time for two types of stochasticity (values of $K_{0.5}$).

Discussion

The numerical analysis of the joint's propagation (initiated by the initial notch) in the 2D region has shown that the fractal dimension of the joint's front is an invariant characteristic parameter of the process. The latter depends on the type of rock stochasticity: an increase in uniformity of spatial distribution of rock properties is linked with the growth of the fractal dimension. Thus, the lattice models for the analysis of fracture should include not only the probability of the presence/absence of links (which is a tradition for percolation approaches) but the law for the spatial distribution either of the material properties or of the zero links. Traditional lattice models exclude, as a rule, the account for the stress intensity factors (in other words, the presence of the initial macroscopic crack/discontinuity). Our analysis has shown that the stress concentration near the crack tip is a sufficient mechanism of stress redistribution.

Introduction of the kinetic equation for damage accumulation makes it possible not only to account for the macroscopic results of the structural processes but also to analyze the temporal effects (change of rate of the joint's propagation, etc.).

The difference of the fractal parameters (fractal dimension and gyration radius) for the samples with the same macroscopic mechanical parameter (integral stiffness of the beams) emphasizes once more that there is no direct correlation of them with traditional mechanical fracture characteristics. This fact was proved by Dauskardt et al. (1990) and Baran et al. (1992) by the treatment of experimental data. Though there are some interesting works on the correlation between fracture toughness and fractal dimension (Mandelbrot et al. 1984, Mecholsky et al. 1989, Main et al. 1990, Davidson 1989) or between fracture toughness and an excess of fractal dimension (Issa et al. 1993) the author agrees with the conclusions of de Archangelis (1990) based on the theoretical analysis that in a general case toughness and fractal dimension are uncorrelated. It is understandable: the fracture toughness is the integral characteristic [as well as other parameters of energy (Ray 1992, Nagahama & Yoshii 1993)] and thus reflects the overall response of material. The fractal dimension, in contrast, is the result of the local events of the fracture process which occur randomly in space and in time.

Acknowledgement. The author is grateful to the Alexander von Humboldt Foundation for the possibility to carry out the research project in Germany.

References

Aharony A, Feder J (eds) (1990) Fractals in Physics. Elsevier, Amsterdam.

de Arcangelis L (1990) Randomness in breaking thresholds. In: Hermann HJ, Roux S (eds) Statistical models for the fracture of disordered media. Elsevier, Amsterdam, pp 159-188.

Atkinson BK (1987) Fracture mechanics of rock. Academic Press, London.

Atluri SN (ed.) (1986) Computational methods in the mechanics of fracture. Elsevier, Amsterdam.

Baran GR, Roques-Carmes C, Wehbi D, Degrange M (1992) Fractal characteristics of fracture surfaces. J Am Ceram Soc 75: 2687-2691.

Cherepanov GP (1979) Mechanics of brittle fracture. McGraw - Hill, New York .

Chudnovsky A, Kachanov M (1983) Interaction of a crack with a field of microcracks. Lett Appl Engng Sci 21: 1009-1018.

Chudnovsky A, Wu S (1992) Evaluation of energy release rate in crack - microcrack interaction problem. Int J Solids Structures 29: 1699-1709.

Dauskardt RH, Haubensak F, Ritchie RO (1990) On the interpretation of the fractal character of fracture surfaces. Acta Metall Mater 38: 143-159.

Davidson DL (1989) Fracture surface roughness as a gauge of fracture toughness: alluminium -particulate SiC composites. J Mater Sci Lett 24: 681-687.

Duxbury PM (1990) Breakdown of diluted and hierarchical systems. In: Hermann HJ, Roux S (eds) Statistical models for the fracture of disordered media. Elsevier, Amsterdam, pp 159-188.

Gorelik M, Cudnovsky A (1992) On strong crack - microcrack interaction problem. Int J Fracture 56: R9-R14.

Herrmann HJ, Roux S (1990) Modelization of fracture in disordered systems. In: Hermann HJ, Roux S (eds) Statistical models for the fracture of disordered media. Elsevier, Amsterdam, pp 159-188.

Hutchinson JW (1987) Crack tip shielding by micro-cracking in brittle solids. Acta Metall 35: 1605-1619.

Issa MA, Hammad AM, Chudnovsky A (1993) Correlation between crack tortuosity and fracture toughness in cementitious material. Int J Fracture 60: 97-105.

Kachanov LM (1986) Introduction to continuum damage mechanics. Martinus Nijhof, Boston.

Knott JF (1973) Fundamentals in fracture mechanics. Butterworth, London.

Lemaitre J, Chaboche J-L (1990) Mechanics of solid materials. Cambridge University Press, Cambridge.

Liebowitz H (ed) (1968) Fracture. Vol.I-VII. Academic Press, New York.

Main IG, Peacock S, Meredith PG (1990) Scattering attenuation and the fractal dimension of fracture systems. PAGEOPH 133: 283-304.

Mandelbrot BB (1977) Fractals: form, chance and dimension. Freeman, San Francisco.

Mandelbrot BB (1982) The fractal geometry of nature. Freeman, San Francisco.

Mandelbrot BB, Passoja DE, Paullay AJ (1984) The fractal character of fracture surfaces in metals. Nature 308: 721-722.

Mecholsky JJ, Passoya DE, Feinberg-Ringel KS (1989) Quantative analysis of brittle fracture surfaces using fractal geometry. J Am Ceram Soc 72: 60-65.

Nagahama H, Yoshii K (1993) Fractal dimension and fracture of brittle rocks. Int J Rock Mech Min Sci Geomech Abstr 30: 173-175.

Naimark OB, Silberschmidt VV (1991) On fracture of solids with microcracks. Europ J Mech A / Solids 10: 607-619.

Peitgen H-O, Jürgens H, Saupe D (1992) Chaos and fractals. New frontiers of science. Springer, Berlin Heidelberg New York.

Ray KK (1992) Study of correlation between fractal dimension and impact energy in a high strength low alloy steels. Acta Metall Mater 40: 463-469.

Rose LRF (1986) Microcrack interaction with a main crack. Int J Fracture 31: 233-242.

Rubinstein AA (1986) Macrocrack-microdefect interaction. J Appl Mech 53: 505-510.

Sih GC (ed) (1973) Methods of analysis and solutions of crack problems. Noordhoff International Publishing, Leyden.

Silberschmidt VG, Silberschmidt VV (1992) Experimental study of salt rock - water interaction. In: Kharaka YK, Maest AS (eds) Water - rock interaction. Balkema, Rotterdam, pp 673-675.

Silberschmidt VG, Silberschmidt VV, Naimark OB (1992) Fracture of salt rocks. Nauka Publishing House, Moscow (in Russian).

Silberschmidt VV (1991) Fractal analysis of propagating cracks. Sverdlovsk (in Russian).

Silberschmidt VV (1992a) Mathematical modelling and fractal analysis of stochastic fracture process. In: Samarsky AA, Sapagovas MP (eds) Mathematical modelling and applied mathematics. Proc IMACS Conf., Elsevier, Amsterdam, pp 399 -406.

Silberschmidt VV (1992b) The fractal characterization of propagating crack. Int J Fracture 56: R33-R38.

Silberschmidt VV (1992c) Damage-crack interaction: constitutive model and fractal approach. In: Owen DRJ, Onate E, Hinton E (eds) Computational plasticity. Fundamentals and Application. Pineridge Press, Swansea, pp 1633-1643.

Silberschmidt VV (1993a) Micro - macro transition for stochastic fracture process: fractal approach. In: MECAMAT 93. International seminar on micromechanics of materials. Moret-sur -Loing, 6-8 July, 1993. Editions Eyrolles, Paris, pp 72-82.

Silberschmidt VV (1993b) Scale invariance in stochastic fracture of rocks. In: Pinto da Cunha A (ed) Scale Effects in Rock Masses 93. Balkema, Rotterdam, pp.49-54.

Silberschmidt VV, Silberschmidt VG (1990a) Fractal approaches in mechanics of jointed rocks. In: Rossmanith HP (ed) Mechanics of jointed and faulted rocks. Balkema, Rotterdam, pp 83-86.

Silberschmidt VV, Silberschmidt VG (1990b) Fractal models in rock fracture analysis, Terra Nova 2: 483-487.

Silberschmidt VV, Silberschmidt VG (1994) Multi-scale model of damage evolution in stochastic rocks: fractal approach. In: Kruhl JH (ed) Fractals and Dynamic Systems in Geoscience. Springer, Berlin Heidelberg New York, pp 53-64 (this volume).

Tada H, Paris P, Irwin G (1973) The stress analysis of cracks handbook. Del Research Corporation, Hellertown, Pennsylvania.

Takayasu H (1989) Fractals in the physical sciences. Manchester University Press, Manchester New York.

Correlation of Fractal Surface Description Parameters with Fracture Toughness

Zuzana Krišťáková[1] & Miriam Kupková[2]
[1] Institute of Geotechnics of SAS,
Watsonova 45, 04001 Košice, Slovak Republic
[2] Institute of Materials Research of SAS,
Watsonova 47, 04001 Košice, Slovak Republic

Abstract. The present chapter investigates the applicability of fractal geometry in fractography. Samples at different times during excavation have been studied to examine fractal characteristics of fracture surfaces. The fractal dimensions of the fracture surfaces were determined by using a vertical section method. Fracture toughness has been obtained by using the three point SENB method. Results show the fracture surfaces to be fractal, but no correlation between the measured quantities has been noticed.

Introduction

Fractal geometry is being used in many fields of material science, physics, chemistry, and engineering because it can describe shapes and processes which are nonlinear and seemingly complex (Mandelbrot 1983 and Mandelbrot et al. 1984). These authors have also shown that fractured surfaces are fractal in nature and that the fractal dimension of surfaces correlates with the toughness of the material. A basic characteristic of a surface or the profile of a surface is its roughness. It is not easy to examine the relationship between fracture surface roughness and fracture toughness because of the difficulty in measuring surface roughness. Quantitative fractography methods have been developed for measuring roughness, but they are seldom applied because of their complexity. The techniques used to measure this variable must have a resolution capable of detecting features of interest at the scale of the material microstructure. It is often necessary to use an empirical approach to determine the resolution needed to acquire data that can be correlated to other material properties. In certain cases, the measured values for the size and area fractions of microstructural features vary with the magnification used to examine the material. These difficulties provide reasons for the use of fractal concepts in describing roughness. Fractal geometry has been recently suggested in the work of Mandelbrot et al. (1984) as a suitable model for fracture surface roughness, but this hypothesis is still being examined.

Experimental studies on the relationship between fractal dimension D of fractured surfaces and fracture toughness K_{IC} are important for the failure analysis of materials and machine parts and several studies have related fractal dimension to other physical or mechanical properties. The aim of this chapter is to report on measurements of fracture surface roughness and relate these data to fracture toughness values measured with the same materials.

Fractal Geometry

Fractal geometry, in comparison to Euclidean geometry, which utilizes a more simplified approach, provides a more appropriate theoretical model for the description of the complicated geometry of some natural objects. It is generally accepted that the fractal dimension is a measure of the roughness of the surface and may be a useful parameter to characterize fracture.

Any characterization of surface roughness using the concept of fractal dimension depends on the accurate determination of the fractal dimension of real data. Mandelbrot (1967) used a step-divider technique to calculate the dimension of a series of natural coastlines. Step dividing can be visualized as measuring the length of an irregular curve by walking a compass along the curve at successively smaller compass opening or step lengths. The length L(r) of an irregular curve measured in this way is a function of the divider length r. If the data follow a fractal model then the fractal dimension D is obtained through the following equation

$$L(r) = F\, r^{(1 - D)} \tag{1}$$

where F is a constant. The two constants F and D fully characterize a fractal model for a surface; fractal dimension D describes how the roughness changes with the scale of observation, while F determines the steepness of the topography, or the total profile variance (Brown 1989, Hough 1989, Power & Tullis 1991).

According to Huang et al. (1992) the fractal dimension D is a measure of the extent that a fractal set fills a Euclidean plane, and it can always be determined using the step divider method if the fractal curve is self-similar or statistically self-similar. A topographic relief of a fracture surface appears to behave as random fractal set (Mandelbrot 1983). Vertical cuts (profiles) through such surfaces are statistically self-affine fractals with a local fractal dimension lying between 1 and 2. However horizontal cuts of these surfaces give rise to coastlines (isorithmic lines) which are indeed statistically self-similar and have a unique fractal dimension D ($1 < D < 2$). Therefore a unique value for the fractal dimension of the surface D_S can be derived as $D_S = D + 1$ (Sakellariou at al. 1991).

On the basis of recent publications and studies (Mandelbrot 1983, Pande et al 1987a), the methods employed to define fractal dimension of fracture surfaces can be roughly categorized as follows:

1. the slit island method (Mandelbrot et al.1984, Pande et al. 1987b);
2. vertical section method (Pande et al. 1987b, Mencholsky et al. 1989);
3. the secondary electron line scanning method (Pande et al. 1987a).

Mandelbrot et al. (1984) studied fracture surfaces of 300 Grade Maraging steel at different heat treatment temperatures and found that fractal dimension correlates inversely with the impact fracture energy (toughness) as measured by the standard Charpy impact test. Similarly, Pande et al. (1987b) studied the fracture behavior of Ti alloys (Ti-6A1-2V) with varying zirconium content. They obtained an approximate inverse correlation between fractal dimension and dynamic tear energy. However, in a study of the fracture of brittle ceramic materials, Mencholsky et al. (1989) found a direct correlation between fractal dimension and K_{IC} for single and multiphase materials. In addition, when fracture surface roughness of certain steels was plotted versus the scale of measurement, Dauskart et al. (1990) found that changes in fractal plot could be correlated with relevant metallurgical and microstructural features.

Experimental Procedure

Material

Magnesite samples from underground magnesite mine openings at Jelsava were selected for this study. Typical representatives of samples were taken at different times during the excavation so that they are with different degrees of disturbance. The term degree of disturbance in this chapter is used as a means of quantifying the quality of rock mass during the excavation process. In this term are included all forms of rock discontinuities such as rock failure or joint, fracturing etc. that alternate with changes in the state of stress as a result of excavation. The samples were taken in different years and they are characterized by physico - mechanical properties of rock as Young's moduli and fracture toughness (Table 2). The samples were marked A, B, C, and D according to the different degree of rock disturbance because they were taken at different times in the process of excavation;

A - the less disturbed material
D - the most disturbed material.

Chemical composition is described in Table 1.

Table 1. Composition of magnesite in %.

SiO_2	MgO	CaO	FeO	Al_2O_3	lost	rez.
0.54	40.85	4.78	3.34	0.45	48.91	0.27

Bars with dimensions of 100 x 20 x 10 mm were cut and polished from the magnesite samples.

Fractal Dimension

Fractal dimensions were obtained by studying profiles of sections through the fracture surfaces. The fracture profiles were generated by the vertical section method. To protect the profile outline from damage, the fractured surfaces were mounted in bakelite. Then the fracture profiles were obtained from three to four fields of vision along the outline at a fixed magnification. The micrographs were taken at magnifications of 25x; 50x; 200x; 400x with a Neophot optical microscope and digitized using a digitizing table.

The data obtained were then entered into the computer for calculation and a special program was created for analysis. The representative pictures of fracture surfaces for samples marked D (the most disturbed material) are shown in Fig. 1. In this study, the vertical section method was used in terms of the structured walk technique (Eq. 1). When equation (1) is normalized by the projected length of the profile L_0, we obtain an expression for lineal roughness parameter $R_L(r)$

$$R_{L(r)} = L(r) / L_0 = R_0 \, r^{(1-D)} \tag{2}$$

a 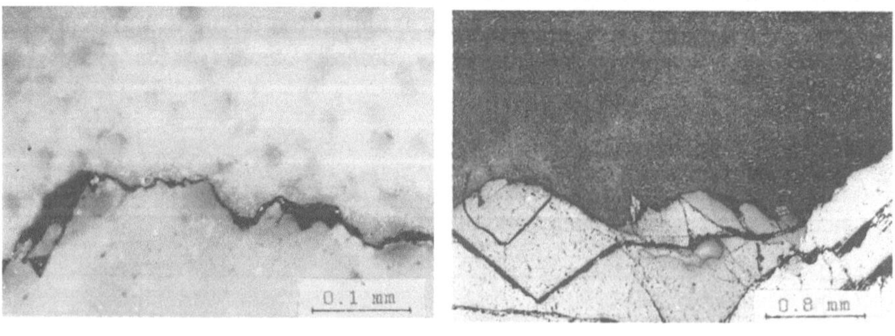 b

Fig. 1 a, b. Representative fracture surfaces of magnesite samples (*A*-the less disturbed) at different magnification **left a**-50x, **right b**-400x.

where R_0 is a constant. The lengths L(r) of those lines were measured with different sizes r of measuring unit. The roughness parameter $R_{L(r)}$ was plotted as a function of r on a log-log scale. Through linear regression for the nearly straight central portion of the graph, the fractal dimension D can be estimated from the slope.

Fracture Toughness

Fracture toughness is a measurement of the work done during fracture. If the microstructure of the material increases the roughness of the fracture surface, then the work done during fracture should be increased because of the increase in newly created surface area. This follows directly from the Griffith theory

$$K_{IC} = \sigma^{1/2}\, a = 2E\gamma \tag{3}$$

where K_{IC} is fracture toughness, σ is stress, a is crack length, E is Young's modulus, and γ is surface energy. The Griffith relationship is strictly true only for brittle materials. For more ductile solids, σ should be considered as being energy expended per unit area of crack surface increase, which would include the contribution of a plastic zone surrounding the growing crack (Davison 1989). Catastrophic failure of nominally brittle material components occurs when a crack propagates to a size sufficient to cause instability at the crack tip, and K_I - the stress intensity factor, exceeds K_{IC} - the fracture toughness of material. The direction of crack propagation is governed on macroscopic length scales by the distribution of stresses within the material, on the microscopic scale by material microstructure, and on the atomic scale by the crystallography of the material. For evaluation of fracture toughness the single edge notched beam method was used.

For the single edge notched beam method (SENB - method), the samples were notched with an 0.1 mm thick diamond wheel resulting in a notch root radius of about 0.05 mm. Tests were performed in 3-point bending with an outer span of L=80 mm. Then the specimens were loaded to failure with a crosshead speed of 0.5 mm/min. and the peak load (F) was recorded to failure. The fracture toughness was computed using this equation

$$K_{IC} = \frac{3}{2}\frac{FLY}{bw^2}\mathrm{a}^{1/2}$$

where a is notch depth, w is specimen high, b is thickness, and Y is a geometrical constant that has the value

$$Y = 1{,}93.3.07\left(\frac{a}{w}\right) + 14.53\left(\frac{a}{w}\right)^2 - 25.11\left(\frac{a}{w}\right)^3 + 25.8\left(\frac{a}{w}\right)^4 \tag{5}$$

under the condition L / w = 4.

Results and Discussion

Values of hardness H, and Young's modulus E (obtained by ultrasonic method as well as in situ seismic method) are presented in Table 3. The values represent the means of 10 tests.

Table 2. Results of the Test.

	A	B	C	D
H 5 /GPa/	2.82±0.30	2.58±0.2	2.62±0.3	2.32±0.4
Young´s modules /10^9.N.m^{-2}/ -ultrasonic	101	100	90	88
Young´s modules /10^9.N.m^{-2}/ -in situ seismic	130	125	100	90

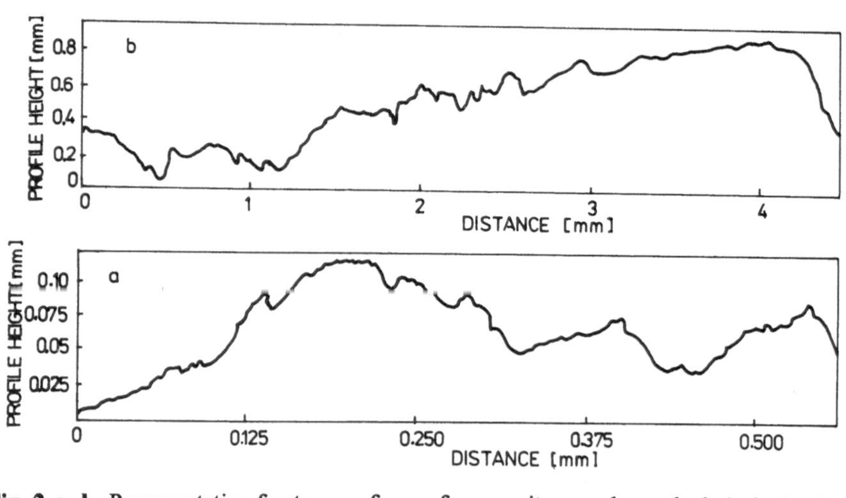

Fig. 2 a, b. Representative fracture surfaces of magnesite sample marked *A* obtained at magnifications of **a** 50x,**b** 400x.

Figure 2 shows representative profiles of fracture surfaces used for calculating fractal dimension.

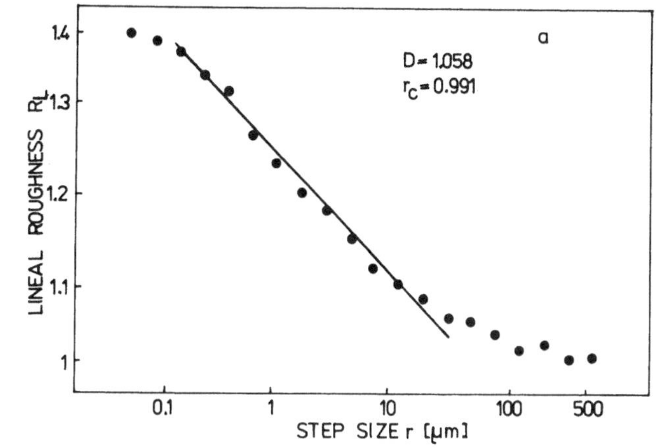

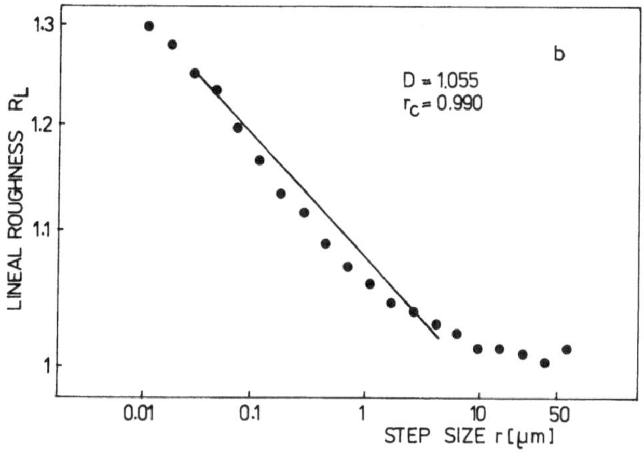

Fig. 3a, b. Log-log plots of the profile's length vs measuring step size marked *A* at magnifications of 50x (**a**), 400x (**b**) *D* is determined from the slope of the linear part of the log-log curve, r_c is corresponding regression coefficient.

Corresponding log - log plots of lineal roughness $R_{L(r)}$ vs measuring step size r for fracture profiles in Fig. 3. The results at different magnifications are presented here in order to verify how the roughness changes with the scale of magnification.

Results show that fractal dimension does not depend on the scale of magnification used.

Figures 4 and 5 show plots of fracture toughness K_{IC} (Fig. 4) determined by SENB method and fractal dimension (Fig. 5) vs degree disturbance. Fig. 6 shows a plot of K_{IC} vs fractal dimension D.

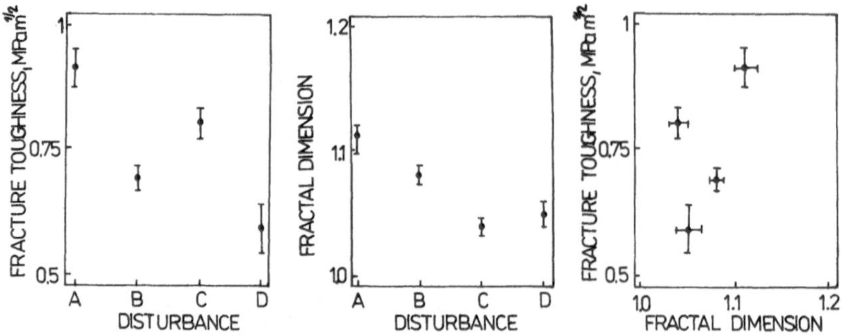

Fig. 4-6. Left Plot of K_{IC} vs degree of disturbance. **Middle.** Plot of D vs degree of disturbance. **Right.** Plot of fracture toughness K_{IC} vs fractal dimension D.

It can be seen from published reports concerning the relationship between fractal dimension and toughness that they do not agree; in the study of Mandelbrot et al. (1984) (using the slit island method) and Pande at al. (1987b) (using the slit island and fracture surface profile method), fractal dimension decreases with an increase in toughness of metals, while Mencholsky et al. (1989) (using slit island and fracture surface profile methods) report an increase in D with an increase in K_{IC} for ceramics. Davison (1989), using the fracture surface profile method, found that fracture toughness did not correlate with any measure of fracture surface roughness in studies of studying aluminum, matrix SiC, and particle reinforced composites.

Conclusions

In this study, the fractal characteristics of profiled surfaces of magnesite samples were investigated to relate them to other physical or mechanical properties. For this reason the fracture toughness (K_{IC}) of these materials was also measured using the three point SENB method.

In summary, we have characterized the fracture surfaces of magnesite samples and found they can be described using fractal techniques. In order to relate the fractal dimension D to measurable mechanical and physical properties, we estimated the toughness K_{IC}, modulus of elasticity E and hardness H. The result shows that there is some correlation between fractal dimension D and other mechanical properties but the precise dependence of the fractal dimension as a measure of fracture toughness remains uncertain and more research is needed in this area. Also there is no agreement in the literature and no theoretical relationship has been obtained.

Finally, this study points out the need for additional investigation of the techniques used to obtain fractal dimension, as well as roughness data for materials. The authors are currently working on projects which attempt to study the relation of the fractal dimension to the microstructure and fracture toughness by means of a large number of fractal dimension measurements of fracture surfaces of magnesite obtained by the other technique of quantitative metallography, the indentation method that is based on measuring the lengths of cracks which extend from the four corners of Vicker's indentations. More detailed analyses relating to other properties and structure is also studied.

Acknowledgements. This study was supported by Grant agency (G-580). We wish to express special thanks to them and many individuals who have discussed aspects of fractal geometry with us. Among them this is Ing. Peter Kmetek who provided the program for calculating fractal dimension.

References

Brown SR (1987) A note on the description of surface roughness using fractal dimension. Geophys Res Lett 14: 1095-1098

Dauskart RH, Haubensak F, Ritchie RO (1990) On the interpretation of the fractal character of fracture surfaces. Acta Metall Mater 38: 143-159.

Davison DL (1989) Fracture surface roughness as a gauge of fracture toughness: aluminum/particulate SiC composites. J materials science 24: 681-687.

Hough SE (1989) On the use of spectral methods for determination of fractal dimension. Geophys Res Lett 16: 673- 676

Huang SL, Oelfke SM, Speck RC (1992) Applicability of fractal characterization and modelling to rock joint profiles. Int J Rock Mech Min Sci Geomech Abstr 29: 89-98.

Mandelbrot BB (1967) How long is the coast of Britain? Statistical self-similarity and the fractional dimension. Science 156: 636-638

Mandelbrot BB (1983) The fractal geometry of Nature. Freeman, San Francisco.

Mandelbrot BB, Passoja DE, Paullay AJ (1984) Fractal character of fracture surfaces of metals. Nature 308: 721-22.

Mencholsky JJ, Passoja DE, Feinberg-Ringel KS (1989) Analysis of Brittle Fracture Surfaces Using Fractal Geometry. J Am Ceramic Soc 72: 60-65.

Pande CS, Richards LE, Smith S (1987a) Fractal characteristic of fractured surfaces. J Mater Sci Lett 6: 295-298.

Pande CS, Richards LE, Louat N, Dempsey BD, Schwoeble AJ (1987b) Fractal Characterization of Fracture Surfaces. Acta Metall 35: 1633-37

Power WL, Tullis TE (1991) Euclidean and Fractal Models for Rock Surface Roughness. J Geophys Res 96: 415-424

Sakellariou M, Nakos B, Mitsakaki C (1991) Technical note on the fractal character of rock surfaces. Int J Rock Mech Min Sci & Geomech Abstr 28: 527-533.

Underwood EE, Banerji K (1986) Fractals in fractography. Mater Sci Eng 80: 1-15

Fractal Dimension of Fracture Patterns - a Computer Simulation

Miriam Kupková[1] & Zuzana Krištáková[2]
[1] Institute of Materials Research of SAS,
Watsonova 45, 04001 Košice, Slovak Republic
[2] Institute of Geotechnics of SAS,
Watsonova 45, 04001 Košice, Slovak Republic

Abstract. We propose a simple model suitable for the computer simulation of fracture phenomena in brittle materials (rocks, ceramics). The real object of interest – polycrystalline sample – is replaced by a mathematical object-lattice of binary variables. Every variable represents a bond between two structural units of the sample and the values of this variable correspond to the two possible states of the bond (intact, broken). So any situation (crack inside the sample) can be described by the distribution of suitable values of variables on a model lattice and the energy of the real state can be associated with a given distribution on the basis of the physical considerations.

For illustration we performed Monte Carlo simulation on a simple model (two-dimensional square lattice of variables and the energy determined according to Griffith's theory). We obtained the shape of the crack and the fractal dimension of the fracture pattern as functions of temperature.

Introduction

Fracture and deformation of materials are wide spread phenomena of great practical importance, so they are the subject of intensive investigation in a variety of rock mechanics and geophysical problems (deformation behaviour, tectonic faults, etc.). Unfortunately, the majority of these phenomena are very complicated because they contain nonequilibrium and nonlinear processes occurring over a wide range of time and size scales and so they are not well understood yet, Meakin (1991).

This is why new procedures and methods are looked for to obtain information about these phenomena. Computer simulation represents one of the possibilities to get a deeper understanding of the processes of interest. The majority of simulations deal with quite simplified models in which the real material is represented by a network of mechanical elements (bonds, springs, beams, etc.) each having its own mechanical properties. Properties assigned to the network elements depend on the processes which we want to study. In the majority of approaches the network element can only exist in two mechanical states (perturbed, unperturbed) and the stress-dependent rate (probability) of transition between these two states controls the kinetics of failure and deformation of the material.

Of course, from the material science point of view one can object that these simplified models provide an oversimplified representation of the deformation and fracture of the real materials. But at present it is impossible to devise a general realistic model applicable to a wide class of materials and conditions, and the results obtained for the simplified models provide us with a heuristic basis for understanding real material behaviour and for development of more realistic models.

Fracture and deformation can be built-in into computer simulations in a number of ways (for instance Herrmann et al. 1989, de Arcangelis et al. 1989, Herrmann 1991). Here we shall concentrate on Monte Carlo methods.

In a typical Monte Carlo simulation there is a probability P assigned to each of the network elements according to some rule, dependent on the concrete model. One element is randomly chosen and with respect to its probability P either it is removed (changed) or it remains unchanged. Then the network relaxes partially or totally to a new mechanical equilibrium, a new probability distribution is calculated for the network elements and the process of transition to the new mechanical state and subsequent relaxation are multiply repeated to simulate the process of fracture and deformation of material.

Monte Carlo methods can be roughly divided into two groups with respect to the rules by which the probability P is assigned to the network elements:

i. Probability P is determined by means of some auxiliary quantity (e.g. scalar electrostatic potential, field of displacements, etc.). This quantity is obtained as a solution of some equation with given boundary conditions (e.g. Laplace equation, Witten and Sander 1981, Navier equation, Louis & Flores 1985, Louis & Guinea 1987, or another, usually partial differential equation of the second order, Taguchi 1989).
ii. Probability P is determined in a phenomenological way in analogy with the kinetics of chemical reactions. Probability of fracture or modification of i-th element is given as

$$P_i \sim \exp(-E_a/k_B T),$$

where E_a is an activation energy, k_B is the Boltzmann constant and T is temperature. Activation energy E_a depends on a concrete investigated model and it contains model parameters, such as external stress, local chemical and structural composition etc. (Meakin 1987, Curtin & Scher 1990).

Definition of the Investigated Model

To study fracture propagation in a brittle material with predominantly intergranular failure we propose a simple model (Kupková & Parilák 1991).

According to the above mentioned classification the presented model belongs to the second group.

Let each bond between two structural units (for example grains, molecules, etc.) is represented by the two-value variable $\xi_{ijk}= 0$ or 1. Subscripts i,j,k determine the spatial position and orientation of the bond represented by $\xi_{i,j,k}$ (e.g. i,j,k are the spatial co-ordinates of the centre of gravity of the boundary between two grains, etc.). $\xi_{ijk}= 0$ corresponds to the situation when the bond is intact, $\xi_{ijk}= 1$ when the bond is broken. The lattice of variables $\{\xi_{ijk}\}$ shows us which bonds are broken, which are not, how crack looks etc. We can attribute the energy of state to each configuration of variables of the model lattice. This energy can be expressed in terms of the variables ξ_{ijk}.

The use of this model is illustrated on a simple example. We assume a two-dimensional structure of identical structural units forming a square lattice. So the lattice with variables x in the lattice sites is also a square. As we are dealing with a two-dimensional case, we need only two co-ordinates i,j to determine the spatial positions of the bonds (e.g. grain bondaries). If measured in the units of the original lattice parameters, these coordinates have the form either $(i, j) = (m, n)$ (for ξ_{ij} representing a horizontal boundary) or $(i, j) = (k + ½, l + ½)$ (for ξ_{ij} representing a vertical boundary) (here m,n,k,l are integers).

We applied the Griffith theory to determine the energy of configuration $\{\xi_{ij}\}$. We consider the case when uni-axial tensile stress is applied in the plane of the model. Every broken bond causes an increase $\gamma(T)$ of energy, $\gamma(T)$ includes the specific surface energy and the specific energy absorbed in the plastic zone or by other mechanism (for instance by phase transformations) at a given temperature T as well as an area of surface originated by the breaking of one bond between structural units.

Fracture propagation is caused by the fact that the fracture removes energy of elastic deformation in the region proportional to the second power of the length of the crack. The constant of proportionality δ depends on the stress acting σ as well as on the modulus of elasticity E_y and the size of the structural unit l, $\delta = \alpha E_y \sigma^2 l^2$, α is form factor. So the energy $E\{\xi_{ij}\}$ corresponding to any configuration $\{\xi_{ij}\}$ can be expressed as

$$E\{\xi_{ij}\}=E_o+\gamma(T)\,[\Sigma_{m,n}\,\xi_{m,n} + \Sigma_{k,l}\,\xi_{k+½,l+½}] - \delta\{\Sigma_m \,\mathrm{sign}(\Sigma_n\, \xi_{mn})\}^2 \quad (1)$$

Here E_o refers to the energy of the configuration with all ξ_{ij} equal to zero, sign $x = 0$ for $x = 0$ and sign $x = 1$ for $x > 0$. We assume that the uni-axial tensile stress σ acts in a vertical direction.

Now we shall perform Monte Carlo simulation based on this model on a square lattice of NxN lattice sites. First, every bond on a square lattice has a corresponding variable $\xi_{ij}= 0$; $i = 1...N$, $j = 1...N$. Next we "break" some of the bonds (i.e. their variable becomes one). We use random numbers to choose a new state and the energy of this state is calculated according to (1). In the case when $E_{new} < E_{old}$ the new state is always accepted. In the case when $E_{old} < E_{new}$ the probability of transition from the old to the new state is given by the Gibbs

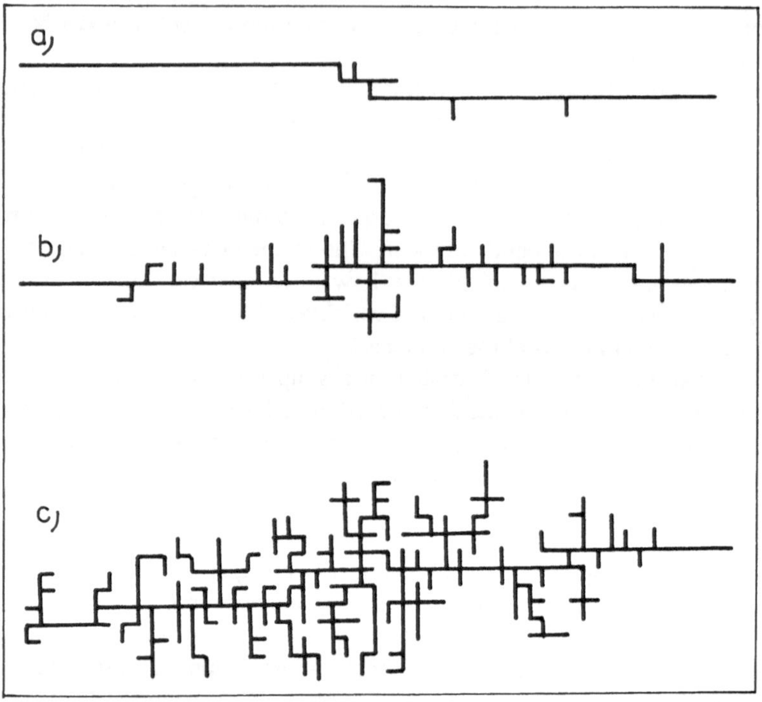

Fig. 1a-c. A typical pattern of broken bonds on 50x50 lattice, **a** for $\tau = 0.2$, **b** $\tau = 0.4$, **c** $\tau = 0.6$. Here $\tau = k_B T/\gamma(T)$ (see text).

probability $Pr = \exp\{-\beta(E_{new} - E_{old})\}$; ($\beta = 1/k_B T$), tested again using a random number generator. (Here k_B is Boltzmann's constant and T is absolute temperature).

Figures 1a, 1b, 1c show a typical fracture on a 50x50 lattice at selected reduced temperatures $\tau = k_B T/\gamma(T)$. We see that with increasing temperature the patterns become more ramified.

The Fractal Dimension of the Fracture Patterns

Next we determine the fractal dimension of the obtained fracture pattern (which can be treated as a fracture profile curve) as a function of temperature and model parameters in this simple case. The relation between the length of a fractal curve l(e) and the size e of the length unit used in the measurement,

$$l(\varepsilon) \propto \varepsilon^{1-D_f},$$

is used for determination of the fractal dimension D_f. So, D_f is determined from the slope of the linear part of the log l(ε) vs. log ε curve as

$$D_f = 1 - \frac{d(\log l(\varepsilon))}{d(\log \varepsilon)}$$

The fractal dimension is a measure of the roughness of a fractured surface (Huang et al. 1989) and it is related to the energy dissipated on creation of new surfaces in directions other than the main direction of fracture propagation.

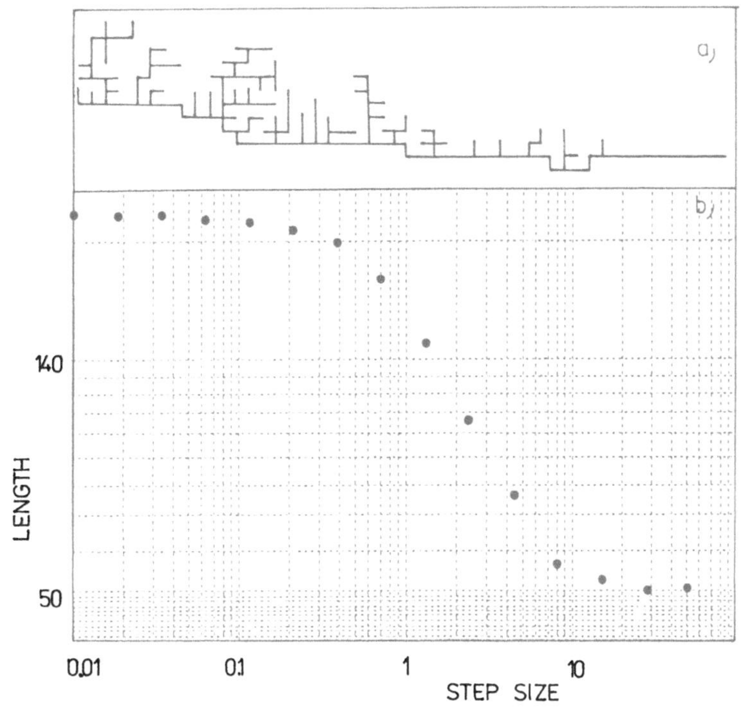

Fig. 2a, b. A simulated fracture curve obtained, **a** for $\tau = 0.5$ and **b** the length of this curve as function of scale size.

Figure 2 shows the fracture curve from one of the Monte Carlo simulations and the length of this curve as function of scale. The fractal dimension is determined from the linear part of this dependence.

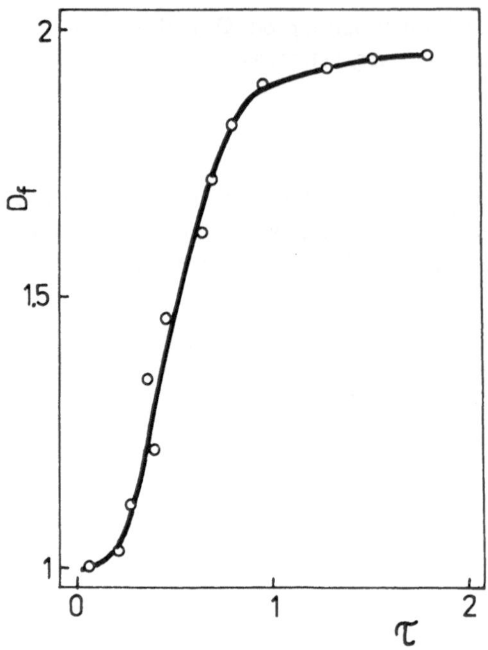

Fig. 3. Dependence of the fractal dimension D_f on τ.

Figure 3 shows dependence of the fractal dimension D_f on τ ($\tau = k_BT/\gamma$ (T)). Here k_BT/γ (T) is the ratio between the characteristic energy k_BT of a thermal fluctuation and the energy γ (T) originated by breaking a simple bond between structural units (because thermal fluctuations afford the necessary energy for creation of the fracture surface except in the main direction of fracture propagation). The transition to the stable state on the fractal dimension versus k_BT/γ (T) curve occurs at about k_BT/γ (T) ~1. In this case the thermal fluctuation energy is sufficient for ramifying of the crack (this process is energetically disadvantageous for low temperatures). If we suppose the applicability of linear fracture mechanics, then the fracture toughness K_{IC} is proportional to the square root of the effective surface energy: $K_{IC} \sim \gamma_{eff}^{1/2}$ where $\gamma_{eff} = (N_b/b)\,\gamma$ (T) in our case (N_b is the total number of broken bonds and b is the system size). Then the temperature dependence of K_{IC} is qualitatively similar to the temperature dependence of the fractal dimension.

Conclusions

In this chapter we proposed a simple mathematical model for investigation of the failure of brittle materials. The basic idea has been to represent each bond between two grains by a binary variable ξ_{ij} ($\xi_{ij} = 0$ corresponds to the situation when the bond is intact, and $\xi_{ij} = 1$ if the bond is broken). Any distribution of broken bonds can be modeled by a corresponding distribution of the values 0 or 1 along the edges of a model lattice, and it is possible to determine an energy of state for any distribution of the ξ_{ij} values according to a chosen physical model for a real brittle sample. Such a model (lattice of variables ξ_{ij} plus the corresponding energy of state) is suitable for computer simulation.

For illustration we used a two-dimensional square lattice of the variables ξ_{ij} and for the determination of the energy of a given configuration we assumed simple Griffith theory (where the energy is determined only by the crack surface energy and the energy of elastic deformation removed by crack).

We used this model for a Monte Carlo simulation of the shape of a crack at various temperatures. The object of interest was the fractal dimension of the fracture patterns as a function of temperature (Fig.3). The shape of the curve fractal dimension vs. temperature is similar to the experimental dependence of the fracture toughness K_{IC} on temperature. If we suppose that the fractal dimension reflects an effective surface energy γ_{eff} , and γ_{eff} is connected with K_{IC} in the sense that an increase of γ_{eff} leads to the increase of K_{IC}, then the similarity of dependences fractal dimension vs. temperature and K_{IC} vs. temperature is an interesting result obtained from such a simple model (Griffith theory).

For further investigations it is necessary to choose a more precise model for the determination of the energy corresponding to the configuration of ξ_{ij} (concentration of stress on the crack tip, plastic deformation, distribution of grain size, etc.) and to use a more sophisticated calculation of probability than in presented paper. By using such an amended model it might be possible to achieve a quantitative agreement between theoretical and experimental results.

References

De Arcangelis L, Hansen A, Herrmann HJ, Roux S (1989) Scaling laws in fracture. Phys Rev B 40:877-880.

Curtin WA, Scher H (1990) Brittle fracture in disordered materials. J Mater Res 5: 535-553.

Herrmann HJ (1991) Patterns and scaling in fracture. Physica Scripta T38: 13-21.

Herrmann HJ, Kertész J, de Arcangelis L (1989) Fractal shapes of deterministic cracks. Europhys Lett 10: 147-152.

Huang ZH, Tian JF, Wang ZG (1989) Analysis of fractal characteristics of fractured surfaces by secondary electron line scanning. Materials Science and Engineering A118: 19-24.

Kupková M, Parilák L (1991) A computer simulation of temperature dependence of fracture propagation in brittle materials. In: Van Mier JGM, Rots JG, Bakker A (eds) Fracture Processes in Concrete, Rock and Ceramics. Spon, London Noordwijk, pp 261-266.

Louis E, Flores F (1985) Fractals in Physics. Pietronero L & Tosatti E (eds), Elsevier, Amsterdam, 177.
Louis E, Guinea F (1987) The fractal nature of fracture. Europhys Lett 3: 871-879.
Meakin P (1987) A simple model for elastic fracture in thin films. Thin Solid Films 151: 165-191.
Meakin P (1991) Simple models for material failure and deformation. In: Van Mier JGM, Rots JG, Bakker A (eds.) Fracture Processes in Concrete, Rock and Ceramics. Spon, London Noordwijk, pp 213-229.
Taguchi Y-H (1989) Fracture propagation governed by the Laplace equation. Physica A 156: 741-755.
Witten TA, Sander LM (1981) Diffusion limited aggregation, a kinetic critical phenomenon. Phys Rev Lett 47: 1400-1402.

The Formation of Extensional Veins: An Application of the Cantor-Dust Model

Jörn H. Kruhl
Geologisch-Paläontologisches Institut, JW Goethe-Universität,
D-60054 Frankfurt/M., Germany

Abstract. The brittle widening of rocks during regional extension follows a power-law relationship and, therefore, is scale-invariant. The development of extensional veins during this process is best described by the Cantor-dust model with a fractal dimension $0.955 < D < 0.970$. D is independent of the rock types, their anisotropies, the amount of widening, the vein type and the vein thickness. The model implies that successively thinner veins are formed reflecting the increase in irregularity of the total stress field with time.

Introduction

Rock extension under brittle conditions produces fractures of different thickness and spacing, which, generally, are filled with different type of material. These extensional veins are often oriented sub-perpendicular to the main elongation direction. The thickness distribution of four different types of extensional veins from different geologic environments and of different scales, from microscopic to macroscopic, and their relationship to the total extension have been studied. Fractal geometry provides tools for characterizing the thickness distribution of the veins and for modelling the mechanism of rock widening.

Measurements

The four different types of extensional veins are from four different localities within the Caledonides, Hercynides and Alps (Fig. 1). Independently of their thickness scale and the type of rock they cut through, the veins are approximately tabular with a length-thickness ratio of 10^3 and more. Therefore, the change of vein thickness with length is negligible and any influence on the thickness distributions, presented later, can be ignored.

Fig. 1. The mountain belts of Europe (after Walter 1992) and the localities *A, B, C* and *D* of the four different types of extensional veins which have been studied.

(A) Within mafites and metapelites of the Hercynian lower crustal segment of Calabria (S.Italy), during mid-crustal shearing under retrograde lower amphibolite and greenschist facies conditions, numerous fractures have been formed approximately perpendicular to the foliation and the direction of tectonic transport (Kruhl 1992). In the metapelites garnet and sillimanite are cut by these fractures, a portion of which are micron-wide and filled with biotite (Fig.2A). The thicknesses of 979 filled fractures have been measured in a foliation-parallel thin section along profiles parallel to the alignment direction of the long axes of garnet and sillimanite. The profiles sum to a total length of 90mm. The accumulated thickness of the fractures, i.e. the total amount of widening by the fractures, is 12.7% of the profile length.

(B) Within the Taunus Quartzite of the Rhenish-Massif (mid-European Hercynides), during probably sub-greenschist facies conditions, a grid of quartz-filled extensional veins has been formed (Fig.2B). These veins show thicknesses in the millimetre range. They are oriented perpendicular to the sedimentary bedding and developed prior to the regional folding, as shown by their displacement along bedding planes near fold crests. The veins probably reflect flattening perpendicular to the bedding planes during an early stage of regional deformation (Anderle 1987, Oncken 1988). The thicknesses of 336 veins have been measured in the outcrop along profiles parallel to bedding and perpendicular to each of the two sets of veins shown in Fig.2B. The total length of the profiles is 10m, and the total amount of widening by the veins 11.5%.

(C) Within the Sesia Zone of the Western Alps, aplitic dykes occur which are related to the intrusion of the late-Hercynian Monte-Rosa granite (Kruhl 1979). Subsequently, these aplites were deformed during prograde greenschist to lower amphibolite facies conditions of the Alpine event (Fig.2C). The thicknesses of 351 dykes have been measured perpendicular to the principal Alpine foliation along a road cut of 42m length. The average flattening of the quartz and feldspar components of the rocks is c. 75% (Kruhl 1979). On the basis of folded aplites (Fig.2C), the average portion of flattening due to pressure solution in the biotite-rich wall rocks has been estimated as 37.5%. The original thicknesses of the aplites before Alpine deformation and the original length of the measurement profile (237m) have been recalculated. The dykes show thicknesses in the centimetre to metre range. The total amount of widening by the aplites before subsequent deformation was 22.4%.

(D) The intrusion of the late Caledonian granites of Scotland into cool country rocks was accompanied by the formation of mainly felsic porphyric dykes. The SW-NE trending dyke swarm of the Etive complex (Fig.2D) intruded older igneous rocks and a sequence of multiply foliated and folded quartzites and metapelites of the Moinian and Dalradian, which they cut without any relation to sedimentary and tectonic structures (Bailey & Maufe 1960). At the eastern end of Loch Leven, along several profiles of a total length of 14 km, perpendicular to the long axis of the dyke swarm, the thicknesses of 184 dykes have been measured. They are in the metre to 10-metre range. The total amount of widening by the dykes is 6.95%. In addition, the average widening by dykes in different areas along a profile strip from the central axis to the margin of the northeastern part of the Etive swarm is set in relationship to the distance to the central dyke swarm axis (Fig.5).

As a summary, the development conditions and the geological environment of the four different types of extensional veins are clearly different. They cut rocks ranging from nearly non-metamorphic sediments to granulites, which are either isotropic or anisotropic parallel to the extension direction. The veins seem to have been formed during one single step of rock widening. There is no indication of a crack-seal mechanism. Independently of the type of the veins, the thicknesses of each vein type range over 1 to 2.5 orders of magnitude and over seven orders in total, starting with the 2.5μm thickness of biotite-filled fractures in garnet and sillimanite and up to the 20m thickness of porphyritic dykes. The upper limits of these magnitudes are clearly defined. The rock widening caused by each vein type varies from 7% to 22%.

The country rocks with vein types A,C and D at least have been lifted from greater depth before formation of the veins. Most probably, the rocks were fractured during uplift. Fig.2A shows that, on the micro scale, numerous fractures are present, many of which have not been changed to extensional veins. This network of fractures was used by the subsequent rock extension.

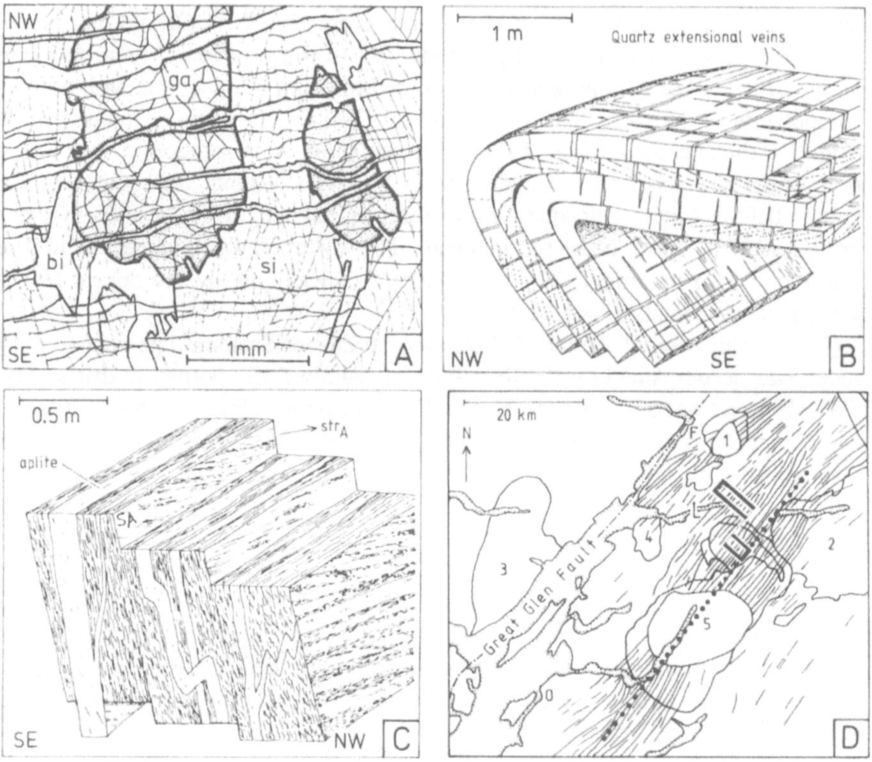

Fig. 2A-D. A Drawing of a thin section parallel to the main foliation of a metapelite from the Hercynian lower crust in Calabria (locality A in Fig. 1). The long axes of sillimanite (*si*) and garnet (*ga*) are aligned with the SE-NW oriented main elongation direction. Biotite (*bi*) filled extensional fractures are oriented approximately perpendicular to this direction. Road cut near S. Vito, northern Serre, Calabria. **B.** Schematic sketch of folded Taunus Quartzite from the Rhensh Massif (mid-European Hercynides) (locality B in Fig. 1). A lattice of quartz-filled extensional veins is oriented perpendicular to and translated along the sedimentary bedding planes and, therefore, has developed prior to the (SE-NW directed) folding. Quarry near Lorch, Rhine valley west of Mainz. **C.** Schematic sketch of aplitic dykes from the Sesia Zone (Alpine "Root Zone"), Western Alps (locality C in Fig. 1). The dykes are formed in relation to the late-Hercynian Monte-Rosa granite and are foliated (S_A) and stretched (str_A) together with their wall rocks during the Alpine event. Those dykes or parts of dykes sub-parallel to the Alpine flattening direction are folded due to the higher amount of pressure-solution related deformation in the biotite-rich wall rocks. Road cut north of Finero/Ivrea Zone. **D.** Late Caledonian granites and dyke swarms (solid lines) in the central Scottish Highlands between Fort William (*F*) and Oban (*O*) (locality D in Fig. 1). Map after Bailey & Maufe (1960). Plutons: (*1*) Ben Nevis, (*2*) Rannoch, (*3*) Strontian, (*4*) Ballachulish, (*5*) Etive. The double bars indicate the profile through the northeastern part of the Etive dyke swarm along which the measurements have been performed. *L*: Loch Leven; dotted line: central axis of the Etive dyke swarm.

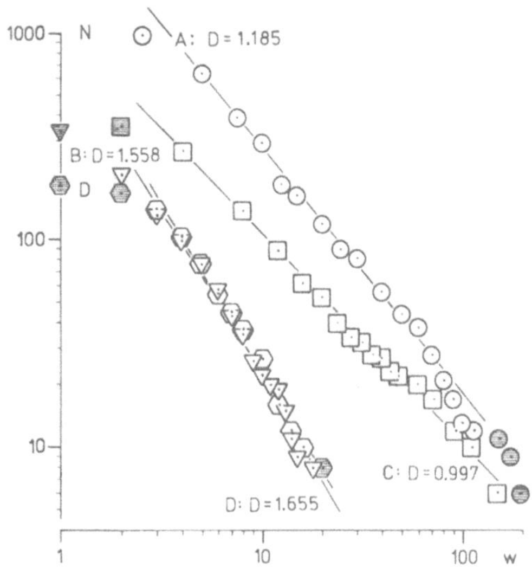

Fig. 3. Log-log plot of vein thickness w versus the number N of veins with a thickness greater than or equal to w. Data from the four different vein types A, B, C and D, as described in Fig.2. Frequencies less than 6 are not presented. Thickness w of vein type *A*: microns, type *B*: millimetre, type *C*: centimetre, type *D*: metre. The four straight lines indicate the best logarithmic fit of the original data, with a correlation coefficient of 0.99. The D-values are the fractal dimensions of the data sets, i.e. the negative slopes of the regression lines. The data represented by hatched symbols have been omitted for regression.

Discussion

Fractals like the Cantor dust, the Sierpinski carpet and the Menger sponge offer opportunities to study the distribution of material in one, two or three dimensions (Mandelbrot 1983). Because the veins presented in this study show a high length-thickness ratio it seems appropriate to investigate their clustering in one dimension along a line perpendicular to the vein extension. The algorithm given by Mandelbrot, which describes the distribution of the Cantor dust along a line, leads to the probability P, with $P = Nd$ or $P = d^{1-D}$, that a certain part of length d of the line has not been removed. N is the total number of the unremoved parts of length d and D is the fractal dimension of the Cantor dust, with $0 < D < 1$. The number N of the parts of the line with a length greater than or equal to d is given by $N = d^{-D}$.

It has been shown that the distribution of vein thicknesses (w) may follow a power-law relationship such as $N = w^{-D}$ (Sanderson et al. 1991). D-values of fault and crack thicknesses given in the literature cover a range of 0.39 to 1.50

(Nagahama 1991). In addition, the spacing of fractures follows the same power-law relationship (Blenkinsop 1993), as well as the two-dimensional distribution of fractures (Skjeltorp 1988, Hirata 1989, Velde et al. 1990, Vignes-Adler et al. 1991). Consequently, fracturing is a scale-invariant process. In Fig.3 the N/w relationship of the four groups of veins are presented in a log-log-diagram. The relationship is more or less linear, with slopes between -0.997 and -1.655. These data support the previously reported fractal relationships for vein thicknesses (Sanderson et al. 1991).

However, as indicated by correlation coefficients of almost exactly 1, a more strictly linear relationship in the log-log diagram is given between the profile length R not covered by veins of thicknesses greater than or equal to a specific value w and this value (Fig.4). Slight deviations in this linearity at small thicknesses can be explained by the possibility that thin veins in outcrops and on

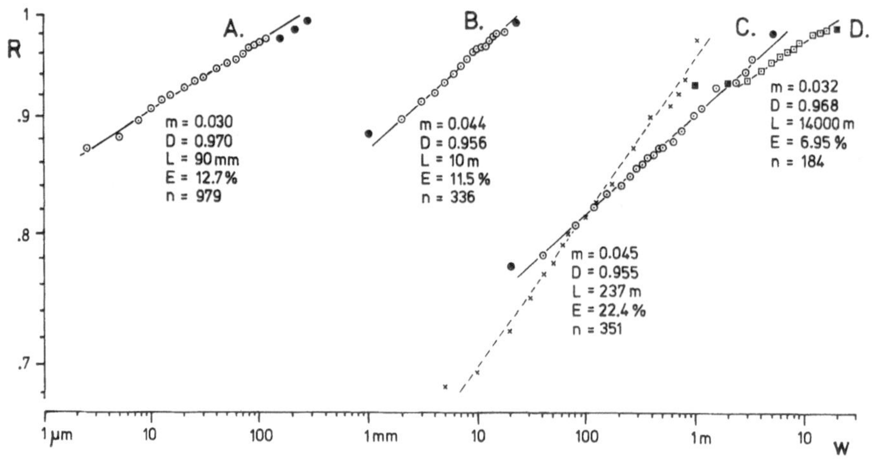

Fig. 4A-D. Relationship between vein thickness and rock widening. W: thicknesses of the different extensional veins from localities A, B, C and D (Fig.1 and 2), measured along profiles perpendicular to the veins. Data of locality D are from the NW-end of the profile strip, outlined in Figure 2D, directly south of the Ben Nevis complex. Data from Kruhl [1974 (D) and 1979 (C)] and this study (A and B). Same data sets as presented in Figure 3. *R*: total profile length minus the accumulated thickness of veins thicker than a specific *w*-value (normalized to the total profile length). Log-log diagram. *m*: slopes of the regression lines of the different data sets, based on a logarithmic regression of the original data, with correlation coefficients of 1.00. The data represented by filled symbols have been omitted for regression. *D*: fractal dimensions of the data sets (D = 1-m); *L*: total length of the measured profiles; *E*: extensional vein portion of the total profile length; *n*: number of measured vein thicknesses. The crosses of data set C represent the present thicknesses of aplites, i.e. without consideration of thickness and profile-length variation by deformation. The circles represent the thicknesses recalculated for the situation before deformation.

the map scale have not been totally registered. This is supported by the fact that the porphyric dykes (data set D), with a worse outcrop situation than the other vein types, show the largest deviation, whereas the smallest fractures on the thin-section scale (data set A), with "100% outcrop", fit the linearity. The deviations at high thicknesses are simply a result of the low number of measurements. The upper limits of vein thickness are clearly defined. For example, at locality A, on the specimen and outcrop scale, there are no biotite-filled extensional veins thicker than some hundred microns. The true lower limits probably do not fall short of the limits presented in Figure 4. For example, thin sections from the host rocks of locality C show that there are no aplites in the millimetre range and below, i.e. before deformation there were no aplites thinner than 2cm. The reason for relating the profile length R to the vein thicknesses and not to the length of the remaining line d is as follows. If some veins remain unresolved the thickness distribution of the veins is only slightly modified. However, the distribution of the vein spacing may be drastically changed.

The slopes m of the linear R/w relationships of the four vein types cover a narrow range of 0.030 to 0.045, which is much smaller than the range of slopes of the linear N/w relationships presented in Figure 3. Only the m-value of the measured aplite thicknesses (crosses in Fig.4) clearly deviates from this range. However, these veins are strongly deformed and, if the vein thicknesses and the profile length before deformation are considered, the m-value is conformable to the m-values of the other three vein types.

At a constant m-value, the total amount of rock extension is governed by the range of vein thicknesses. The lowest amount of extension (vein type D) correlates with the smallest range of vein thickness, the highest amount of extension (type C) with the largest range of vein thickness. Because perpendicular to the fractures the tensile strength of the rock is drastically decreased the total amount of extension by the veins can be considered proportional to the total amount of stress perpendicular to the veins.

A possible mechanism to extend rock would be to successively produce veins of the same or different thickness. The algorithm leading to the Cantor dust follows the mechanism of producing veins of successively lower thickness. Figure 5 shows the development of a Cantor dust of fractal dimension 0.960, which approximately models the thickness distribution of the studied extensional veins. Mandelbrot (1983) introduced the term "trema" for the gaps which are cut out of the profile line. Figure 4 represents the increase of the total length of the trema (vein thickness), i.e. the decrease of the unveined profile line, with decreasing trema size. The slope m of this linear relationship defines the fractal dimension of the unveined rock, $D = 1\text{-}m$. D is independent of (i) the rock type, (ii) the anisotropy of the rocks, (iii) the amount of widening, (iv) the vein types, and (v) their ranges of thickness. It may be speculated that, therefore, D is governed by a more general process. It has been suggested that the spacing between joints may be ruled by the perturbation and/or the relief of the stress field during formation of the joints (Pollard & Segall 1987, Blenkinsop 1993). In the same way, the formation of veins of different thickness could be related to the local decrease of the stress field around these veins. A continuous formation and subsequent broadening of veins would lead to successively thinner veins and to an increasing irregularity of the total stress field, as has been suggested for the fracturing process of thin films (Skjeltorp 1988). With regard to the Cantor-dust model, D is related to the probability that a dyke of a certain thickness occurs. In addition, it may be considered as a measure how "fast" the initial bar "disappears" (Fig.5). A high D-value indicates a low "rate of disappearance" and contrary. Consequently, in relation to the dyke formation, D would be a measure how "fast" the dyke thickness decreases with successive formation of dykes or of how "fast" the irregularity of the stress field increases. However, the relationship to the different types of stress acting upon a rock and to other parameters, or to a combination of them, is not evident.

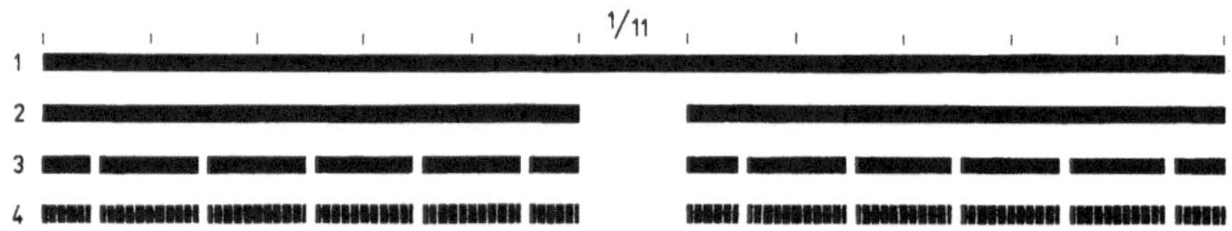

Fig. 5. Construction algorithm for forming a Cantor dust (Mandelbrot 1983) of fractal dimension D = 0.960. A bar is divided into 11 segments of equal length. One of the segments is cut away. Each of the remaining 10 segments is again divided into 11 segments of equal length, one of which, again, is cut away, and so on. The distribution of the interstices may be random or not. They represent the thickness and distribution of extensional veins along a cross profile. For each stage (2-4) of the Cantor-dust construction, in the log-log diagram any point on the x-axis represents the maximum thickness of the removed segments. Along the y-axis the remaining (black) portion of the bar is figured as percentage of the total bar length.

In addition, the regional decrease of rock widening can be described by a logarithmic relationship. Fig.6 shows the length R of the "rest-profile" (profile length minus amount of rock widening) versus the distance s from the centre of a dyke swarm, which is considered as the locality of maximum rock widening. This power-law relationship, with a D-value smaller than that of the local rock widening, probably reflects the general regional change of the stress field.

Conclusions

Independently of the different rock types, anisotropies, amounts of widening, vein types and ranges of thickness, the thickness distribution of the studied extension-

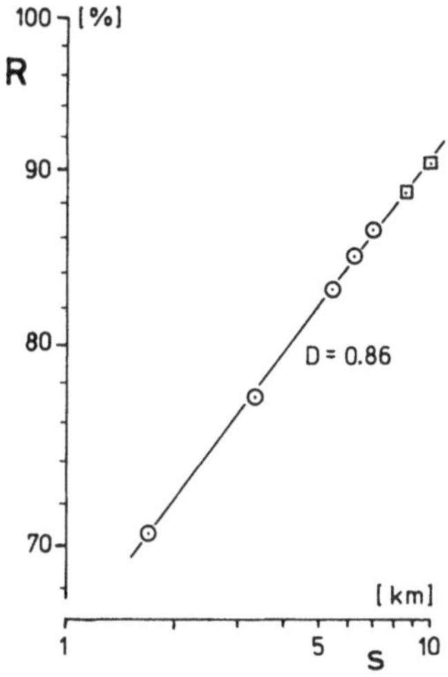

Fig. 6. Relationship between rock widening and distance to the central axis of the Etive dyke swarm. *R*: lengths of the different profile parts minus the total dyke extension, expressed as percentage of the total length of the different profile parts; *s*: distances of the centres of the different profile parts to the dyke swarm axis. *D*: fractal dimension of the data set. The straight line of best fit has been calculated by logarithmic regression, with a correlation coefficient of almost exactly 1. Measurements from the profile drawn in Figure 2D. Data from Bailey & Maufe (1960) (circles) and Kruhl (1974) (squares).

al veins and its relationship to the rock widening follow a power law $R = w^{1-D}$, with the vein thickness w, the profile length R not covered by veins of thicknesses greater than or equal to w, and with the fractal dimension D ranging from 0.955 to 0.970. Consequently, the probability P that a certain point of the profile across the veins is covered by a vein of the thickness greater than or equal to w is given by $P = 1\text{-}R = 1\text{-}w^{1-D}$, with w normalized to the maximum occuring vein thickness and R normalized to the total profile length.

The power-law relationship $R = w^{1-D}$ does not describe the formation of fractures but the brittle extension of the country rocks, which is a scale-invariant process. The mechanism can be modelled by the algorithm leading to the Cantor dust, as given by Mandelbrot (1983). The Cantor-dust model describes the increase in irregularity of the stress field in time.

Acknowledgments. Thanks are due to G.Zulauf for comments on an early draft of this chapter and to T.Blenkinsop and G.Mandl for two critical reviews.

References

Anderle HJ (1987) The evolution of the South Hunsrück and Taunus borderzone. Tectonophysics 137: 101-114.

Bailey EB, Maufe HB (1960) The Geology of Ben Nevis and Glen Coe. Mem Geol Survey Scotl, Edinburgh.

Blenkinsop TG (1993) Fracture spacing distributions in rock. Int Symp on Fractals and Dynamic Systems in Geoscience, Book of Abstracts, Frankfurt/M.

Hirata T (1989) Fractal dimension of fault systems in Japan: Fractal structure in rock fracture geometry at various scales. PAGEOPH 131: 157-170.

Kruhl JH (1974) Structural mapping of the Moine series at the eastern end of Loch Leven, Argyllshire, Scotland. Unpubl Diploma thesis, University of Hamburg. (in German)

Kruhl JH (1979) Deformation and metamorphism of the southwestern Finero Complex (Ivrea Zone) and the northerly adjacent gneiss zone. Unpubl PhD thesis, University of Bonn. (in German)

Kruhl JH (1992) The structural history of a lower crustal section (Calabria, S.Italy). 29th Int Geol Congr, Kyoto, Symp II-6-6, Book of Abstracts.

Mandelbrot BB (1983) The fractal geometry of nature. 2nd ed Freeman, New York.

Nagahama H (1991) Fracturing in the solid earth. Sci Rep Tohoku Univ, 2nd ser 61: 103-126.

Oncken O (1988) Aspects of the reconstruction of the paleostress history of a fold and thrust belt (Rhenish Massif, FRG). Tectonophysics 152: 19-40.

Pollard DD, Segall P (1987) Theoretical displacements and stresses near fractures in rock: with applications to faults, joints, veins, dikes, and solution surfaces. In: Atkinson (ed), Fracture Mechanics of Rock. Academic Press Geol Ser, London, pp 277-349.

Sanderson D, Roberts S, Gumiel R (1991) Power-law distribution of vein thickness: a useful model in mineral exploration and reserve estimation. Tectonic Studies Group / Joint Association for Geophysics, Meeting "Fractals in Structural Geology and Seismology", Book of Abstracts.

Skjeltorp AT (1988) Fracture experiments on monolayers of microspheres. In: Stanley HE, Ostrowsky N (eds), Random Fluctuations and Pattern Growth: Experiments and Models. NATO ASI Ser E 157, Kluwer, Dordrecht, pp 170-173.

Velde B, Dubois J, Touchard G, Badri A (1990) Fractal analysis of fractures in rocks: the Cantor's Dust method. Tectonophysics 179: 345-352.

Vignes-Adler M, Le Page A, Adler PM (1991) Fractal analysis of fracturing in two African regions, from satellite imagery to ground scale. Tectonophysics 196: 69-86.

Walter R (1992) Geologie von Mitteleuropa. 5th edition, Schweizerbart, Stuttgart.

Fractal Fault Displacements: A Case Study from the Moray Firth, Scotland

Giles Pickering[1], Jonathan M. Bull[1], David J. Sanderson[1] & Paul V. Harrison[2]
[1] Dept. of Geology, University of Southampton, Highfield, Southampton, SO9 5NH, UK
[2] Mobil North Sea Ltd, Mobil Court, Clements Inn, London WC 2A, UK

Abstract. Displacement on faults can be described using a power-law distribution with exponent D. To justify using a power-law, the relationship needs to be proved over several orders of magnitude. Offshore, faults with displacement >20m can be measured within Triassic rocks from seismic reflection profiles. Around the Moray Firth there is limited exposure of Triassic rocks which are continuous with the offshore basin. Faulted sections from Triassic sandstones were measured, with fault displacements varying from 1 to 500 mm. These seismic and field data are limited in both size and scale range.

Computer simulations were run where a population with a known power-law was sampled to see how the measured distributions matched the original. The samples gave a variable and biased power-law exponent. Strategies were designed to find the underlying value from these samples. The simulations were also used to estimate the confidence intervals for the D-values calculated.

A combined graph of the field and seismic data showed that a single power-law with a D-value of ~0.8 can describe the fault population in this area.

Introduction

Displacement on faults has been found to be power-law distributed in several recent papers, for example Kakimi (1980), Childs et al. (1990), Heffer & Bevan (1990), Walsh et al. (1991) and Jackson & Sanderson (1992). A power-law distribution is defined as:

$$N \propto U^{-D} \qquad (1)$$

where N is the number of objects with size > U and D is the exponent. Thus a graph of logN against logU is a straight line with a slope of -D. To justify using a power-law, for extension estimates or reservoir studies, the relationship needs to be established over several orders of magnitude.

Sources of geological information from an offshore area are limited to remote geophysical methods and the analysis of well-data. Of the geophysical methods, only normal incidence seismic reflection allows detailed measurement of fault displacement. As such surveys are bandwidth limited due to the attenuation of the

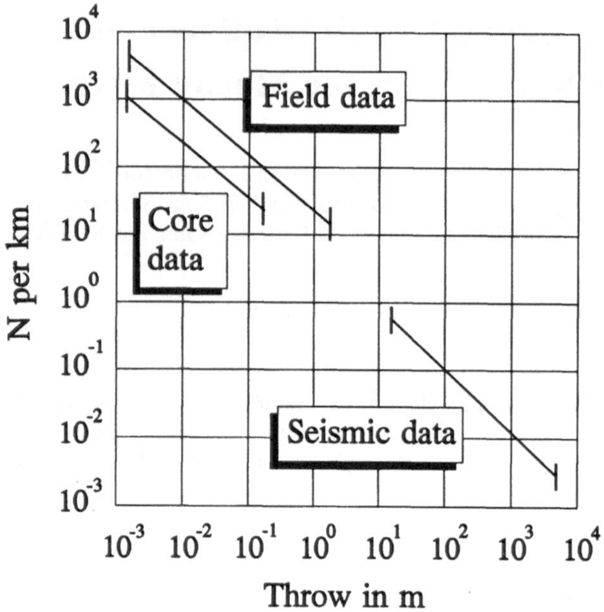

Fig. 1. Schematic size and frequency ranges of fault throw based on various sources of data. Seismic resolution is controlled by the value of a quarter of the dominant wavelength, while core data is limited by the dimensions of the sample taken. Field data is limited by the extent of the outcrop and the difficulty in correlating marker beds for large faults.

high frequencies, only faults with displacements >~20 m can be measured. Structural information from well-core is limited to estimates of small scale displacement (<100 mm). Therefore, below the resolution of seismic surveys there is a gap in the data range (Fig. 1). Estimates of fault density from core can be unreliable. This is due to the possible inclusion of fractures introduced by the drilling and extraction process, and movement of original fractures caused by drag on the core margins (Blackbourn 1990). There is, therefore, a lack of good small-scale data to compare to the large scale. Field data are more reliable and cover a larger range of fault sizes than is possible from core (Fig. 1), but clearly can only be collected onshore. This chapter presents a case study, where fault data collected in the field are compared to data from an offshore seismic survey in order to test the power-law model. As both data sets are limited in size and scale range, sampling effects can introduce errors into the calculation of D-values. These effects were investigated using simulations to find the best way of calculating D-values and confidence intervals for them.

Sampling Effects

Two important effects have been recognised in sampling from power-law distributions; these are known as truncation and censoring (e.g. Jackson & Sanderson 1992). Both lead to deviations from the expected linear relationship on the log-log plot between cumulative number (or rank) and value.

Truncation

Truncation is caused by a lower limit to the resolution imposed by the sampling method. This limit is often unavoidable, for example the width of a marker bed used to measure fault displacements at outcrop, or the limitations of seismic reflection data due to the bandwidth of the seismic signal. If every fault with displacement above this limit were measured then there would be an abrupt cut-off in the data set. However, in most cases some faults will be missed as the limit is approached, causing a curve on the log-log plot at the small-scale, which looks like a deviation from a power-law.

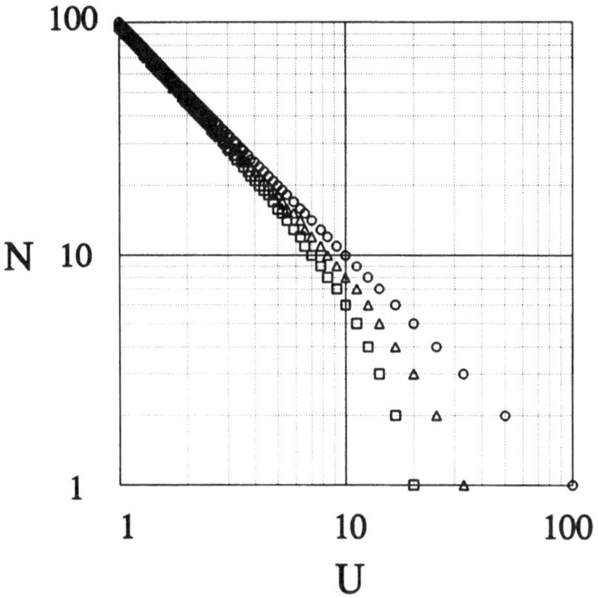

Fig. 2. Graph of *log N* against *log U*, where *N* is rank or cumulative number and *U* are values from artificial data following an exact power-law with a D-value of 1.0. ○ = complete data set, Δ = largest two values removed, □ = largest four values removed. As the censoring is introduced by removing the large values, the graph becomes more curved and the "linear" part steepens (after Jackson & Sanderson 1992).

Censoring

Any set of measurements will be made over a limited area and may not contain the largest values. There are two, to some extent independent effects. Firstly, the largest values in the entire population may have a low probability of being sampled over a finite area or section. Secondly, sampling constraints may specifically exclude the largest values, for example a seismic section may not include a basin bounding fault, or outcrop sections may end at a large unmeasurable fault. This under-sampling of the large values reduces the rank of all the values in the data set, which causes a curve at the large scale when plotted on a log-log graph and increases the slope, hence the D-value. This effect is illustrated in Fig. 2 where artificial data sets with and without the largest values are shown for comparison.

Simulations

Clearly these sampling effects could lead to an inaccurate estimation of the D-value, especially the censoring effect, which will cause an overestimation. In order to test how sampling affects D-values, a controlled computer simulation was conducted. The program randomly samples from an exact power-law distribution, for which both the D-value and a maximum value have been set, using the Inverse Transform Method (Rubinstein 1981). The user can define the scale range and size of the sample in order to simulate the size and scale limitations of the actual data. A large series of data sets are produced in order to determine the distribution of D-values for samples of the specified size and scale range. In this paper we are primarily interested in the application of the results from the simulation to the Moray Firth data sets, therefore we have confined our discussion to this aim. A full discussion of the simulation and its' general application to samples from power-law distributions is contained in Pickering et al. (in review).

Results from these simulations show that truncation presents few problems; provided data from the affected area are not used to define the linear part of the graph it does not change the slope of this portion and thus the D-value found. This is best achieved using a robust algorithm to fit a line only to the majority linear trend. The method used minimises the total absolute deviation of the data points rather than the sum of the squares. Further details of robust statistical methods can be found in Press et al. (1986).

The simulations of censored data sets show that the D-value found from fitting a line to such data is always an overestimate of the true value. This will cause a large error in any application of the D-value. The appropriate solution to this problem for the Moray Firth data sets is to try and estimate the number of "missing" fault measurements and add this number to the cumulative number N.

If N_T is the number of faults with size greater than U_{MIN}, and N_C is the number of faults with size greater than U_{MAX} , where U_{MIN} and U_{MAX} are the size range of the sampled feature, in this case fault throw, then the number of faults > U_{MAX} from a power-law distribution can be obtained as follows. Taking logs of (1) and substituting into the equation for a straight line gives:

$$\log N_T - \log N_C = D_T(\log U_{MAX} - \log U_{MIN}) \tag{2}$$

where D_T is the D-value for the distribution. From (2) the number of missing (censored) faults N_C can be found:

$$\log N_C = \log N_T - D_T(\log U_{MAX} - \log U_{MIN}) \tag{3}$$

As D_T is unknown it must be estimated, usually by fitting a line to the censored sample to give an approximate estimate D_E. N_T is equal to the size of the sample (N_E) plus the correction N_C which is unknown. N_T can be estimated as N_E, as long as $N_E \gg N_C$. Therefore, N_C can be calculated:

$$\log N_C \approx \log N_E - D_E(\log U_{MAX} - \log U_{MIN}) \tag{4}$$

Both approximations will lead to an underestimate of N_C and therefore the correction is always conservative. With this correction, the D-value found from the simulated samples are much closer to the original value, even when the scale range is narrow.

The simulation can also be used to estimate the confidence intervals for D-values. This approach is common in experimental physics, where a good simulation of the measurement process gives an accurate estimation of the confidence which can be attached to the measured parameters (Rubinstein 1981). The synthesis of the results from 1000 simulations for each of a series of sample sizes gave the following empirically derived equations:

$$68\%\ \text{Interval} \approx \pm 0.075\sqrt{\frac{200}{\textit{Sample size}}} \times D$$

$$95\%\ \text{Interval} \approx \pm 0.15\sqrt{\frac{200}{\textit{Sample size}}} \times D \tag{5}$$

Fault Displacement Data

The data used in this study are measurements of vertical separation or throw, made on dip sections through the fault set. Throw is used in this study because it is the most accurately determined parameter from seismic sections. Where displacement is in the plane of the section and the fault dips are reasonably const-

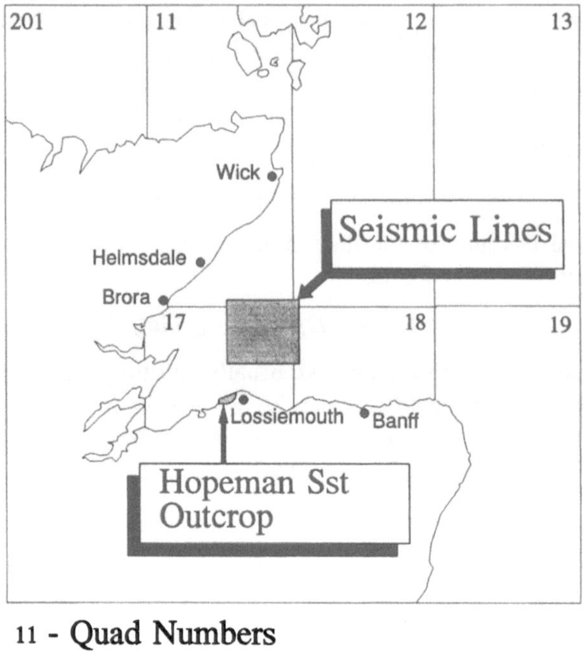

Fig. 3. Outline map of NE Scotland showing the location of the SSL-MF89 seismic lines used in relation to the Hopeman Sandstone outcrop.

ant, throw is directly proportional to displacement. Although there has been a belief in a significant strike-slip component to the displacement of all of the faults in the Moray Firth (e.g. Roberts et al. 1990), it is now accepted that those faults which developed predominantly in the Mesozoic are normal (Underhill 1991). The use of sections, rather than sampling from maps or other sources, has several advantages. Firstly, as they represent a projection of the fractal set, the D-value found is directly related to the D-value describing the full set in three dimensions (Marret & Allmendinger 1991). Secondly, the use of sections avoids the problems associated with fault maps required for sampling in higher dimensions (Yielding et al. 1992). The data are drawn from two sources, seismic sections and field outcrops.

Fault Measurements from the Seismic Survey

The large scale measurements for this study come from a set of seismic lines shot in the Inner Moray Firth, just offshore from Lossiemouth (Fig. 3). They are good

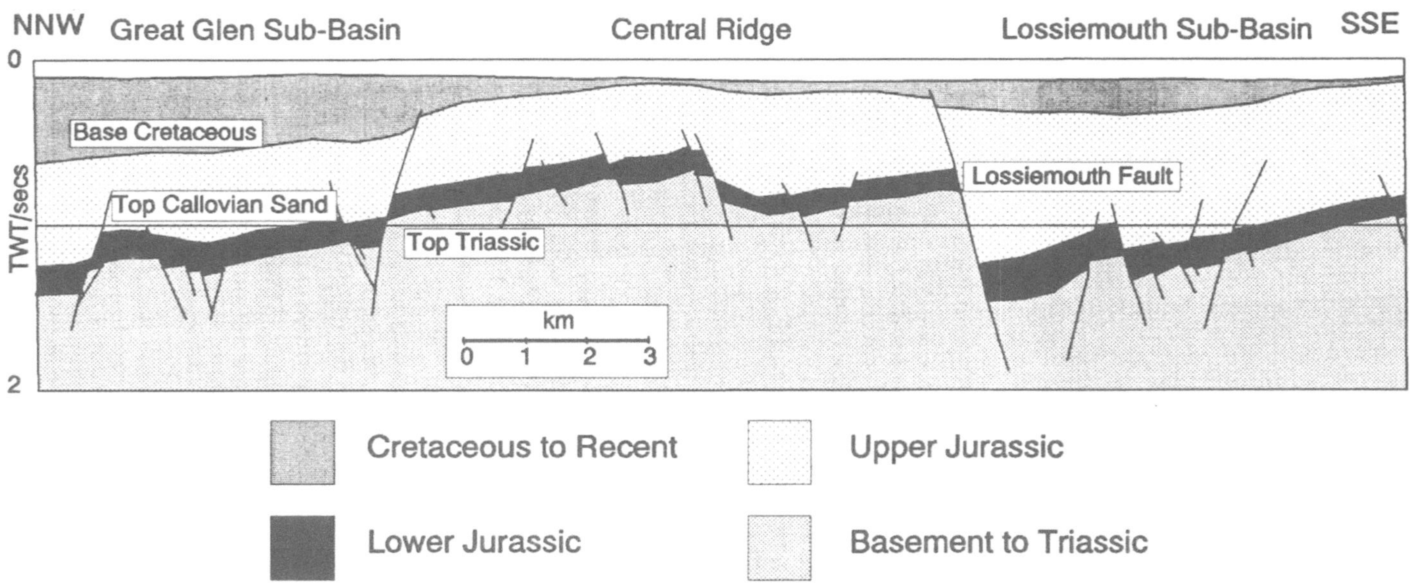

Fig. 4. Line-drawing of one of the seismic sections showing the major horizons and faults. Fault measurements from the Top Triassic horizon were used in this study. The faults are primarily steeply dipping normal faults. The general location of the sections is shown in Fig. 2.

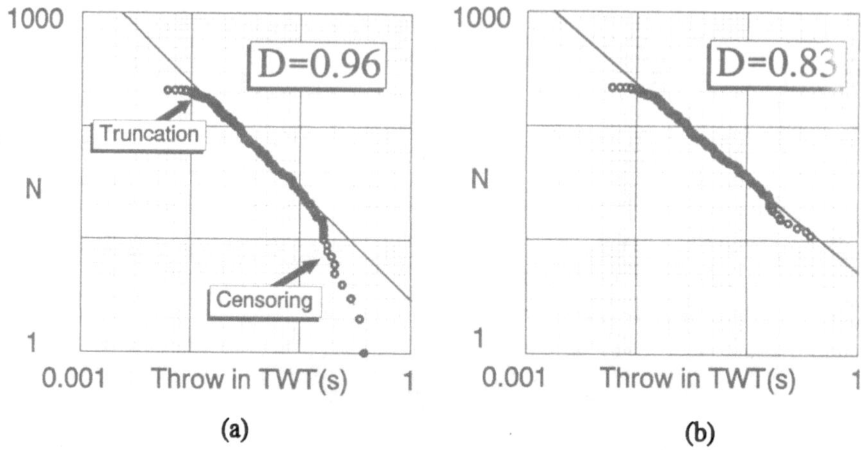

Fig. 5. a Log-log plot of cumulative number (N) against throw for the faults from the seismic sections. The deviation from a power-law at the large scale is due to censoring, that at the small scale is due to truncation. The line fitted using a robust line fitting algorithm gives a D-value of 0.96. **b** A censoring correction ($N_C = 10$) has been applied to the data shown in (a) which has extended the linear portion up to the large scale. Note that the D-value from the fitted line has now been reduced to 0.83, which is more likely to represent the true value. The truncation at the small scale remains, but this does not affect the measurement of the D-value.

quality, migrated sections; an example line-drawing is shown in Fig. 4. The horizon chosen for the study was the top Triassic, as the base Triassic is not well defined. The faults penetrate the entire Triassic section, so comparison with field data from the early Triassic Hopeman Sandstone is valid. Measurements from eleven sections were combined and graphed on a log-log plot (Fig. 5a). The data show a deviation from a power-law distribution at both the small and large scales. These deviations are caused by truncation and censoring respectively.

Given the conclusions from the simulations, the truncation will not affect the measurement of D as long as the line is fitted only to the remainder of the data. However the censoring will bias the D-value. A censor correction of 10 has been calculated using equation (4) and a D_E of 0.96, which has been applied to the data to give the graph shown in Fig. 5b. A line fitted to this graph using a robust algorithm gives a D-value of 0.83. From equation (5) with a sample size of 212, the confidence intervals for this value are: $D = 0.83 \pm 0.06$ at 68% and $D = 0.83 \pm 0.12$ at 95%.

Fault Measurements from the Field

Many of the rocks that outcrop along the coast of the Moray Firth are of Triassic or Jurassic age and represent the on-shore equivalents of rocks buried under

Cretaceous and Tertiary sediments below the Inner Moray Firth. On the southern coast of the Moray Firth, near Lossiemouth (Fig. 6a), the Triassic Hopeman Sandstone is exposed in accessible cliff sections. These dune-bedded sandstones are highly faulted; most of the faults are high-angle normal faults. They generally strike E-W (Figs. 6a & 6b). Five N-S transects of the cliff sections were made, the fine laminations allowed measurement of fault displacements down to 1mm. The position, throw and angle to layer of each fault were noted; where possible dip & dip direction were also recorded.

Of the five sections measured, section #1 contributed the majority of the data and is considered independently from the combined set. Initial visual inspection of Fig. 7a would suggest that a power-law model was inappropriate; statistically, a log-normal or exponential distribution would give a better fit. This deviation from the expected linear trend is in fact due to censoring. Using an estimated $D_E = 1.1$ in equation (4) gives a calculated correction of $N_C = 6$. If this correction is applied to the data (Fig. 7b), a good linear trend is found with a D-value of 0.86. The combined data set is shown in Fig. 6c and gives a D-value of 0.79. The wider scale-range removes the need for correction. Using equation (5) the estimated confidence intervals for this value are $D = 0.79 \pm 0.08$ at 68% and $D = 0.79 \pm 0.16$ at 95%.

Comparing the Field and Seismic Data

One application of the power-law model is to predict fault density at scale ranges outside that which can be directly observed. Individually, the data sets do fit the model, but such estimations will only be valid if a single power-law describes the entire scale range. The faults seen in both the seismic and field data are in rocks of Triassic age and have similar orientations. It is possible, therefore, to test whether one power-law distribution describes both sets. Two procedures are necessary to achieve this comparison. Firstly, the number of faults must be normalised by dividing through by the length of the section used, to provide number per km. Secondly, the throw measurements from the seismic data must be converted into depth from two-way time. This was done using an interval velocity of 3500 ms^{-1}. As the log of the throw is graphed the choice of velocity is not crucial, and even a large lateral velocity change along the section, say 500 ms^{-1}, would have little effect. Fig. 8 shows the corrected seismic data set and the combined field data on the same log-log plot.

Both data sets fit a single power law with a D value of 0.84. As the field data was measured only from sections that showed significant faulting, it is possible that the number density found at this scale is higher than the average value. If the true density is lower, then this would tend to lower the D-value found from the combined graph. However, halving this density would still leave the D-value at

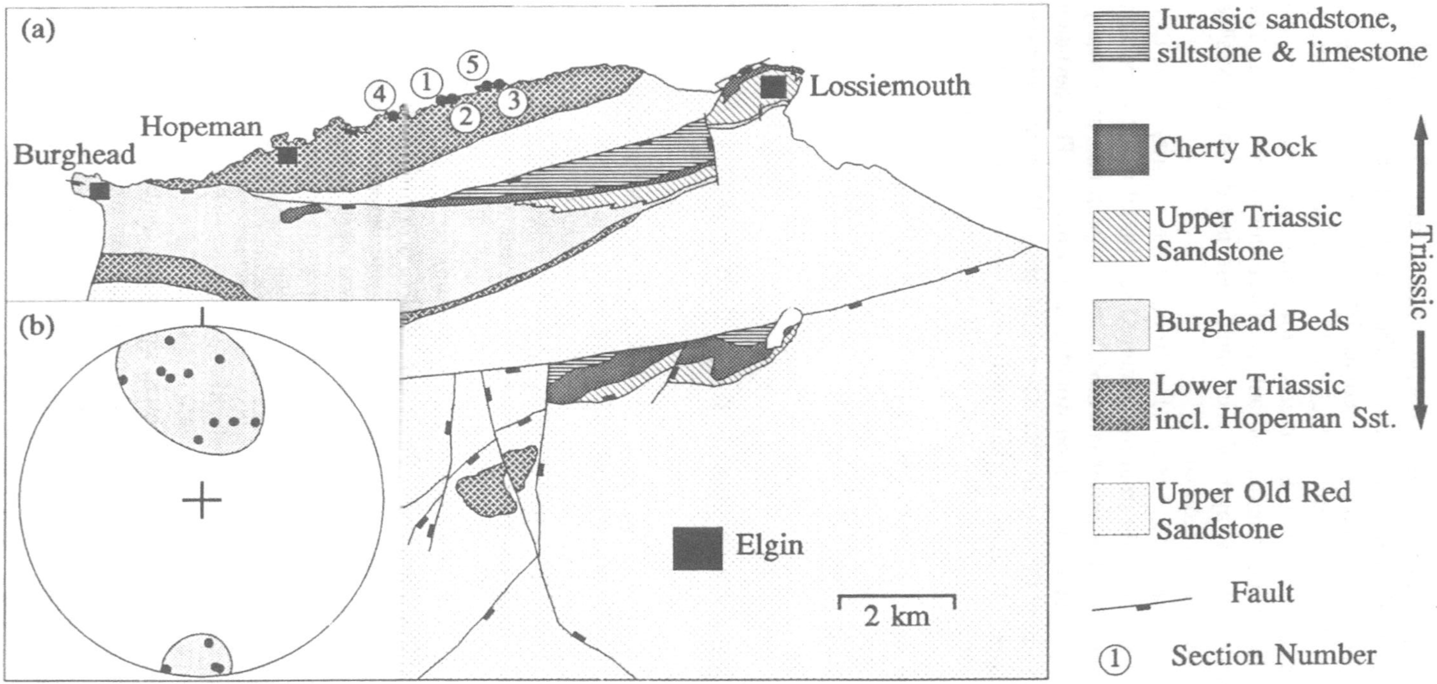

Fig. 6. a Map showing the Mesozoic cutcrops around Lossiemouth and Elgin (after BGS Elgin Sheet - number 95). The Hopeman Sandstone fault data were taken from cliff sections to the east of Hopeman village. **b** Equal-area stereographic projection of poles to fault planes. The faults are hig angle and strike ~E-W. This str ke is close to the regional trend for this area and the strike of the faults seen on the seismic sections.

~ 0.8. Clearly the fault population distribution is power-law over at least six orders of magnitude, with a D value of ~ 0.8.

In addition to confirming the model, the advantage of having two data sets covering different scale ranges is to improve the confidence in the D-values calculated. The 95% confidence limits of the D-value from the seismic are $0.71 < D < 0.95$. From inspection of Fig. 7, a line with a gradient of -0.95 drawn from the seismic data would overestimate the fault density at the field scale by two or three times, similarly a D-value of 0.71 would under-estimate the density. Therefore, confidence in the D-value is much greater than would be predicted from each individual data set, or one of their combined size but only covering one of the scale ranges.

Discussion

Papers which have looked at other areas in the North Sea have found D-values similar to 0.8 (0.8 - 1.0). Walsh, Watterson & Yielding (1991) and Marrett & Allmendinger (1992) used seismic and well core data from the Viking Graben but made no rigorous analysis of sampling effects. Jackson & Sanderson (1992) carried out a similar comparison of map and field data in SW Spain and show their general compatibility. However, although they recognise the truncation and censoring effects there was also no rigorous attempt to correct either data set prior to comparison. They also obtained different D values from the data from each source, whereas in this study $D_{seismic} \sim D_{field} \sim D_{total} \sim 0.8$. Therefore, this study represents the most complete analysis of scaling of fault displacements to date.

The importance of estimating the confidence interval can be illustrated by a calculation developed in Scholz & Cowie (1990) for the total displacement of a fault system, or the fraction contained in different scale ranges:

$$\int_{U_{min}}^{U_{max}} DCu^{-D} du = \left[\frac{DC}{(1-D)} U^{1-D} \right]_{U_{min}}^{U_{max}} \qquad (6)$$

Equation (6) can be used to calculate the fraction of the total extension (along a line) which is developed by faults beneath the resolution of the seismic section. In this case the displacement measure needed is heave. As this is also proportional to displacement for normal faults, it will scale with the same D-value as for throw. Therefore if we estimate a maximum fault heave, say 5km, and take the resolution limit of the seismic sections as 20m, the fraction of the extension below the resolution will be ~ 35% for a D-value of 0.8. The 95% confidence limits for a sample size of 250 and a D-value of 0.8 would be $0.7 < D < 0.9$. These values substituted into equation 6 would give fractions of 20% and 58% respectively. This error is mainly controlled by the D-value rather than the

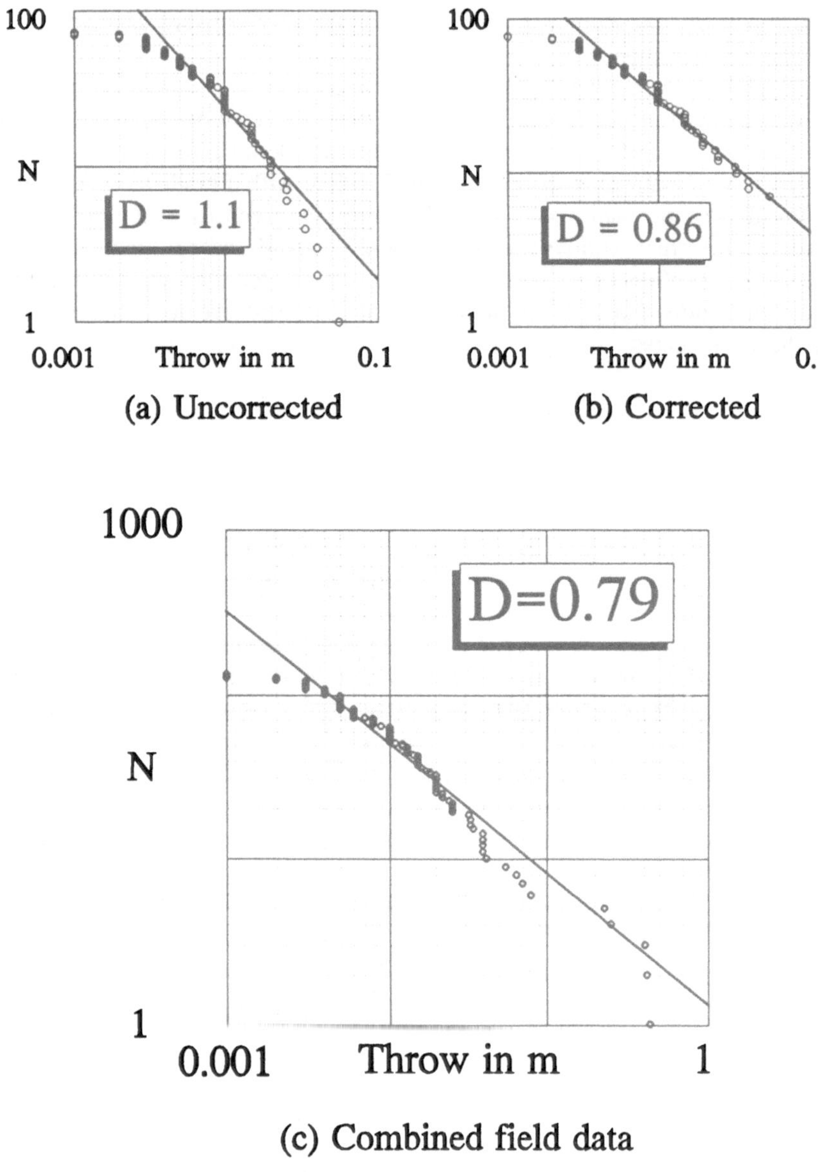

Fig. 7a-c. **a** Graph of the original data set from field section #1 (for location see Fig. 6a). The graph is heavily censored due to the narrow scale range, and does not appear to fit a power-law distribution. **b** Censor corrected data, using a N_C of 6. This graph shows a good linear trend giving a D-value of 0.86, similar to that of the seismic data. **c** Graph of the combined data from all the field sections, which gives a D-value of 0.79. The wider scale range gained by combining the data sets has removed the need for any censoring correction.

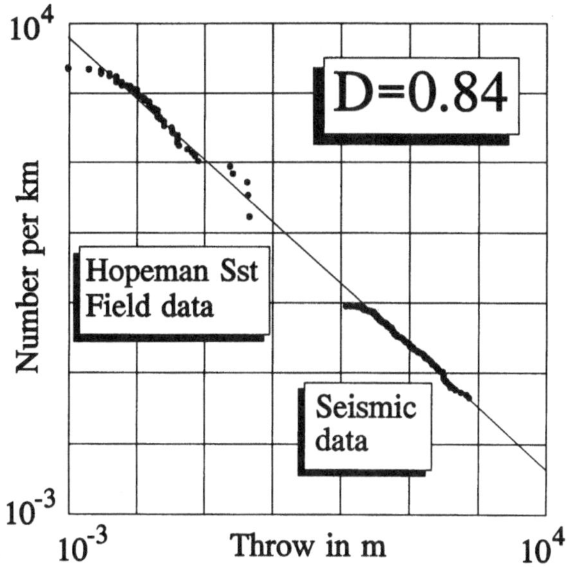

Fig. 8. Graph showing both the field and seismic data, normalised to cumulative number per kilometre. The seismic data were converted into throw in metres using a velocity of 3500 m/s. A single power-law distribution with a D-value of 0.84 fits the combined data over six orders of magnitude of displacement.

estimated maximum heave; for example halving the maximum heave to 2.5km would only increase the estimated fraction to 38% in the case of $D = 0.8$. The possible error is therefore quite high, and would need to be considered when drawing conclusions from the extension estimate. This example also illustrates the need to improve confidence in the D-value, either by increasing the sample size or by confirmation using a data set covering a different scale range.

Conclusions

To fit a distribution function to a fault population, data from a wide scale-range are needed. Offshore, core data are all that is available at small scale. Core data are only a small vertical sample of the section, from which horizontal (line) density has to be inferred. The field data used in this chapter gives a much more direct and accurate measurement of fault density at the small scale and an estimate of the D-value of the population. Seismic sections were used to provide measurements of faults with large displacements. Both of these data sets were affected by sampling problems due to their scale and size limitations. Simulations were used to develop strategies to find the underlying D-value and to estimate the confidence intervals for the values calculated. D-values found for the field data and the seismic data were all ~ 0.8. The combined graph of field and seismic data shows that a single power-law with a D-value of 0.84 can describe the Inner Moray Firth fault population over at least six orders of magnitude. The additional advantage of having data covering such a wide scale range is the improved confidence this gives in the calculated D-value, as compared to that estimated for data sets only covering a smaller scale range. This allows reliable estimation of extension and fault density below seismic resolution.

Acknowledgements. The authors are grateful to Seismograph Service Ltd & SHELF for permission to use the SSL-MF89 Seismic Survey. This research is funded by Mobil North Sea Ltd.

References

Andrews IJ, Long D, Richards PC, Thomson AR, Brown S, Chesher JA, McCormac M (1990) The geology of the Moray Firth. BGS United Kingdom Offshore Regional Report. HMSO.

Blackbourn GA (1990) Cores and core logging for geologists. Whittles Publishing Services, Scotland.

Childs C, Walsh JJ, Watterson J (1990) A method for estimation of the density of fault displacements below the limits of seismic resolution in reservoir formations. In North Sea Oil and Gas Reservoirs II. Edited by A. T. Buller. 309-318. Graham & Trotman London.

Heffer K, Bevan T (1990) Scaling relationships in natural fractures - data, theory and applications. Proc European Petrol Conf 2: 367-376 (SPE paper No. 20981).

Jackson P, Sanderson DJ (1992) Scaling of fault displacements from the Badajoz-Cordoba shear zone, SW Spain. Tectonophysics 210: 179-190.

Kakimi T (1980) Magnitude-frequency relation for displacement of minor faults and its significance in crustal deformation. Bull Geol Surv Japan 31: 467-487.

Marrett R, Allmendinger RW (1991) Estimates of strain due to brittle faulting: sampling of fault populations. J Struct Geol 13: 735-738.

Marrett R, Allmendinger RW (1992) Amount of extension on "small faults": an example from the Viking Graben. Geology 20: 47-50.

Pickering G, Bull JM, Sanderson DJ (in review) Sampling power-law distributions. submitted to Journal of Geophysical Research.

Press WH, Flannery BP, Teukolsky SA, Vetterling WT (1986) Numerical Recipes: The Art of Scientific Computing. Cambridge University Press, Cambridge.

Roberts AM, Badley ME, Price JD, Huck W (1990) The structural history of a transtensional basin: Inner Moray Firth, NE Scotland. J Geol Soc 147: 87-103.

Rubinstein RY (1981) Simulation and the Monte Carlo method. John Wiley & Sons, New York.

Scholz CH, Cowie PA (1990) Determination of total strain from faulting using slip measurements. Nature 346: 837-839.

Underhill JR (1991) Implications of Mesozoic-Recent basin development in the western Inner Moray Firth, UK. Mar Petrol Geol 8: 359-369.

Walsh JJ, Watterson J (1992) Populations of faults and fault displacements and their effects on estmates of fault-related regional extension. J Struct Geol 14: 701-712.

Walsh JJ, Watterson J, Yielding G (1991) The importance of small scale faulting in regional extension. Nature 351: 391-393.

Yielding G, Walsh J, Watterson J (1992) The prediction of small-scale faulting in reservoirs. First Break 10: 449-460.

High-Temperature Viscoelastic Behaviour and Long Time Tail of Rocks

Hiroyuki Nagahama
Institute of Geosciences, School of Science, Shizuoka University,
836 Oya, Shizuoka 422, Japan

Abstract. The relaxation function $\phi(\xi)$ for high temperature flow of rocks can be represented by $\phi(\xi) \propto \xi^{-\gamma}$ where ξ is the temperature reduced time. This shows the relaxation behavior has a temporal fractal property called a long time tail. This relaxation function can be regarded as the result of the superposition of standard exponential decay functions with different mean lifes, τ. Here, a universal relaxation function of rocks is introduced and the relaxation of high temperature flow of rocks is discussed from the viewpoints of fractal concepts.

Introduction

Among other fields of study, experimental rock deformation has contributed to the general concept of plate tectonics. Several researchers have studied the flow laws of high temperature rocks and proposed empirical power-law creep equations given by

$$\dot{\varepsilon} = A\sigma^n \exp(-Q / RT) \qquad (1)$$

where $\dot{\varepsilon}$ is the strain rate, σ is the applied stress, Q is the activation energy, R is the universal gas constant, T is the absolute temperature and A and n are so called "constants". This empirical equation is called Dorn's equation of steady-state flow of rocks. However, previous research has not established a sufficient description of transient behaviors.

Shimamoto (1987) described high temperature flow of rocks including their transient behaviors (relaxation behaviors) from the viewpoint of thermorheology (Figs. 1, 2) and proposed a viscoelastic constitutive equation given by

$$\dot{\varepsilon} = \left\{g(\varepsilon)E'\right\}^{1/m}(\varepsilon / C)\sigma^{1/m} \exp(-Q / RT) \qquad (2)$$

where ε is the strain, $g(\varepsilon)$ is the strain-dependent function, $1/m$ is the stress sensitivity of strain rate, E' and C are constants. This viscoelastic constitutive equation for high temperature rocks is nonlinear and provides a linkage to the

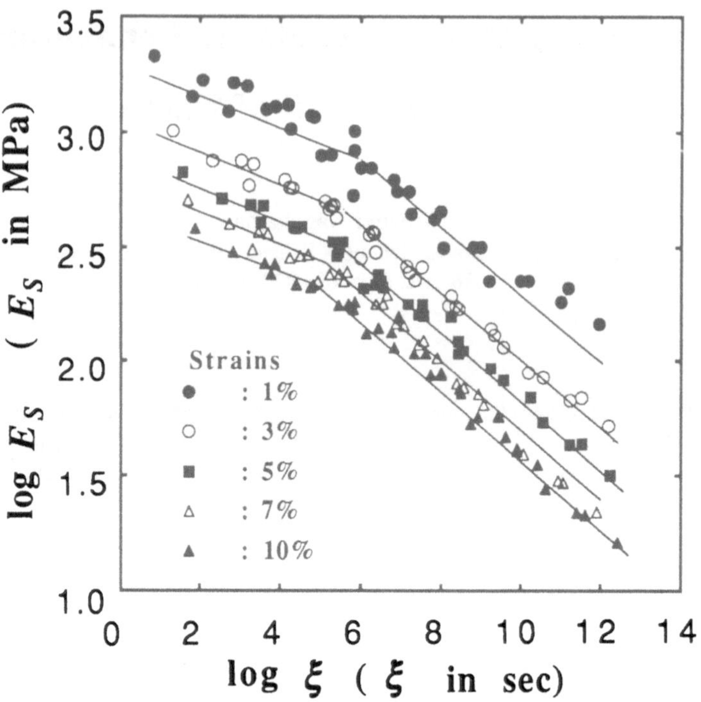

Fig. 1. Secant modulus (E_S) vs. temperature reduced time (ξ) master curve at 100 °C reference temperature under different strain levels (partly modified from Shimamoto 1987, Fig.3; data from Heard 1972). Solid line can be expressed by eq.(6). If all curves are shifted parallel to the secant modulus axis so that they are superimposed, they will be found to lie along a single line, as shown in Fig. 2.

empirical mechanical equation (Dorn's equation) for the steady-state flow of rocks.

Moreover, eq.(2) is generalized by the relaxation function

$$\sigma = h(\varepsilon)\int_{o}^{t} E(\xi - \xi')\frac{d\varepsilon}{d\tau}d\tau, \quad h(0) = 1 \tag{3}$$

where $E(\xi - \xi')$ is the relaxation modulus at the interval of temperature reduced time between two different temperatures, $\xi - \xi'$ (Shimamoto 1987). This nonlinear constitutive equation is a special form of Smith (1962) and Schapery's (1969, 1974, 1982) nonlinear viscoelastic constitutive equation. However, the present status of the theory for the relaxation function of this equation is extremely unsatisfactory, because the basic underlying mechanical concept is unknown. Here I will discuss a viscoelastic equation for high temperature flow of rocks from the viewpoint of fractal concepts.

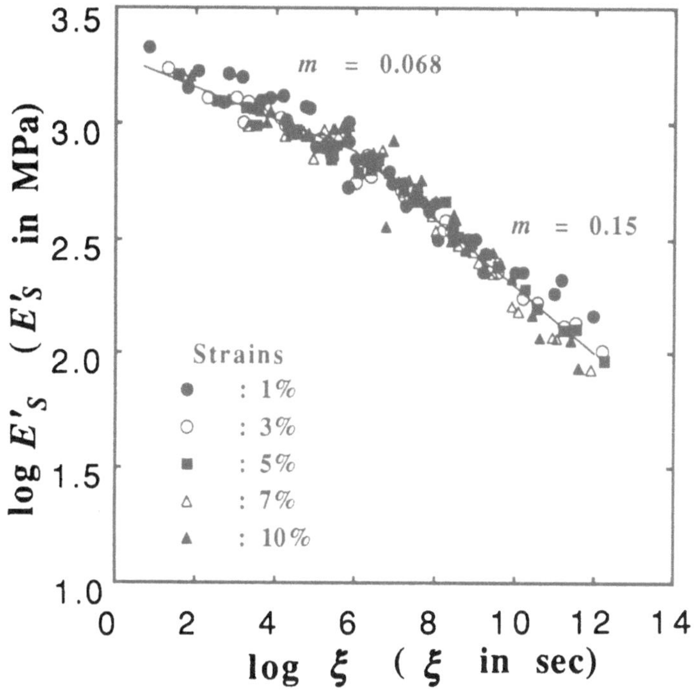

Fig. 2. Strain-reduced secant modulus (E'_S) vs. temperature reduced time (ξ) master curve at 1% reference strain and 100 °C reference temperature, data from Fig. 1 (partly modified from Shimamoto 1987, Fig. 5; data from Heard 1972). Solid line can be expressed by eq.(6).

Relaxation Modulus of Rocks

Geometrically, the relaxation modulus can be interpreted as being the slope of a tangent line of stress-strain curve at a given finite strain ε and the secant modulus can be interpreted as being the slope of a straight line between a point at a finite strain and a point at an initial condition on the stress-strain curve. Now if we define the secant modulus E_S (t) by the ratio of a stress to a finite strain ε, then the secant modulus can be represented mathematically by

$$E_S(t) = \frac{\sigma}{\varepsilon} = \frac{1}{t}\int_0^t E(\nu)d\nu \tag{4}$$

where t is the deformation time and $E(\nu)$ is the relaxation modulus at the deformation time ν. Furthermore, the relation between the secant modulus and relaxation modulus $E(t)$ at the deformation time t,

$$E(t) = E_S(t)\left\{1 + \frac{d\log E_S(t)}{d\log t}\right\} \tag{5}$$

can be obtained from eq.(4) (Schapery 1974).

From Figs. 1 and 2, the strain-reduced secant modulus E_S (ξ) at a temperature reduced time ξ can be expressed by

$$\mathrm{E_S}\,(\xi) = h(\varepsilon)\,E'\,\xi^{-m}\,,\ h(\varepsilon) = 1/g(\varepsilon),\ h(0) = 1 \tag{6}$$

Moreover, the relaxation modulus $E(\xi)$ at a temperature reduced time ξ can be written as

$$E(\xi) = (1-m)E'\xi^{-m} \tag{7}$$

This equation says that the relaxation modulus of the nonlinear viscoelastic constitutive equation (eq. 2) is a power-law equation of the temperature reduced time. In other words, this indicates that the relaxation of high temperature rocks shows a fractal property with respect to the temperature reduced time.

Time-Temperature Superposition Principle

In eqs.(6) and (7), the temperature reduced time ξ is given by

$$t = a_T\,\xi \tag{8}$$

where a_T is the temperature shift factor (Shimamoto 1987). This equation indicates that the deformation time can be changed into temperature. This principle is called the time-temperature superposition principle and expresses that the effect of temperature on time dependent mechanical behavior is equivalent to a shifting of real time for temperatures above (or below) the reference temperature (for example, Tobolsky 1958, Findley et al. 1976). In other words, the behaviour of rocks at high temperature and high strain rate is similar to that at low temperature and low strain rate. Materials exhibiting this property are called "thermorheologically simple". Therefore, rocks (e.g., halites) at high temperature behave thermorheologically simply. In this case, the temperature shift factor a_T can be represented by

$$a_T = \exp\left(\frac{Q}{RT}\right) \tag{9}$$

Moreover, the Zener-Hollomon factor $Z(\sigma)$ can be written as

$$\frac{d\varepsilon}{d\xi} = \frac{d\varepsilon}{dt}a_T = Z(\sigma) \tag{10}$$

Stress Sensitive Exponents of Rocks

Several researchers have studied the stress sensitive exponents n of several rock materials at different experimental conditions. These exponents range from 1.0 to about 4.0 (Kirby & Kronenberg 1987) and show various values at different deformation modes (Table 1). For example, exponent values in the diffusion creep mode are equal to 1 and exponent values in the dislocation creep mode are equal to 3 or 5. Moreover, those in the recovery creep mode range from 4.5 to 6.

Table 1. Stress sensitivity of strain rate and creep modes.

Creep type	Exponent (n)
Diffusion creep	
Nabarro-Herring creep	1
Cobble creep	1
Dislocation Creep	
Subgrain creep	3
Nabarro creep	
Lattice diffusional model	3
Dislocation core diffusion model	5
High temperature recovery creep	
Edge climbing model	4.5
Jog-dragging screw model	6
Micro-creep (Dislocation glide creep)	3

Comparing Dorn's equation with the nonlinear constitutive equation (eq. 2), the stress sensitive exponent n and the constant A are given by

$$n = \frac{1}{m}, \quad A = \left\{ \frac{g(\varepsilon)}{\varepsilon E'} \right\}^{\frac{1}{m}} \left(\frac{\varepsilon}{C} \right) \tag{11}$$

In Fig. 2, the exponent value of small reduced times is equal to $n = 15$ ($m = 0.068$) and the exponent value of large reduced times is equal to $n = 6.7$ ($m = 0.15$) (Shimamoto 1987). However, we cannot derive high exponent values from the existing creep models. Thus, next I will discuss the relationship between the relaxation of rocks and stress sensitive exponents from the viewpoint of fractal concepts.

Universal Relaxation Functions of Rocks

From eq.(8), the relaxation function $\phi(t)$ for eq.(7), which is equal to $E(t)$, is generalized by

$$\phi(t) = \alpha t^{-m} \tag{12}$$

where α is a constant. This equation says that the relaxation function for high temperature flows of rocks shows a temporal fractal property. A similar equation has also been proposed for other materials by Chua & Henderson (1991). Moreover, this equation is also a special form of the Nutting equation (Nutting 1921, Scott Blair & Caffyn 1942, Dingle 1949, Scott Blair & Reiner 1950, Scott Blair 1967).

Relaxation similar to that expressed by eq.(12) occurs not only in the high temperature flow of rocks but also in the mechanical relaxation of other materials (Ferry & Williams 1952, Scher et al. 1991). It was first found by Weber and Gauss about 150 years ago and is called a long time tail (Takayasu 1990).

A simple explanation for this empirical function is to regard a decay function $\phi(t)$ as the result of the superposition of the usual exponential decay functions with different relaxation times. Saito & Maruyama (1987) represented this relaxation function by the Laplace transform of the distribution function of the life time τ_l as follows:

$$\phi(t) = \int_0^{\infty} p(\omega)\omega \cdot \exp(-\omega t)\,d\omega, \quad \omega = 1/\tau_l \tag{13}$$

The tables of Laplace transforms and the considerable literature on their application to quite different sorts of problems may find some application to the analysis of non-exponential relaxation functions. See for example, Carslaw and Jaeger (1941), Doetsch (1947) and Maguire et al. (1952). This superposition principle is similar to that of Wiechert's viscoelastic model which is the general Maxwell model in a linear viscoelastic model (Wiechert 1893 a, b, Kuhn 1939). For this relaxation function $\phi(t)$, the distribution function $p(\omega)$ can be easily obtained by the inverse Laplace transformation of Eq.(13). Therefore, we can get the distribution $p(\omega)$ by

$$p(\omega) = \frac{1}{2\pi i\,\omega}\int_{-i\infty}^{i\infty} \phi(t) \cdot \exp(\omega t)\,dt = \frac{C\omega^{-(m+2)}}{\Gamma(m)} \tag{14}$$

where $\Gamma(m)$ is a gamma function and C is a constant. If we rewrite the exponent value of eq.(14) as

$$D = m + 1 \tag{15}$$

then Eq.(14) can be rewritten as

$$p(\nu) \propto \nu^{-(D+1)} \tag{16}$$

This equation indicates that the distribution function $p(\omega)$ expresses the fractal size distribution of material elements with different life times τ_l, where the D-value is the fractal dimension for the size distributions of different life-times.

Discussion and Summary

In this chapter, a universal distribution function of the relaxation time was proposed for the power-law relaxation function usually observed in relaxation mechanical phenomena. This distribution function $p(\omega)$ is one in the group of Lévy distribution functions (Takayasu 1987). The Lévy distribution function has often been discussed in connection with fractal concepts (Mandelbrot 1982, Schroeder 1991). It is quite likely that the relaxation function (eq.12) or the universal relaxation function (eq.13) originates in the fractal geometry or the fractal time stochastic process (Shlesinger 1984, Klafter & Shlesinger 1986, Ito et al. 1992) caused by disordered structures (defects) in materials. The exponent value n $(=1/m)$ is not universal and is related to the fractal structure, which depends on the creep mode of rocks. Therefore, we can draw the interesting conclusion that the stress sensitivity of the strain rate is characterized by a fractal texture with different relaxation rates in different rocks. Further analysis of material structures (micro-texture) and their relaxation behaviours in various rock materials will be required to better understand their fractal characteristics.

Acknowledgements. I thank J. H. Kruhl and two anonymous reviewers for insightful comments which improved the manuscript. The author would like to thank Robert M. Ross for improving the English from an earlier version of this manuscript.

References

Carslaw HS, Jaeger JC (1941) Operational methods in applied mathematics. Clarendon Press, Oxford.

Chua SM, Henderson PT (1991) Changes in microhardness and creep modulus during the ageing of polypropylene. J Mater Sci Lett 10: 1379-1380.

Dingle H (1949) On the dimension of physical magnitudes (Seventh paper; a paradox in dimensional theory). Phil Mag, Seven ser 40: 94-99 .

Doetsch G (1947) Tabellen zur Laplace Transformation und Anleitung zum Gebrauch. Springer, Berlin.

Ferry JD, Williams ML (1952) Second approximation methods for determining the relaxation time spectrum of a viscoelastic material. J Colloid Sci 7: 347-353.

Findley WN, Lai JS, Onaran K (1976) Creep and relaxation of nonlinear viscoelastic materials with an introduction to linear viscoelasticity. North-Holland Publishing Company.

Heard HC (1972) Steady-state flow in polycrystalline halite at pressures of 2 kilobars. Am Geophys Union Monograph 16: 191-210.

Ito HM, Ogura Y, Tomisaki M (1992) Stretched-exponential decay laws of general defect diffusion models. J Stat Phys 66: 563-582.

Kirby SH, Kranenberg AK (1987) Rheology of the lithosphere: selected topics. Rev Geophys 25: 1219-1244.

Klafter J, Shlesinger MF (1986) On the relationship among three theories of relaxation in disordered systems. Proc Nat Acad Sci USA 83: 848-851.

Kuhn W (1939) Beziehungen zwischen Viscostät und elastischen Eigenschaften amorpher Stoffe. Zeit Phys Chem 42: 1-38.

Maguire BA, Pearson ES, Wynn AHA (1952) The time intervals between industrial accidents. Biometrika 39: 168-181.

Mandelbrot BB (1982) The fractal geometry of nature. Freeman, San Francisco.

Nutting PG (1921) A study of elastic viscous deformation. Proc Am Soc Testing Mater 21: 1162-1171.

Saito R, Murayama K (1987) A universal distribution function of relaxation in amorphous materials. Solid State Commun 63 (7): 625-627.

Schapery RA (1969) On the characterization of nonlinear viscoelastic materials. Polym Eng Sci 9: 295-310.

Schapery RA (1974) Viscoelasticity of solids and structures. Lecture Notes at Texas A & M University, Texas.

Schapery RA (1982) Development of cyclic nonlinear vicoelastic constitutive equations for marine sediment. In: Dungar R, Pande GH, Studer JA (eds.) Proc Int Symp on Numerical Models in Geomechanics. Balkema Rotterdam Zürich, pp 13-17.

Scher H, Shlesinger MF, Bendler J (1991) Time-scale invariance in transport and relaxation. Physics Today 44: 26-34.

Schroeder M (1991) Fractals, chaos, power laws, Freeman, San Francisco.

Scott Blair GW (1967) A model to describe the flow curves of concentrated suspensions of spherical particles. Rheol Acta 6: 201-202.

Scott Blair GW, Caffyn J (1942) The classification of the rheological properties of industrial materials in the light of power-law relations between stress, strain and time. J Scientific Instruments 19: 88-93.

Scott Blair GW, Coppen FMV (1939) The subjective judgement of the elastic and plastic properties of soft bodies; the "differential thresholds" for viscosities and compression model. Proc Roy Soc London, Series B, Biol Sci 128: 109-125.

Scott Blair GW, Reiner M (1950) The rheological law underlying the Nutting equation. Appl Sci Res. A mechanics, heat chemical engineering mathematical method 2: 225-234.

Shlesinger MF (1984) Williams-Watts dielectric relaxation: A fractal time stochastic. J Stat Phys 36: 639-648.

Shimamoto T (1987) High temperature viscoelastic behavior of rocks. Proc 7th Japan symp rock mech 467-472. (in Japanese with English abstract)

Smith TL (1962) Nonlinear viscoelastic response of amorphous elastomers to constant strain rates. Trans Soc Rheol 6: 61-80.

Takayasu H (1984) Stable distribution and Lévy process in fractal turbulence. Prog Theoret Phys 72, 471-479.
Takayasu H (1987) $f^{-\beta}$ power spectrum and stable distribution. J Phys Soc Japan 56: 1257-1260.
Takayasu H (1990) Fractals in the physical sciences. Manchester Univ Press, Manchester New York.
Tobolosky AV (1958) Stress relaxation studies of the viscoelastic properties of polymers. In: Eirich FR (ed.) Rheology, theory and applications. Academic Press Inc., New York (Volume II).
von Wiechert E (1893a) Gesetze der elastischen Nachwirkung für constante Temperatur. Wied Ann Phys Chem 50: 336-348.
von Wiechert E (1893b) Gesetze der elastischen Nachwirkung für constante Temperatur. Wied Ann Phys Chem 50: 546-570.

The Fractal Geometry of Patterned Structures in Numerical Models of Rock Deformation

Alison Ord
CSIRO Division of Exploration and Mining,
Private Bag, PO Wembley, WA 6014, Australia

'... and I doubt that we are founding a new science, but at least we are having fun.'
Nonlinear dynamics, chaos and mechanics. P. Holmes, 1990

Abstract. Geologic structures are repetitive in a quasi-periodic or erratic manner. The geometries of these structures are also manifestations of the mechanical behaviour of a deforming rock mass. Can we therefore obtain information on the *dynamics* of a complex geologic system from an examination of the *geometry* of geologic structures? We examine this question from the point of view of non-linear dynamics.

We investigate the variability in space of the velocity of growth of crenulations in a model rock mass which is undergoing a (numerical) simple shearing deformation. That is, we follow the distribution and evolution of the velocity in space, rather than in time. We investigate the thesis that the behaviour of this one variable reflects the presence of all other variables participating in the dynamics and, by use of increasing multiples of a fixed space lag, discretize the system, and unfold the system's dynamics into a multidimensional phase space. The trajectories within this phase space of the system converge to a subspace which is the geometrical attractor for the system. We infer from this that our deforming model rock can be described by a set of deterministic laws. The dimension of this attractor is about 2.5; that is, the system may be completely represented by a fractal attractor. Further, this fractal attractor embeds in a phase space of three so that at least three variables must be considered in the description of the underlying dynamics. These are the variables involved in the three independent differential equations of the numerical model: the stress equations of motion, the yield criterion and the flow rule. We conclude that geological systems may be successfully modelled on the basis of such a system of equations, and analysed using the concepts of fractal geometry.

This new application of nonlinear dynamics to the spatially erratic structures of deformed rocks results in an improved understanding of rock deformation behaviour and in an improved prediction of the distributions of structurally-controlled phenomena.

Introduction

The patterns observed in geological structures such as crenulations, crenulation cleavages, and shear zones are typically arranged in ways which are close to, but not strictly, periodic. In this chapter, we shall refer to these patterns as quasi-periodic. Chaos is an attribute of any system which is extremely sensitive to

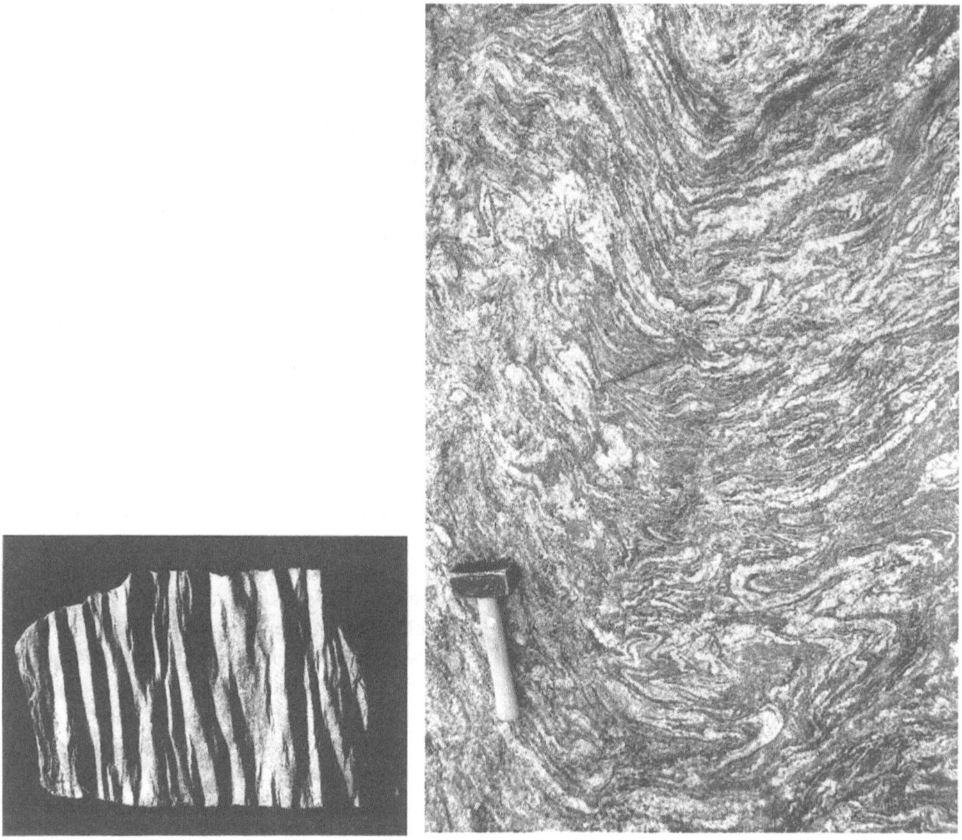

a b

Fig. 1. a. Deformed, crenulated rock with quasi-periodic spatial distribution of fold axes. Maximum dimensions are 13 x 8 cm. **b.** Multiple pegmatite intrusion and folding during progressive deformation. Albany Mobile Belt, Southern Domain migmatitic gneiss, Western Australia. Hammer length is 30 cm. Photograph by Lyal Harris.

initial conditions. In general, this means that adjacent points behave quite differently in time and space. In the quasi-periodic structures characteristic of most deformed rocks, strict chaos, in either time or space, is unlikely because adjacent points in the deforming body commonly follow closely-related paths in time and space throughout their history in order to preserve the continuity of the deforming rock mass. Although rock masses such as that shown in Fig. 1a are irregular, they are not chaotic because adjacent points have followed closely-related deformation paths. However, the geometric structures observed in multiply-deformed terrains associated with migmatites (Fig. 1b) are of the highly irregular and multiply-stretched and folded nature of chaotic systems. The patterning of these quasi-periodic structures in space is analogous to the patterning in time of, for example, electroencephalograms (Albano et al. 1986) and the responses in some electrical systems (Van der Pol 1927). More erratic or chaotic phenomena include the weather (Lorenz 1963, 1979) and Couette flow (Gollub & Swinney 1975). Geological phenomena which have been observed and analysed to be patterned in time include earthquakes (Hobbs 1990), volcanic tremor and gas piston events (Chouet & Shaw 1991) and volcanic eruptions (Sornette et al. 1991).

Quasi-periodicity and chaos may be described as natural phenomena associated with the behaviour of a dynamic, self-organising system. The thesis here is that a deforming rock is as much such a dynamic, self-organising system as is the weather or the more controlled example of Couette flow and is therefore amenable to the same analytic techniques. However, in geology as in all other disciplines, sets of linear differential equations are simpler to solve than sets of non-linear differential equations, so that until now, geological structures have been analysed in terms of strictly periodic or regular patterns (e.g. Biot 1961, Ramberg 1964).

It is therefore considered feasible to analyse geological structures using the mathematical techniques developed in discussions of theories describing temporally patterned phenomena, and in the practical analysis of the resultant data. We describe here the data resulting from a simple numerical model for a deforming rock mass, and construct a spatial attractor for the spatially-disposed pattern in a manner analogous to that in which Packard et al. (1980, see also Takens 1981, Crutchfield et al. 1986) obtained the attractor for a temporally-disposed system.

We need to quantify the information contained in the attractor. Here, we do this by determining the dimension of the spatial attractor for the system through description of the attractor in different embedding dimensions. A collection of points produced by a random process will always tend to fill space so that, for n-dimensional space, the dimension of this collection of points will be close to n. However, for a structured system, as n increases, the attractor of the system will not increasingly fill space, and its dimension will become independent of n. This saturated dimension is then the dimension of the system. If this dimension is fractal (Mandelbrot 1977), then, by comparison with Lorenz's (1963) discovery of a 'strange attractor', the phenomenon is worth further investigation. Its dimension may then indicate the minimum state space needed to reproduce the attractor, and further investigation of its topological behaviour may lead to significant new theoretical results, as well as applications; for example, prediction of the distributions of vein-hosted ore deposits. In particular, the importance of the true dimension of the attractor is that it reflects the number of degrees of freedom of the system, and this number is indicative of the number of independent first-order, nonlinear, differential equations which govern the system.

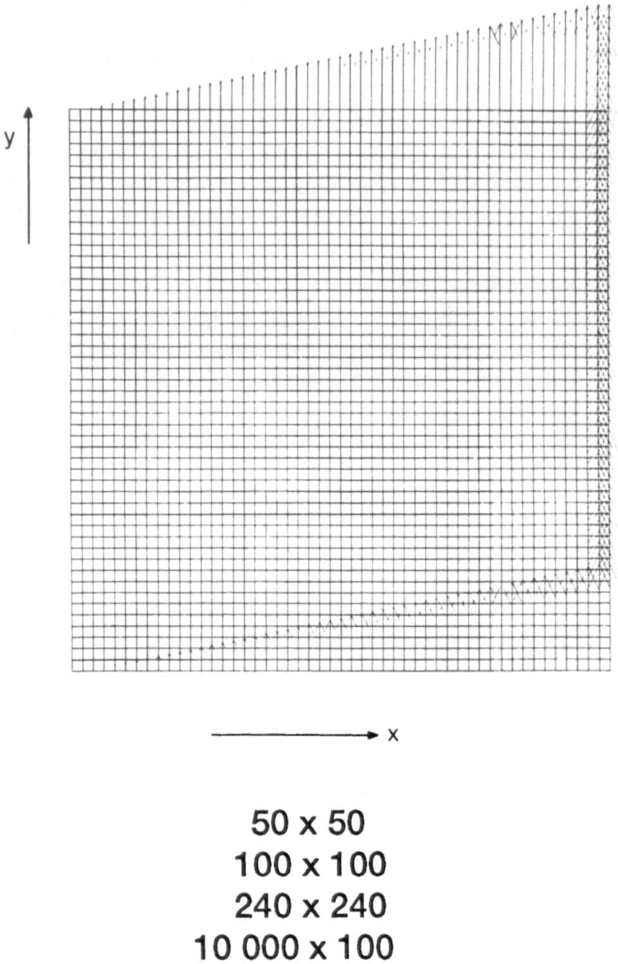

Fig. 2. Finite difference grid of 50 x 50 zones. The arrows represent the direction and relative magnitudes of the velocities applied to the external nodes throughout the deformation.

The aim of this work is therefore to determine and improve information on the constitutive behaviour of rock masses and on the governing equations for rock deformation, from observations of the geometry of geological structures. For example, for description of a simply deforming, non-hardening rock mass, we know that we have interaction between the stress equations of motion, the yield criterion, and the flow rule. Here the dimension of the resulting spatial attractor is expected to be between 1 and 3. Further relationships are required to model more sophisticated behaviour such as strain-hardening or softening, volume change with deformation, coupling of volume change and permeability, fluid flow coupled with deformation, stress dependence of grain size, and so on. Incorporation of the equations describing these additional phenomena may be required to increase the dimension of the attractor established for the numerical models up to that of the attractor observed for natural systems.

The recognition that the state of such a system may be measured by its fractal dimension provides a new, exciting and rigorous framework within which to improve our understanding of how rock deforms and to understand and to predict the patterning of geological structures.

We first describe the development of a quasi-periodic crenulated structure using a numerical model, and indicate the input and output of this model. Second, we describe how we construct a spatial attractor for the model. A method for calculation of the dimension of the spatial attractor is then summarised. Third, we apply this methodology to analysis of the numerical data, and fourth, we discuss the results in terms of the interaction of the equations describing the numerical model, and consider the application of this method to naturally deformed rocks.

The Model

Numerical modelling of a non-hardening, frictional-dilatant, Mohr-Coulomb material undergoing a simple shearing deformation history results in patterned shear band formation exhibiting both temporally and spatially weakly chaotic or quasi-periodic behaviour (Ord 1990). Similar behaviour is observed also by Cundall (1989, 1991) and by Hobbs & Ord (1989) for constant velocity, plane strain shortening.

The computer code FLAC (Fast Lagrangian Analysis of Continua, Cundall & Board 1988) is used for simulation of rock deformation behaviour. FLAC is a plane-stress or plane-strain explicit finite difference code based upon a Lagrangian calculation scheme suitable for modelling the non-linear behaviour and the large deformation characteristic of geologic structures. It incorporates the basic governing equations for a solid body, that is, the stress equations of motion and the constitutive relations. Further details may be obtained from Board (1989) and FLAC (1991). Physical conditions required for localization in geological materials are discussed and reviewed by Hobbs et al. (1991) who also describe some examples modelled by FLAC.

In FLAC (FLAC 1991) a dynamic relaxation type scheme is employed. In any one time step, new velocities and displacements are derived from stresses and forces through the equations of motion. Strain rates are then derived from the velocities, and new stresses from these strain-rates according to the stress/strain rate relationship (constitutive law) for the deforming material.

The basic equation of motion used is:

$$m\dot{v} = F,$$

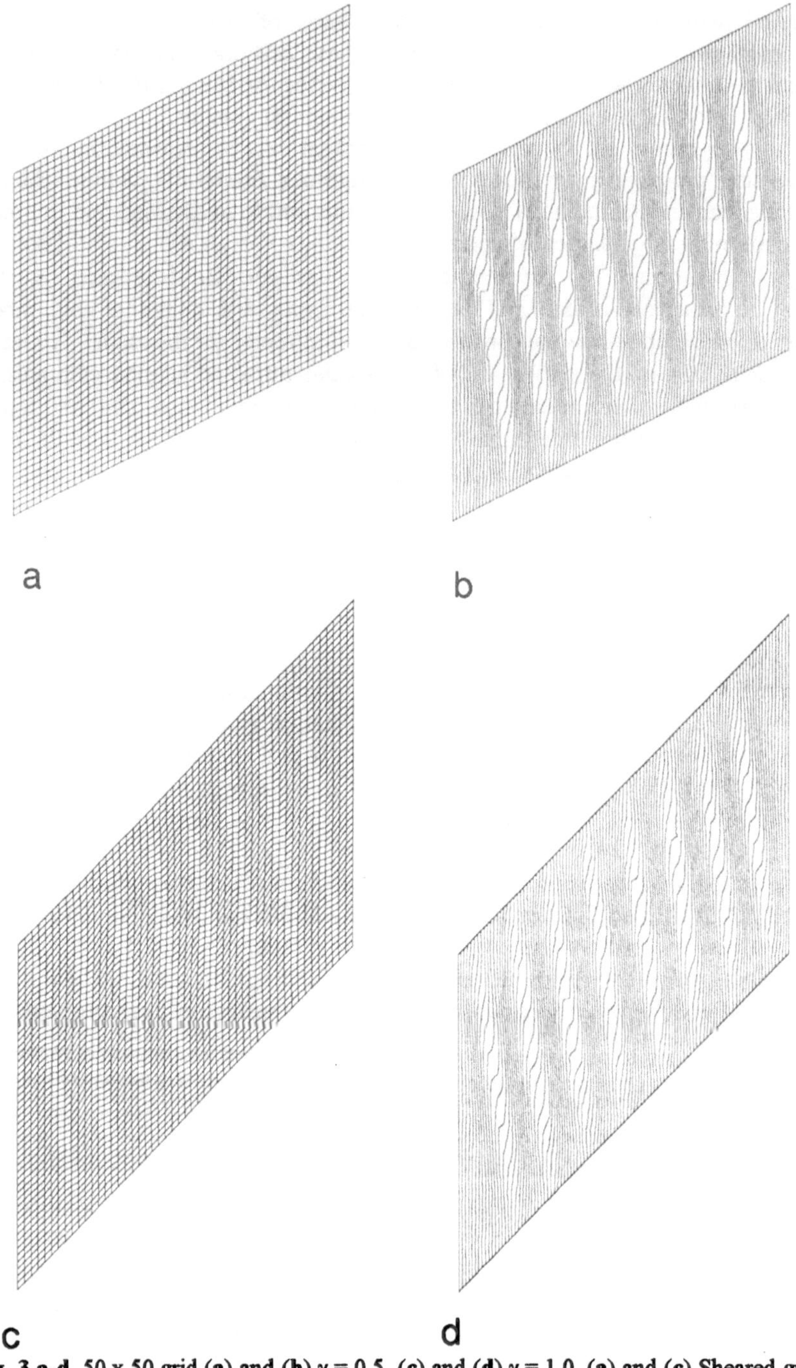

Fig. 3 a-d. 50 x 50 grid (**a**) and (**b**) $\gamma = 0.5$. (**c**) and (**d**) $\gamma = 1.0$. (**a**) and (**c**) Sheared grids. (**b**) and (**d**) Contours of the instantaneous y-velocity. The contour interval is 2.5×10^{-5} units per time step in each case.

which describes a mass, m, which is moved at an acceleration, $\dot{v}$ by a force, F, which may vary with time, t.

The generalised stress equations of motion are:

$$\rho \dot{v}_i = \frac{\partial \sigma_{ij}}{\partial x_j} + \rho g_i$$

for ρ the mass density, x_i the components of the position vector, g_i the components of gravitational acceleration, and σ_{ij} the components of the stress tensor. Indices i,j denote components in a Cartesian coordinate frame, and summation is implied for repeated indices in an expression.

The strain rate is then derived from the velocity gradient according to

$$D_{ij} = \frac{1}{2}\left[\frac{\partial v_i}{\partial x_j} + \frac{\partial v_j}{\partial x_i}\right],$$

for D_{ij} the stretching components and v_i the velocity components.

Finally, the stress is derived from the strain rate according to a constitutive law which is of the form (FLAC, 1991, Eq. 3-4)

$$\sigma_{ij} := M\,(\sigma_{ij}\,,\, D_{ij}\,,\, \kappa)$$

where M(...) is the functional form of the constitutive law, κ is an optional history parameter or set of parameters, and : = means 'defined by'.

The external loading is increased sufficiently slowly so that at each time step, the artificial dynamics of the relaxation scheme can be considered as decayed or subsided.

The interest here is the simplicity and known basis of the numerical model in contrast to the complexity and unknown basis of geological structures.

Geometry and Boundary Conditions

The finite difference grid chosen for most of these simple shearing deformation history experiments contains n x n zones. n needs to be large for any analysis of the data to be of use in terms of fractal geometry. In this case, we investigate the effects of having n equal to 50, 100 and 240. In order to bridge the gap with analysis of a time series, we also investigate a grid of 10 000 x 10 zones (n x m). The corners of each zone are defined by nodes. The left hand column of nodes in Fig. 2 was given zero velocity while the most extreme right hand column of nodes was given a velocity of n x 10^{-4} units per time step, parallel to y, where a unit is the length of an initial element. The rows of external nodes between the outermost columns were given velocities varying from 1 to (n - 1) x 10^{-4} units per time step so that, together with the condition of plane strain, the bulk deformation was constrained to be isochoric (constant volume). The remaining

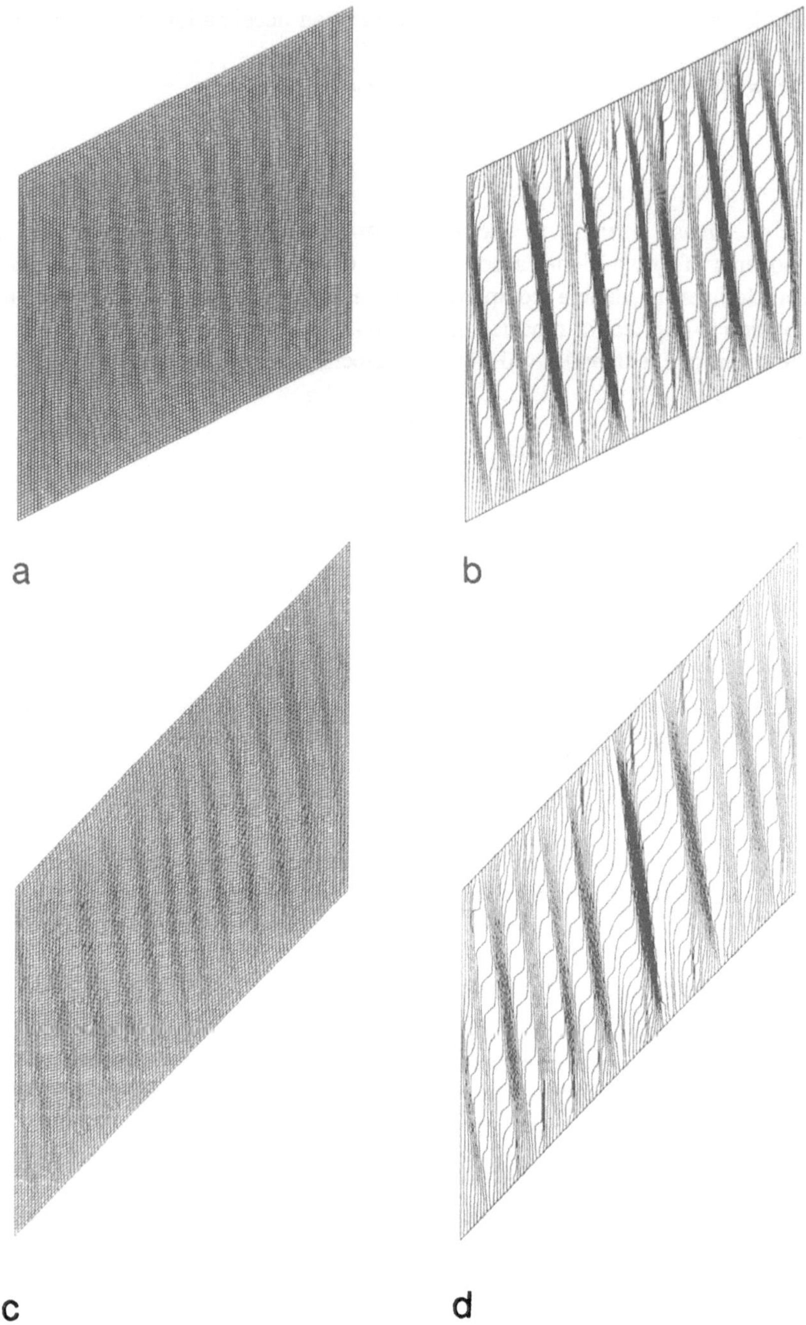

Fig. 4 a-d. 100 x 100 grid. (**a**) and (**b**) $\gamma = 0.5$. (**c**) and (**d**) $\gamma = 1.0$. (**a**) and (**c**) Sheared grids. (**b**) and (**d**) Contours of the instantaneous y-velocity. The contour interval is 5×10^{-5} units per time step in each case.

nodes were given velocities similarly, but only for the first computational step, resulting initially in an homogeneous simple shearing of the material. Thereafter, the n^2 or n x m deforming zones may dilate or contract, and undergo whatever deformation is consistent with the local kinematics and dynamics so long as the overall boundary conditions are satisfied. A total of 10 000 time steps results in a shear strain of 1.

Material Properties

The material is equivalent to that described by Hobbs & Ord (1989) in that it is isotropic and elasto-plastic with no hardening. The material follows a non-associated Coulomb constitutive law (Vermeer & de Borst 1984, Hobbs et al. 1991) as a result of the non-equality of the friction and dilation angles. The magnitudes of these angles are the same as those used in Ord (1990, 1991).

All elements have the same elastic shear modulus (1 GPa) Poisson's ratio (0.125), cohesion (10 MPa), friction angle (30°) and dilation angle (10°). One or two elements were, in some experiments, given higher elastic moduli (shear modulus of 10 GPa), but the same plastic properties. They behave as an elastically hard inclusion within a deforming mass. The shear band formation appears to be triggered by the inherent instability of the non-associated constitutive law as well as by the presence of this initial material heterogeneity, as described also by Cundall (1989, 1991). However, different positions of these elements within the grid made no difference to the stress/strain curve or to the results of the following analysis.

Results

Geometrical results only are presented. Mechanical results are described in Ord (1990), Hobbs et al. (1991) and Ord (1991).

We wish to examine the spatial distribution of one component of the system. In this instance, the component of the system is the spatial distribution of the instantaneous velocity of growth of the crenulations. The velocity of growth of the crenulations is represented by determining the velocity in the y-direction (Fig. 2), parallel to the shearing direction, for each node. The patterning of the grids, which represents a finite displacement, together with the contours of instantaneous y-velocity, are shown for shear strains of 0.5 and 1.0 in Fig.s 3 (50 x 50 grid), 4 (100 x 100 grid) and 5 (240 x 240 grid). 'Cross-sections' through these contours are shown in Fig. 6 for comparison with the results of the 10 000 x 100 grid.

The patterning appears periodic for the 50 x 50 grid, and quasi-periodic for the 100 x 100, 240 x 240 and 10 000 x 100 grids. This interpretation is confirmed by plots of both the Fourier transform and the autocorrelation of the data, as shown

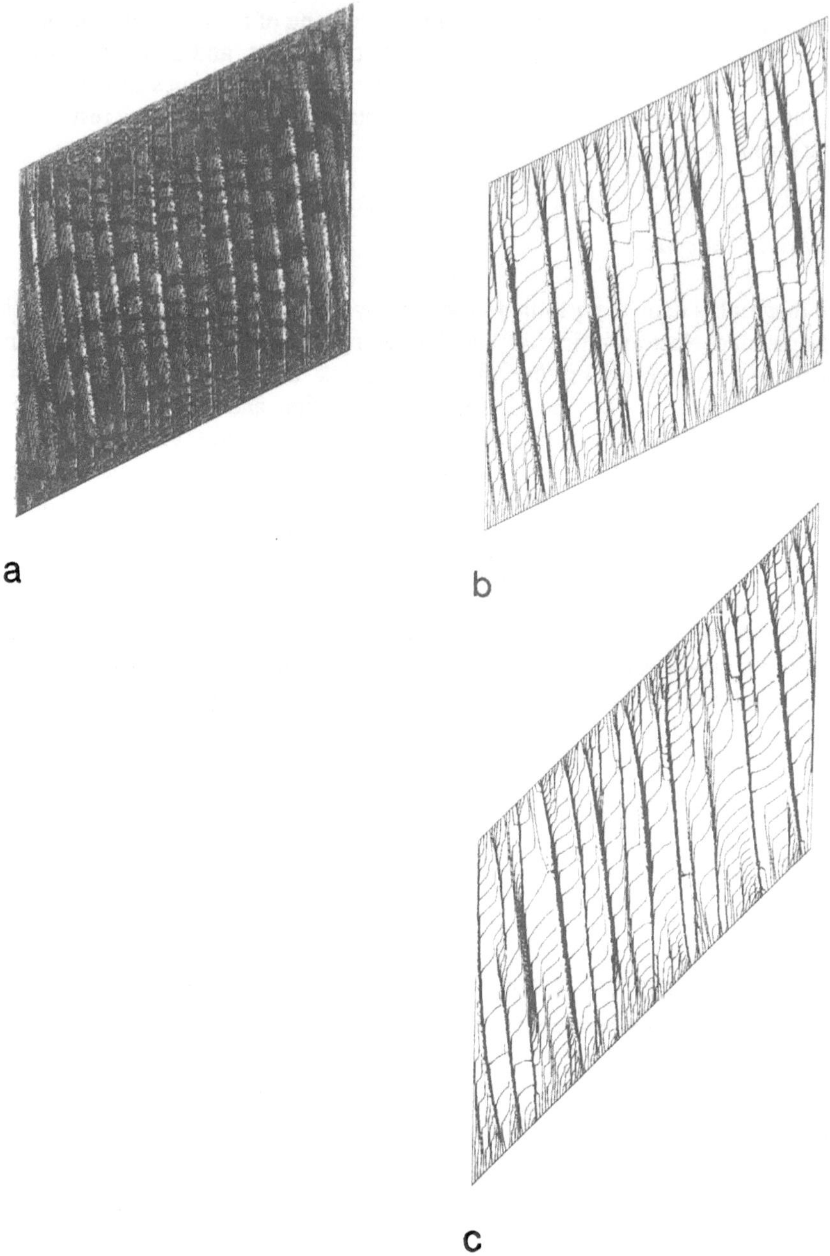

Fig. 5 a-c. 240 x 240 grid. (**a**) and (**b**) $\gamma = 0.5$. (**c**) $\gamma = 1.0$. (**a**) Sheared grid. (**b**) and (**c**) Contours of the instantaneous y-velocity. The contour interval is 1×10^{-4} units per time step in each case.

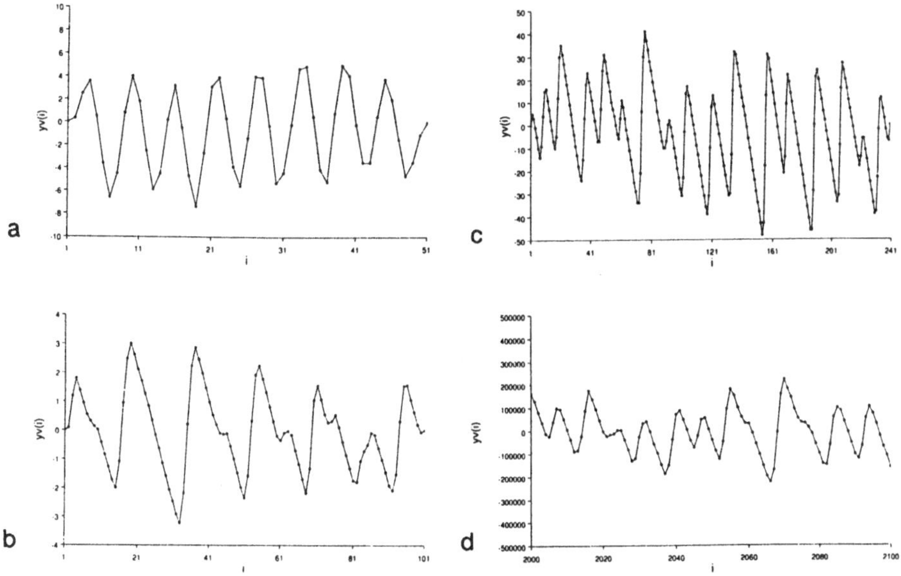

Fig. 6 a-d. Instantaneous y-velocity yv(i) versus node i for one row of each grid for a finite shear strain of 0.5. (**a**) 50 x 50 grid. row 26. (**b**) 100 x 100 grid. row 51. (**c**) 240 x 240 grid. row 121. (**d**) 10 000 x 100 grid. row 51.

in Fig. 7. Figure 8 displays variety within one row. Both the broad-banded spectrum underlying the periodic spikes of the Fourier transform and the spike at x = 0 of the autocorrelation increase in intensity as the grid length increases indicating an increase in wide band noise. The strong periodicity displayed in Fig. 7a-c degenerates to responses indicative of wide band noise superimposed by narrow band noise. The dominant microstructural wavelength varies from about 6 to 12 units.

Phenomena such as possible period doubling require further investigation.

We wish to determine first, if this behaviour does indeed represent the behaviour of a dynamic, self-organising system, and if it does, then how many degrees of freedom has the system?

Methodology of Dimensional Analysis

Temporally erratic behaviour has been analysed for a variety of non-linear dynamic systems; here we investigate spatially erratic behaviour. Techniques for analysis of spatiotemporal chaos have been described by, for example, Mayer-Kress & Kaneko (1989) for coupled map lattices. We construct an attractor for the spatially disposed pattern in the same way that Packard et al. (1980, see also Takens 1980, Crutchfield et al. 1986) obtained the attractor for a temporally-

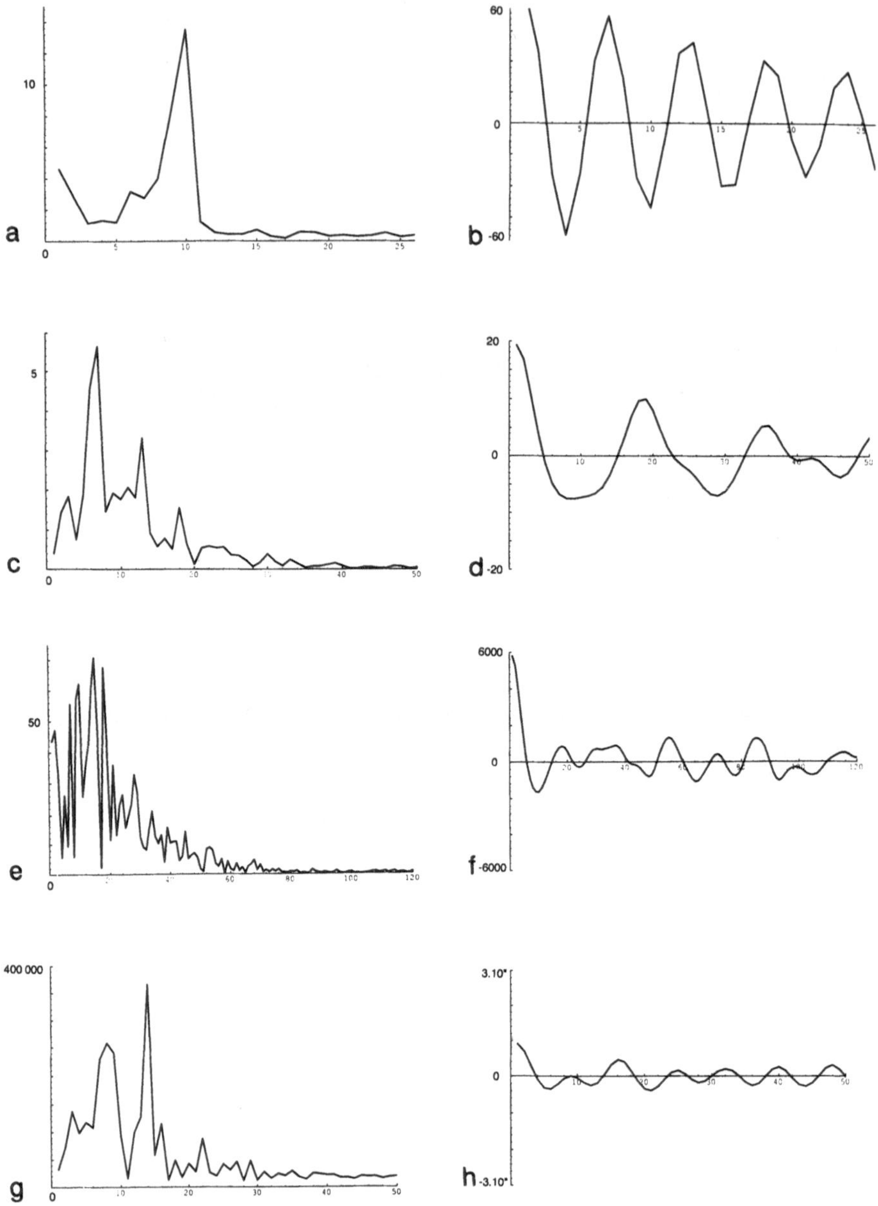

Fig. 7 a-h. Fourier transforms (**a**,**c**,**e**,**g**) and autocorrelations (**b**,**d**,**f**,**h**). (**a**,**b**) 50 x 50 grid. row 26. (**c**,**d**) 100 x 100 grid. row 51. (**e**,**f**) 240 x 240 grid, row 121. (**g**,**h**) 10 000 x 100 grid. row 51.

disposed pattern. We recognise that the evolution of one component of a system is determined by its interaction with other components, and reconstruct an equivalent' state space by examining the measured values at fixed space (rather than fixed time) delays with respect to a single component as though these space delays were new dimensions in this 'equivalent' state space. In this instance, the component of the system is the velocity of growth of the crenulations in the numerical modelling experiments. We are then able to examine the dimension of this spatial attractor within different embedding dimensions, and infer the dimension of the system (for a succinct review, see Ruelle 1990). The dimension of the system is representative of the number of degrees of freedom of the system, and therefore of the number of independent, non-linear differential equations required to describe the system.

This methodology is described with respect to the finite difference grid of the numerical model (Fig. 9). The magnitude of the y-velocity is provided for each grid point at each finite difference step. The identification of grid point (i, j) indices is as described in FLAC (1991, Fig. 4-3b). In this instance the space delay calculations are calculated separately for each row; that is, in terms of i, not in terms of j. However, the information obtained for all values of j is incorporated when describing the spatial phase portraits or attractors for the different embedding dimensions. The calculations may therefore be described as follows.

A single trajectory in p-dimensional 'equivalent' state space is described for $i = 1, 2, \ldots m$ and $j = n$ by constructing a space series $yv(i,j)$, $i = 1, 2, \ldots N$; $N = m - (p - 1) S$ by grouping p values of the space series into p-dimensional vectors

$$yv_{i,j}^{(p)} = (yv\,(i,j), yv(i+ S,j), \ldots , yv\,(i + (p - 1)\,S,j)) \qquad (1)$$

where S is the space delay, and represents an integral number of grid points, and $yv(i,j)$ is the y component of the instantaneous growth velocity at the grid point (i,j) after subtraction of the overall homogeneous shearing. The dimension p is called the embedding dimension. The attractor for the entire model is described by allowing j to range from 1 to n.

The formation of such a spatial attractor is shown in Figure 10 for a grid size of 100 x 100 zones (101 x101 grid points). The maximum value of N is dependent on m and on p and S so that $N = m - (p-1)S$. Figure 10a shows the form of a single trajectory for $j = 51$ and $S = 10$, in 2 dimensions. Let the embedding dimension be 2, then $N = 101-(2-1)10 = 91$ so that $i = 1, 91$ and the axes are $yv(i,51)$ and $yv(i+10,51)$ or $yv_{i,51}^{(2)} = [yv(i,51), yv(i+10,51]$. The 101 trajectories for the entire grid ($i=1, 91$; $j=1, 101$) are shown in Fig. 10b. Figure 10c may be considered as a 2D section for $S = 20$ and $p = 2$, or for $S = 10$ and $p \geq 3$. Similarly, Figure 10d may be considered as a 2D section for $S = 30$ and $p = 3$, or for $S = 10$ and $p \geq 4$. The only difference would be in the allowed value for N. Further, Figures 10e, 10f and 11 show 3D sections, in this case unambiguously for $S = 10$. Again, unless N is provided, the embedding dimension can only be inferred to be at least 3 in Figure 10e and at least 4 in Figure 10f. Figure 12a shows the attractor for the 240 x 240 grid also for $S = 10$, and from the same viewpoint as Figure 10e.

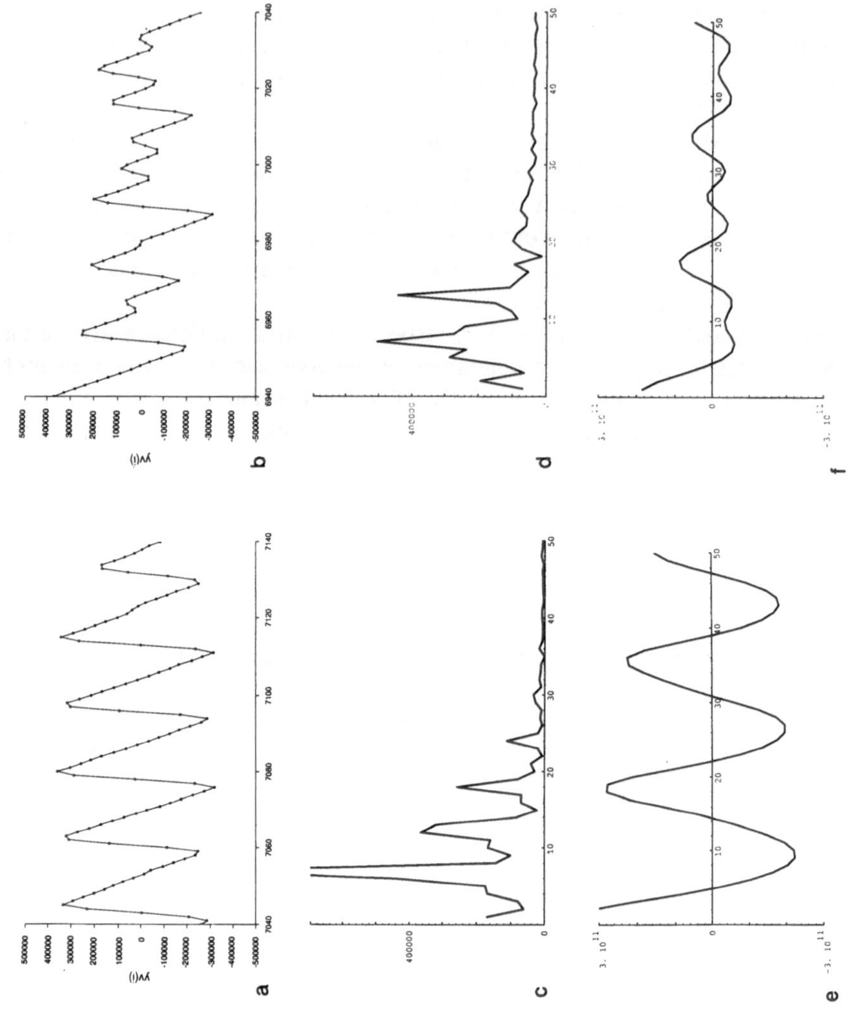

Fig. 8 e-f. 10 000 x 100 grid. Row 51 (**a,c,e**) nodes 7040 to 7140 (**b,d,f**) nodes 6940 to 7040 (**a,b**) Instantaneous y-velocity yv(i) versus node i (**c,d**) Fourier transform. (**e,f**) Autocorrelation.

A random array of 240 x 240 points is shown in Figure 12b for visual comparison of an array which fills space with one which does not fill space (Fig. 12a).

The spatial attractors for different parts of row 51 of the 10 000 x 100 grid are shown in Figure 13. There is a strong similarity between Figure 10a and Figure 13a, which should be expected from examination of Figure 10. Comparison of Figures 13a, b and c shows graphically that the attractor for a single row has

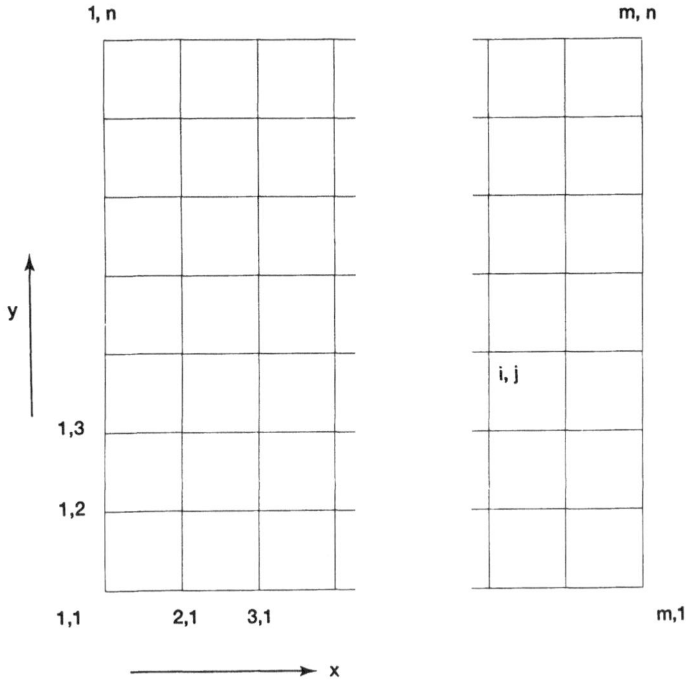

Fig. 9. Numbering system for the nodes of the finite difference grid. For the square grids m = n, and m = 50, 100 or 240. i = j = 1, m. Otherwise, m = 10 000 and n = 100. i = 1, m. j = 1, n.

structure. However, it is obviously not as well-defined as the superimposed attractors from juxtaposed rows.

These attractors are at least as aesthetically pleasing as, for example, the classic Lorenz attractor obtained from the simple equations

$$\left.\begin{aligned} \dot{x} &= \sigma(y-x), \\ \dot{y} &= \rho x - y - xz, \\ \dot{z} &= -\beta z + xy, \end{aligned}\right\} \quad \begin{aligned} &(x,y,z) \in R^3 \\ &\sigma,\rho,\beta > 0 \end{aligned}$$

These equations are used to represent fluid convection and contain three parameters, the Prandtl number, σ, the Rayleigh number, ρ, and an aspect ratio, β, as summarised by Guckenheimer & Holmes (1983). x, y and z are Cartesian coordinates and a dot denotes differentiation with respect to time. Similarities and comparisons with systems such as these provide at least some of the motivation for investigating rock deformation behaviour in terms of self-organisation and non-linear dynamics.

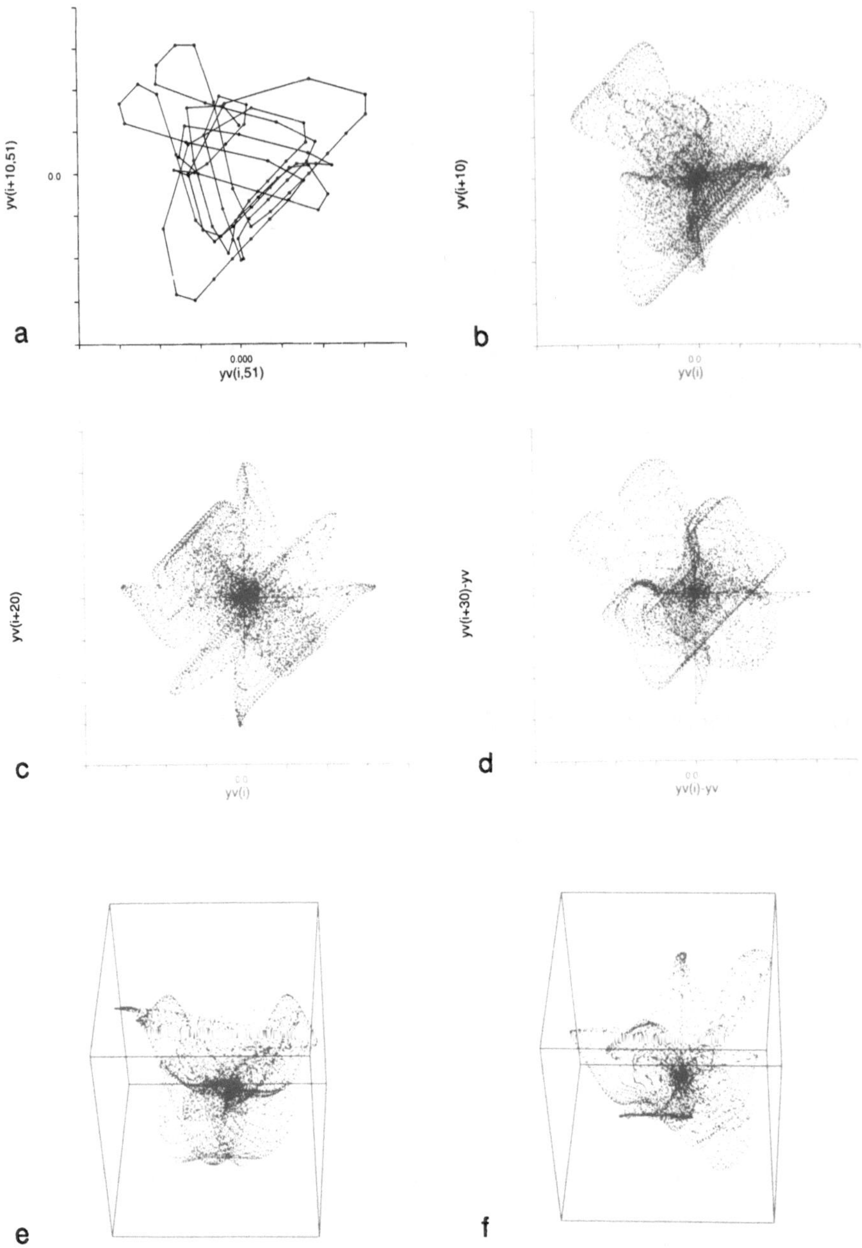

Fig. 10 a-f. 100 x 100 grid. Homogeneous shearing velocity has been subtracted so as to demonstrate only variations in y-velocity with respect to the homogeneous state. **a** single trajectory of the magnitudes of the y-velocities for j = 51 and S = 10. **b** Multiple trajectories of the magnitudes of the y-velocities for j = 1, 101 and S = 10. **c** Multiple trajectories of the magnitudes of the y-velocities for j = 1, 101 and S = 20.

d Multiple trajectories of the magnitudes of the y-velocities for j = 1, 101 and S = 30. **e** and **f** 3D sections of multiple trajectories of the magnitudes of the y-velocities for j = 1, 101 and S = 10. Axes are yv(i), yv(i+10), and yv(i+20) for **e**, and yv(i), yv(i+10), and yv(i+30) for **f**.

The aim therefore is to use the information contained in the attractors presented here as the basis for refining our present models for the description of rock deformation behaviour.

We choose initially to determine the dimension of the spatial attractors described here by a box-counting algorithm. Different techniques (see, for example, Rasband, 1990, Chap. 4.1) may be used if the results of this initial study show that they are required.

The fractal dimension D (Mandelbrot 1977) is given by

$$D = \lim_{\varepsilon \to 0} \{-\ln N(\varepsilon) / \ln \varepsilon\},$$

where N(ε) is the number of boxes of side ε required to cover the attractor in phase space.

This reduces to D = - ln N(ε) /ln ε, for sufficiently small ε so that the slope of a best fit straight line to ln N versus -ln ε gives D.

The numerical code follows the logic described by McGuinness (1983). The variables $yv_{i,j}^{(p)}$ are normalised so that the bounds of the attractor are 0,1 in all directions. The coordinates of the current box entered are given by dividing the $yv_{i,j}^{(p)}$ by the box side e and taking the integer part, and a "1" bit is entered when a box is entered by a trajectory on the attractor.

A CONVEX 220 with 2 processors was used for computations involving full vectorisation, and 790 Mb of real memory (Young 1991).

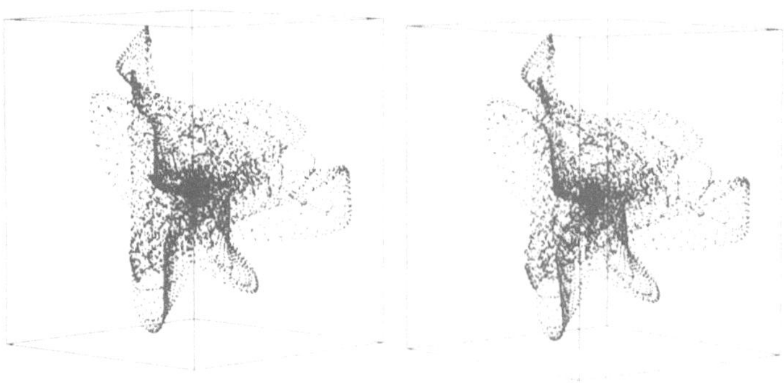

Fig. 11. Stereoscopic images of the phase portrait described in Figures 9e and 9f. Axes are yv (i), yv (i + 10), and yv (i + 20).

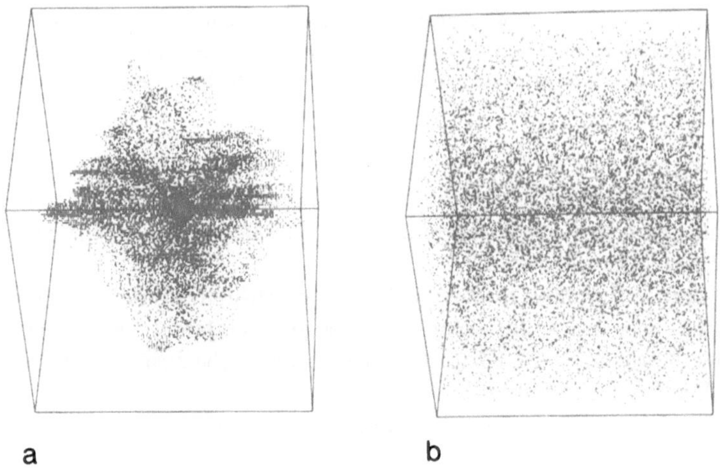

Fig. 12 a,b. a Phase portrait for the 240 x 240 grid for j = 1, 241 and S = 10. **b** Random array of 240 x 240 points. Axes are yv (i), yv (i + 10), and yv (i + 20).

For this initial investigation, the best-fit straight line to ln N versus -ln ε was chosen by eye.

The dimension of the attractor may be determined in this manner for any embedding dimension. The fractal dimension of the attractor for the entire system is the maximum limit to the dimension for the attractor as the embedding dimension is increased.

Results of Dimensional Analysis

We used the above methodology in analysis of the numerical model for different grid sizes deformed to different shear strains. The dimension of the spatial attractor for the system varies from about 2.3 to slightly less than 3.

Figure 14 demonstrates the steps involved in determination of the dimension of the system for the 100 x 100 grid. Figure 14a shows ln N versus ln ε for 10201 points of the attractor reconstructed in embedding spaces with dimensions ranging from p = 2 to p = 7 for S = 1. The magnitudes of the gradients of the best-fit straight lines sketched in Figure 14a are plotted in Figure 14b as fractal dimension D versus embedding dimension p.

Similarly, Figure 15a shows ln N versus ln ε for the 58 081 points of the 240 x 240 grid for p = 2 to p = 7 and for S = 1, and Figure 15b presents the fractal dimensions derived from the gradients of these data versus the embedding dimension.

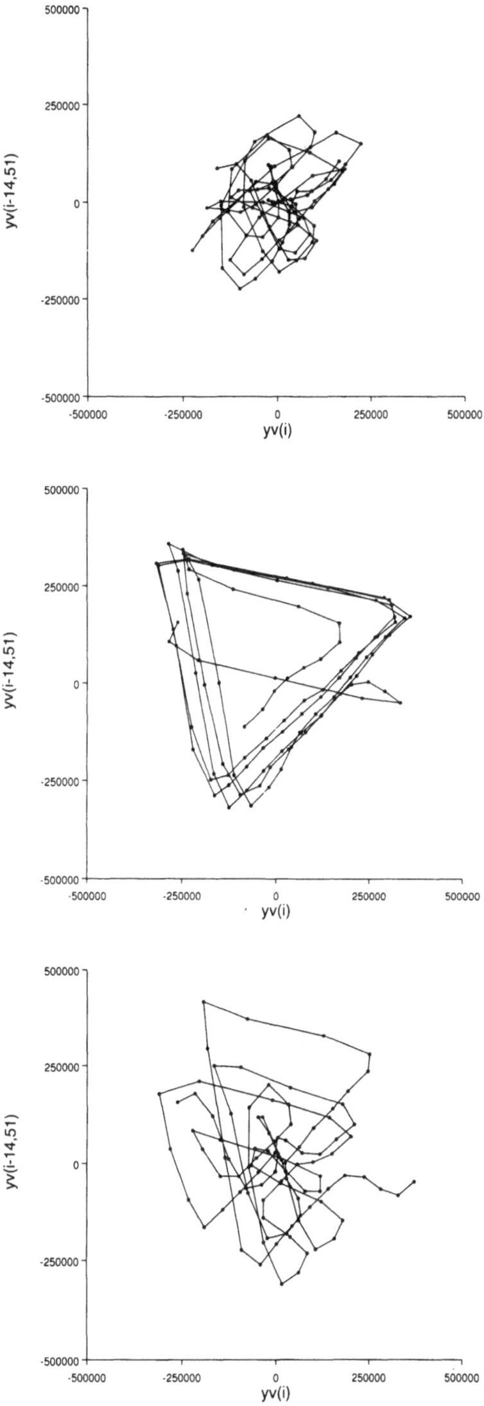

Fig. 13 a-c. Phase portraits for the 10 000 x 100 grid for j = 51. Space lag = 14.
a nodes 2000 to 2100 (Fig. 6d)
b nodes 7040 to 7140 (Fig. 8a)
c nodes 6940 to 7040 (Fig. 8b)

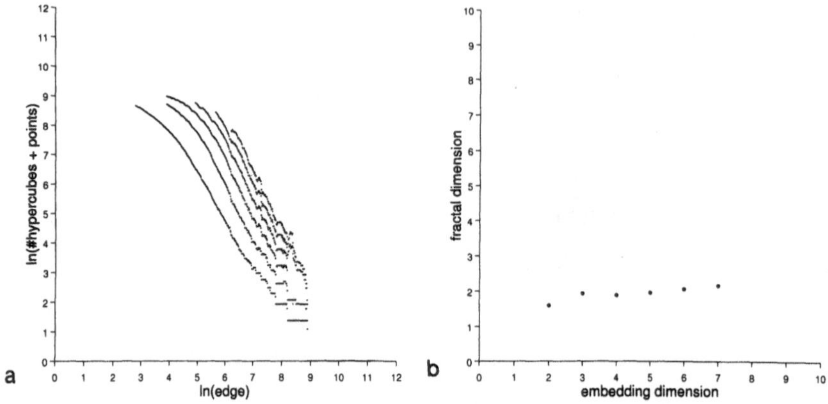

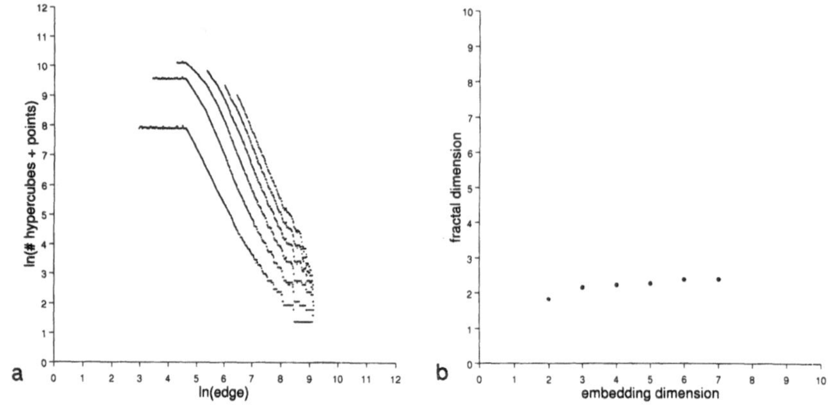

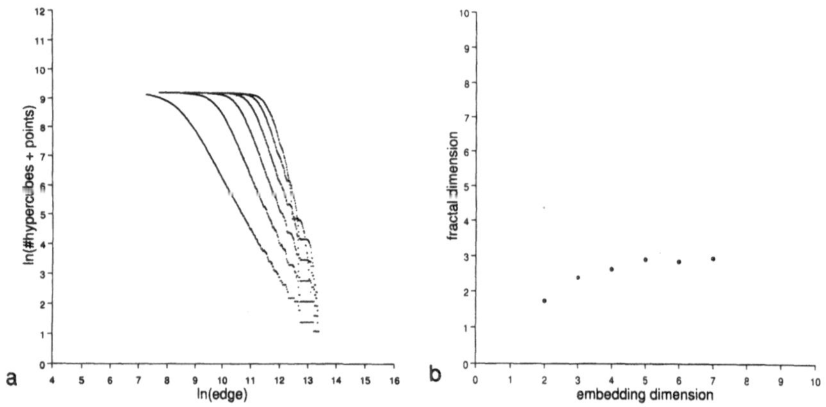

Fig. 14 a,b. (upper): 100 x 100 grid. S = 1. **a** ln N(ε) versus ln ε. **b** D versus p.
Fig. 15 a,b. (middle): 240 x 240 grid. S = 1. **a** ln N(ε) versus ln ε. **b** D versus p.
Fig. 16 a,b. (lower): 10 000 x 100 grid. row 51 S = 10. **a** ln N (ε) versus ln ε. **b** D versus p.

For comparison, Figure 16a shows ln N versus ln ε for the 10 001 points of row 51 of the 10 000 x 100 grid for p = 2 to p = 7 for S = 10, and Figure 16b presents the fractal dimensions derived from the gradients of these data versus the embedding dimension.

Figure 17 is a synoptic diagram which presents the data from Figures 14b, 15b and 16b, together with the results for S = 10 for the 100 x 100 grid and the 240 x 240 grid, and for a collection of points produced by a random process. Changing S makes no difference to the space-filling properties of the data produced by the random process. S = 10 for both the 100 x 100 and 240 x 240 grids results in D higher by at most 0.4 units than for S = 1. The grid point spacing is therefore not so close that the associated instantaneous y-velocities represent the same information. S = 10 was chosen as approximating the wavelength of the crenulations (as determined from Figs. 6 and 7) and maximising the information on the attractor. Also, there are not enough points to investigate the system for values of S greater than 10. D is also slightly higher for the 240 x 240 grid than for the 100 x 100 grid. This appears to be a result of the larger number of points and the more clearly defined region of constant slope.

The following points arise from this analysis:

a) For large values of ε the slope of ln N versus ln ε tends to 0 since when ε is larger than the size of the attractor, all the points of the attractor are contained in the target hypercube.

b) As ε approaches zero, the attractor progressively loses resolution as ε approaches the numerical precision with which the position of individual points is known.

c) The fractal dimension of the attractor in any embedding dimension must be determined over some intermediate range of ln ε that shows a stable constant slope (the scaling region), preferably over 2 or 3 orders of magnitude. Ruelle (1990) notes that gradients for a slope determined over less than a decade seem unreasonable. Figure 14b is then a plot of the slope of ln N versus ln ε over the range of ln N between roughly 3 and 8.

d) As described above, the fractal dimension of the system is represented by the constant value to which the dimension of the attractor converges as the embedding dimension is increased. This value varies from about 2.3 almost to 3 as shown in Figure 17. This satisfies the requirement described by Takens (1981) that the fractal dimension F of a system must be obtained by investigation into an embedding space of at least 2F+1, i.e. $p \geq 2F+1$. For $F = 3.0$, $p \geq 7.0$, and we have investigated p up to a value of 7.

e) Number of samples needed to define the attractor adequately. As the number of points is decreased, the population of points at small scales becomes increasingly diffuse, resulting in large fluctuations of ln N which can completely mask the scaling region. In analysis of the reliability of results for the Grassberger-Procaccia algorithm, Ruelle (1990) infers that the total number of points examined, N, should be greater than D according to $2\log_{10}N \geq D$. Applying this to our results, our maximum D is 3.0. For N only

5000, $2\log_{10}N = 7.4$. So the results from row 51 (10001 points) of the 10000 x 100 grid and from the 100 x 100 (10261 points) and 240 x 240 (58081 points) grids may be worth considering.
Further, as the calculation of N shows, embedding a fixed number of points into higher and higher dimensions effectively reduces the number of points for which the fractal dimension of each embedding space is calculated. This also results in a poorly defined scaling region.

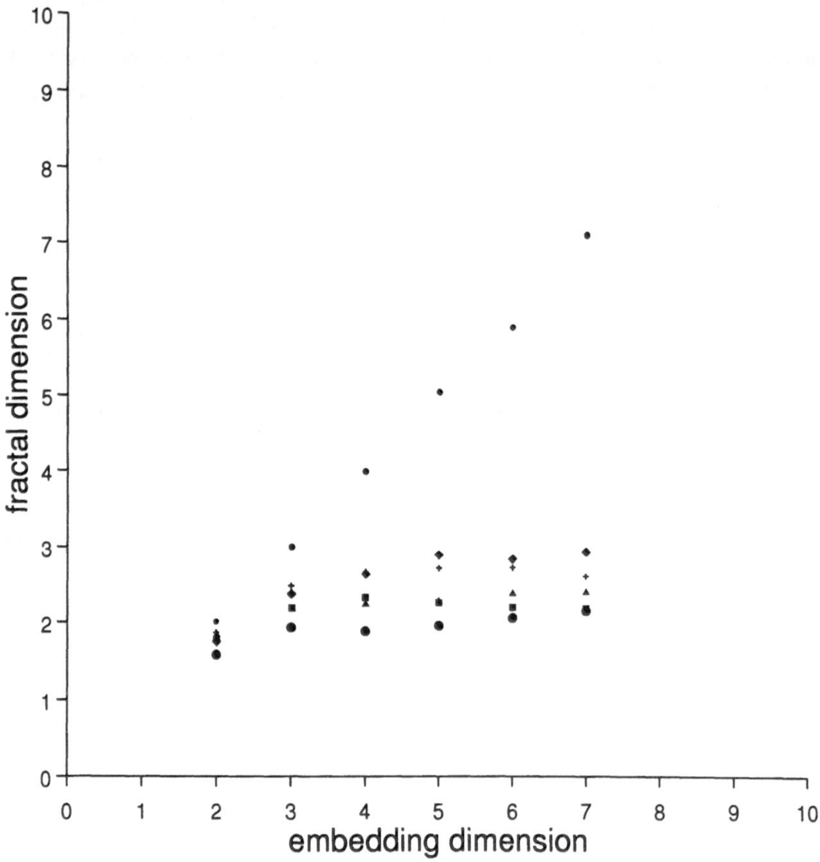

Fig. 17. Synoptic diagram Random points. Small circles. S = 1 and S = 10. 100 x 100 grid. Filled circles S = 1. Filled squares S = 10. 240 x 240 grid. Filled triangles S = 1. Crosses S = 10. 10000 x 100 grid. row 51. Filled diamonds S = 10.

Conclusions

i) A clearly-defined attractor has been described for the first time for a natural system with spatially quasi-periodic behaviour.
ii) A fractal dimension may be derived for this attractor using a box-counting algorithm.
iii) This fractal dimension varies from about 2.3 up to 3, determined for a maximum embedding dimension of 7, and for data sets of 10201 and 58081 points.
iv) It is inferred from this analysis that three first order, non-linear, independent, ordinary differential equations are required to describe this system.
v) We infer, to be consistent with the numerical model, that these are the stress equations of motion, the yield criterion and the flow rule. Although the analysis of the full non-linear problem for folding has not yet been carried out, we infer that it is the feedback relations between these three non-linear equations that leads to the observed quasi-periodicity in the velocity of growth of the crenulations described here and to the resultant quasi-periodicity in the resulting geometries. We conclude that geological systems may be analysed using the concepts of fractal geometry thus providing the basis for inferences as to the number of ordinary differential equations involved in the development of the system The treatment presented here forms the basis for a new way of viewing the irregularity inherent in geological structures and relating this irregularity to the kinematics and dynamics involved in the development of those structures.

Acknowledgements. I thank Bruce Hobbs for his continuing support of this work, ITASCA Consulting Group for continuing access to FLAC, and Bruce Hobbs, Hans Mülhaus and Frank Horowitz for their critical and positive comments on this manuscript.

References

Albano AM, Abraham NB, Guzman GC de, Tarroja MFH, Bandy DK, Gioggia RS, Rapp PE, Zimmerman ID, Greenbam NN, Bashore TR (1986). Lasers and brains: complex systems with low-dimensional attractors. In: Mayer-Kress G (ed) Dimensions and Entropies in Chaotic Systems . Springer, Berlin Heidelberg New York, pp. 231-240.

Biot MA (1961). Theory of folding of stratified viscoelastic media and its implication in tectonics and orogenesis. Geol Soc Am Bull **72**: 1595-1620.

Board M (1989). FLAC (Fast Lagrangian Analysis of Continua) Version 2.20. Vol.1, Software Summary. Prepared for Division of High-Level Waste Management, Office of Nuclear Material Safety and Safeguards, NUREG/CR-5430. Washington, D.C., USA.

Chouet B, Shaw HR (1991). Fractal properties of tremor and gas piston events observed at Kilauea Volcano, Hawaii. J Geophys Res 96: 10177-10 189.

Crutchfield JP, Farmer JD, Packard NH, Shaw RS (1986). Chaos. Scientific American, 255 (6): 38-49.

Cundall PA (1989). Numerical experiments on localization in frictional material. Ingenieur-Archiv 59: 148-159.

Cundall PA (1991). Shear band initiation and evolution in frictional materials. In: Mechanics Computing in 1990's and Beyond (Proceedings of the Conference, Columbus, Ohio, May 1991), Vol. 2: Structural and Material Mechanics, pp. 1279-1289, ASME New York.

Cundall PA, Board M (1988). A microcomputer program for modelling large-strain plasticity problems. In: Numerical Methods in Geomechanics (Innsbruck, 1988), Vol 3, Balkema, Rotterdam, pp. 2101-2108.

FLAC (1991). Fast Lagrangian Analysis of Continua. Version 3.0. Volume I : User's Manual. Volume II : Verification Problems and Example Applications. ITASCA Consulting Group, Inc. Minnesota, USA.

Gollub JP, Swinney HL (1975). Onset of turbulence in a rotating fluid. Phys Rev Lett 35: 927-930.

Guckenheimer J, Holmes P (1983). Nonlinear oscillations, dynamical systems and bifurcations of vector fields. Springer, New York Berlin Heidelberg.

Hobbs BE (1990). Chaotic behaviour of frictional shear instabilities. Proc. 2nd Int. Symp. on Rockbursts and Seismicity in Mines. Minneapolis, 8-10 June 1988. Ed. C. Fairhurst, Balkema, Rotterdam.

Hobbs BE, Ord A (1989). Numerical simulation of shear band formation in a frictional-dilational material. Ingenieur-Archiv 59: 209-220.

Hobbs BE, Mühlhaus H-B, Ord A (1991). Instability, softening and localization of deformation. In: Knipe RJ & Rutter EH (eds) Deformation Mechanisms, Rheology and Tectonics. Geol Soc Spec Pub 54, The Geological Society, London, pp. 143-165.

Holmes P (1990). Nonlinear dynamics, chaos, and mechanics. Appl Mech Rev 43: S23-S39.

Lorenz EN (1963). Deterministic non-periodic flow. J Atmos Sciences 20: 130-141.

Lorenz EN (1979). On the prevalence of aperiodicity in simple systems. In: Global Analysis (Eds. M. Grmela and J.E. Marsden). Lect Notes Math **755**, Springer, New York, pp. 53-75.

Mandelbrot BB (1977) Fractals: form, chance, and dimension. Freeman, San Francisco.

Mayer-Kress G, Kaneko K (1989). Spatiotemporal chaos and noise. J Stat Phys 54: 1489-1508.

McGuinness MJ (1983). The fractal dimension of the Lorenz attractor. Phys Lett 99A: 5-9.

Ord A (1990). Mechanical controls on dilatant shear zones. In: Knipe RJ & Rutter EH (eds) Deformation Mechanisms, Rheology and Tectonics. Geol Soc Spec Pub 54, The Geological Society, London, pp. 183-192.

Ord A (1991). Fluid flow through patterned shear zones. In: Beer G Booker JR & CarterJP (eds) Computer Methods and Advances in Geomechanics Proc. 7th Intl Conf on Computer Methods and Advances in Geomechanics, Balkema, Rotterdam, pp. 393-398.

Packard NH, Crutchfield JP, Farmer JD, Shaw RS (1980). Geometry from a time series. Phys Rev Lett 45: 712-716.

Ramberg H (1964). Selective buckling of composite layers with contrasted rheological properties; a theory for simultaneous formation of several orders of folds. Tectonophysics 1: 307-341.
Rasband SN (1990). Chaotic Dynamics of Nonlinear Systems. John Wiley and Sons, New York.
Ruelle D (1990). Deterministic chaos: the science and the fiction. The Claude Bernard Lecture, 1989. Proc R Soc Lond A427: 241-248.
Sornette A, Dubois J, Cheminee JL, Sornette D (1991). Are sequences of volcanic eruptions deterministically chaotic? J Geophys Res 96: 11931-11945.
Takens F (1981). Detecting strange attractors in turbulence. In: Rand DA & Young L-S (eds). Dynamical Systems and Turbulence. Lecture Notes in Mathematics 898, Springer, New York, pp. 366-381.
Van der Pol B (1927). Forced oscillations in a circuit with nonlinear resistance (receptance with reactive triode). London, Edinburgh, and Dublin, Phil Mag 3: 65-80.
Vermeer PA, de Borst R (1984). Non-associated plasticity for soils, concrete and rock. Heron, 29(3): 1-64.
Young P (1991). Convex 200 humbled by CSIRO program. Computerworld Australia 14: No.23, 1, 6.

Part II
Physical Features and Behaviour of the Earth

Non-Linear Processes in Earthquake Prediction Research, a Review

Rolf Meissner,
Institut für Geophysik, Christian-Albrechts-Universität zu Kiel,
Olshausenstr. 40, D-24098 Kiel, Germany

Abstract. Two sources of non-linearity in the initiation and continuation of rupture processes are the build-up of critical stresses and their propagation along inhomogeneous rupture zones including the underlying non-seismogenic crust. The most important sources for a non-linear build-up of critical stresses seem to be dilatancy processes, transfer of stresses from nearby foreshocks, and possibly interactions with the lower crust. Propagation of rupture processes are dependent on inhomogeneities of the dynamic friction, asperities, fault gouge, geometry, and possibly complex interaction with the non-seismogenic lower crust. Models based on chaos theory are shortly reviewed. They can explain several statistically observed phenomena. A deterministic approach to earthquake prediction requires a dense network as well as short-and long-time monitoring of any deformation in the neighbourhood of the suspected rupture area. In accordance with weakly chaotic systems a limited prediction might be possible.

Introduction

Most earthquakes occur repetitively on pre-existing faults. The repetition rate is often connected with a period of quiescence before large ruptures. It is rather regular in some areas (Kanamori 1981), but irregular or with a tendency of earthquake clustering in others (Kagan & Jackson 1991, Nishenko & Sykes 1993). The observations are often used for assessing times of an increased probability (Wyss & Habermann 1988) and not for a deterministic prediction of an imminent earthquake. Some earthquakes have a pronounced foreshock activity or show certain clustering of regional seismicity (Keilis-Borok & Rotwain 1990). Large earthquakes do not show correlations with the earth's tides (Knopoff 1964), an observation which will form the main argument for assuming non-linearity in the build-up of critical stresses.

Another major source of non-linearity seems to be the rupture process. It is well known that dynamic frictional stresses σ_{df} of rupture are much below those of static friction σ_{sf}, but they must be highly variable, depending on the fault gouges' thickness and composition, on the velocity of rupture, and on the geometry of the fault zone. The particle size distribution of the fragmented gouge material seems to have a fractal distribution which is related to the fragmentation by previous rupture processes (Blenkinsop 1991). It follows that neither σ_{df} nor σ_{sf} can be assumed to be constants in space or time, implying that even the more refined slider-block models (Huang & Turcotte 1990, Bak & Tang 1989) are only poor replicae of natural processes. If the critical stress σ_{sf} is reached in one place, rupture might start, but where might it end? Will it fill a whole seismic gap or only the distance to the next asperity or bend ? What role does dilatancy play in the accumulation and release of stresses ? It is certainly not involved in the slider-block models.

Local or regional processes have to be monitored or controlled before earthquake prediction can be more than an assessment of the "time of increased probability" (Turcotte 1992). In the following two main sources of non-linearity will be described.

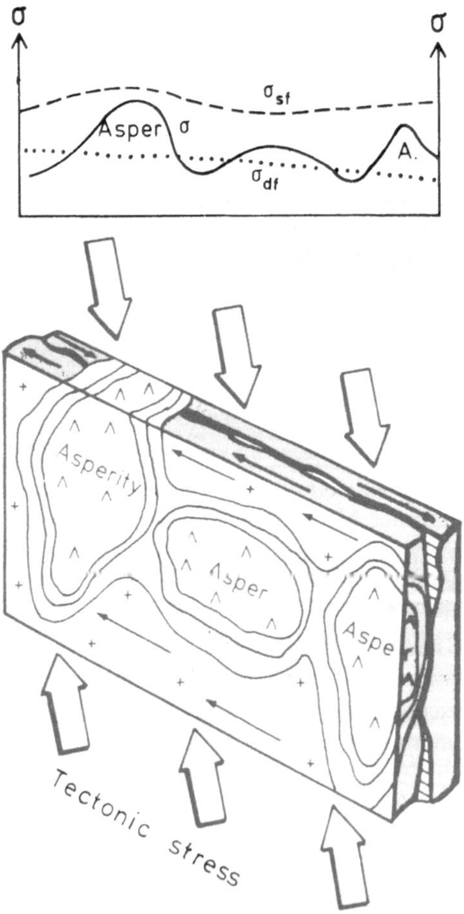

Fig. 1. Schematic view of a vertical strike-slip fault with asperities shortly before rupture
Upper picture: static friction σ_{sf}, not yet reached by growing stress (σ); dynamic friction σ_{df}, relevant for the subsequent rupture process; asperity = A = patches of difficult slip
Lower picture: Fault zone with current slip (thin arrows) and tectonic stresses (thick arrows). Some parts of the fault zone have already slipped, either by creep or foreshocks.

The Build-Up of Critical Stresses

Most large earthquakes seem to take place along plate boundaries or/and along prominent fault zones. The sequence of earthquakes along a single fault zone is quite arbitrary but a network of fault zones, which often seems to be fractally distributed, produces a fractal relationship in the time domain, such as the Gutenberg-Richter law (Gutenberg & Richter 1954, Turcotte 1992, Korvin 1991).

In the following the stress situation along a single fault will be considered. Although the average strength and the frictional stress along a prominent fault zone is supposed to be much weaker than in the surrounding because of repeated creep and rupture, there are significant inhomogeneities also along the fault zone.

Figure 1 is a sketch of a section from a near-vertical strike-slip fault like the San Andreas or the North Anatolian Fault Zone. The stress situation with stress approaching the static frictional stress σ_{sf} is shown in the upper part of the figure. Along the fault there are several inhomogeneities like asperities, i.e. rock units which can hardly break and concentrate relatively high stresses, while the weaker rocks around them have a smaller critical stress, and many already have slipped by creep or small ruptures giving rise to a certain foreshock activity. Little is known about the size of these patches of easy or difficult slip. These simplified pictures of a fault zone (Figure 1) will show a – heterogenous – stick-slip behaviour which might lead to an overall linear or non-linear accumulation of stress, finally reaching the critical stress where rupture begins. Non-linearity seems to dominate in the last critical phase of stress accumulation.

Another source of non-linearity is provided by the dilatancy theory, which has been studied by many field and laboratory investigations (Nerzerov et al. 1969, Spetzler et al. 1982, Rummel & Frohn 1982). Also in this theory after a non-elastic volume increase, the final stage of stress accumulation is described by a coalescence of microcracks and/or by an intrusion of fluids into the cracks. Both processes are fast and strongly non-linear. Especially in laboratory measurements coagulation and coalescence of microcracks into one big crack with rupture is so fast that special feed-back systems had to be developed in order to study crack propagation (Rummel & Frohn 1982).

While the above mentioned sources of non-linearity give us some positive evidence for fast accumulating critical stresses, another observation provides a negative evidence for linearity. It is the absence of a general correlation between the period of the solid earth tides and large earthquakes, as has been demonstrated frequently from catalogues of global or large-scale seismicity of the earth, beginning with the early work of Knopoff (1964), Simpson (1967), Shlien (1972), up to Shirley (1988), and Rydelek et al. (1993). Only from regional and local catalogues some correlations with varying degree of significance have been reported, e.g. by Shlien (1972), Young & Zürn (1979), Kilston & Knopoff (1983), Shirley (1988) and others. Research on this topic had accelerated after all deep moonquakes were shown to be triggered by terrestrial tidal forces (Meissner et al. 1973), but on earth even the low-level seismicity near rifts and volcanoes could not conclusively demonstrate a general tidal triggering.

The solid earth tides represent the largest short-period oscillatory strain in the earth, about 10^{-7} for homogeneous media, inducing stresses of 5 mbar (Shlien 1972). Strong inhomogeneities and an activation of pore fluids might increase stresses considerably. The expectation that tides should trigger earthquakes is based on three assumptions:

(1) tidal and tectonic stresses can be superimposed
(2) rupture occurs when a critical level of stress (such as σ_{sf} in Figs. 1 and 2) has been reached
(3) the tectonic stress increases more or less linearly and smoothly, even shortly before rupture.

It is easy to see that assumption (3) is rather vague. Dilatancy or a stepwise stress increase from adjacent earthquakes certainly show strong deviations from linearity. If these induced (tectonic) stresses have a steeper gradient than the tidal forces have, then the critical stress σ_{sf} is reached at any time and not supported by tidal stresses.

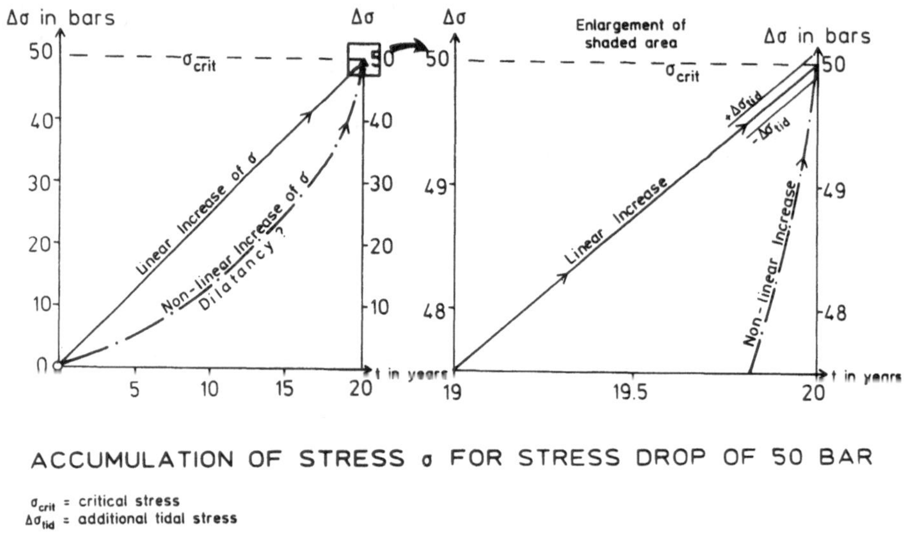

Fig. 2. Example for a linear and a non-linear accumulation of stresses, longtime (**left**) and shorttime, i.e. pre-critical, behaviour (**right**). Modulation of the linear increase by tidal forces causes triggering (3 months earlier for a 50 mb, 10 days earlier for a 5 mb tidal amplitude). Modulation of the non-linear increase would not be expected to trigger earthquakes, if only its gradient is steeper than that of the tidal forces. σ_{sf} = level of static friction (critical stress). $\Delta\sigma_{tid}$ = amplitude of additional tidal stress.

It is suggested therefore that a non-linear increase of stress in the critical pre-rupture phase is responsible for the general absence of tidal triggering. Fig. 2 shows as an example the situation for an earthquake with an average stress drop of 10 bars. While a linear stress increase, modulated by tidal forces, would generate tidal triggering once the critical stress σ_{sf} is reached, a non-linear increase may reach σ_{sf} at any time.

Another argument for non-linearity of stress increase comes from the interaction between the seismogenic, brittle upper crust and the ductile lower crust. The lower crust is suspected of taking an active part in the transfer of stresses from creeping lower crust to rupturing upper crust. For the San Andreas Fault this seems to be the only serious candidate for loading, because maximum horizontal stresses near the surface are all nearly perpendicular to the fault (Zoback & Zoback 1981, Zoback & Healy 1992), and cannot load the fault zone. In several areas there is no stress component in the direction of the observed strike-slip movement.

In conclusion, there are several strong arguments for non-linearity in the accumulation of stresses. Dilatancy, microcrack coagulation, general absence of tidal triggering, and the loading mechanism by creep provide the main arguments for non-linearity.

The Rupture Process

If the critical stress overcomes static friction σ_{sf} , rupture starts and progresses under the much smaller dynamic friction σ_{df}. If σ_{df} were constant along the fault, rupture would be linear and show a constant velocity σ_{Ru}. The non-linearity in this case concentrates on the beginning and termination of rupture, similar to the slider-block models of Burridge & Knopoff (1967) or more refined models of Huang & Turcotte (1988) or Bak & Tang (1989). The latter two models show already chaotic behaviour for different spring constants for the two-block model (Turcotte 1992), or a more general unpredictable behaviour with self-organized criticality in the multiple block model (Brown et al. 1991, McCloutskey & Bean 1992). Figure 3 shows a sketch of the two block- and the multiple block-slider model and the corresponding non-linear equations of motion. Many similarities with natural behaviour, mainly in a statistical approach, are observed. The (now "fractal") Gutenberg-Richter relationship (Gutenberg & Richter 1954) and even its gradient b can be matched by multiple block models, and certain patterns of quiescence and rather periodic recurrence times of large (multiple) events are observed. This means that a "time period of increased probability" might be understood and predicted. But examples from nature, e.g. Parkfield (Michael & Langbein 1993) or from the North Anatolian Fault (Zschau & Ergünay 1989; Berckhemer et al. 1991) show that even quasi-regular periods of recurrence times might break up and are useless for a determi-

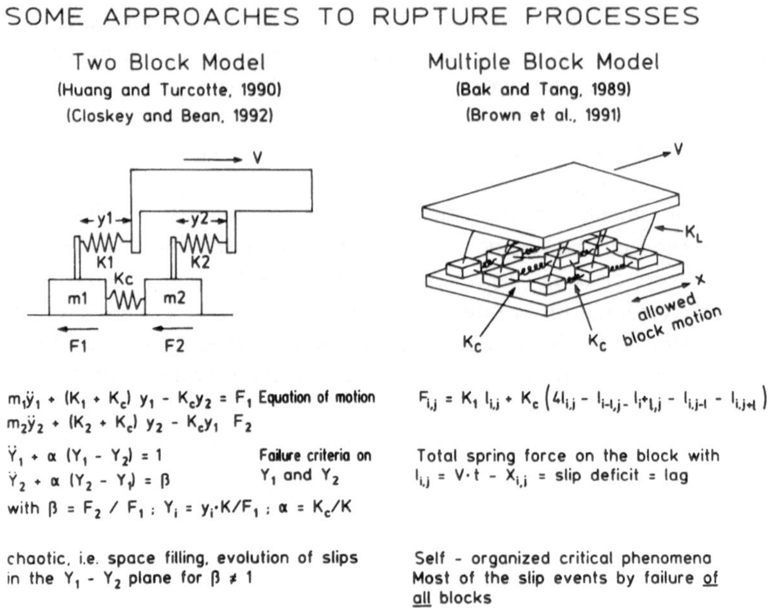
SOME APPROACHES TO RUPTURE PROCESSES
Two Block Model
(Huang and Turcotte, 1990)
(Closkey and Bean, 1992)
Multiple Block Model
(Bak and Tang, 1989)
(Brown et al., 1991)
V
y1
y2
K1
K2
Kc
m1
m2
F1
F2
V
K_L
x
allowed block motion
K_c
K_c
$m_1\ddot{y}_1 + (K_1 + K_c)\, y_1 - K_c y_2 = F_1$ Equation of motion
$m_2\ddot{y}_2 + (K_2 + K_c)\, y_2 - K_c y_1 \;\; F_2$
$\dot{Y}_1 + \alpha\, (Y_1 - Y_2) = 1$
$\dot{Y}_2 + \alpha\, (Y_2 - Y_1) = \beta$
Failure criteria on Y_1 and Y_2
with $\beta = F_2 / F_1$; $Y_i = y_i \cdot K/F_1$; $\alpha = K_c/K$
$F_{i,j} = K_1\, l_{i,j} + K_c\,(4l_{i,j} - l_{i-1,j} - l_{i+1,j} - l_{i,j-1} - l_{i,j+1})$
Total spring force on the block with
$l_{i,j} = V \cdot t - X_{i,j}$ = slip deficit = lag
chaotic, i.e. space filling, evolution of slips in the Y_1 - Y_2 plane for $\beta \neq 1$
Self - organized critical phenomena
Most of the slip events by failure of all blocks

Fig. 3. Two models to simulate slip processes. The two-block model leads to space-filling, i.e. chaotic, behaviour for different frictional stress F_1,F_2. The multiple-block model simulates the Gutenberg-Richter relationship (1954) and large slip events after periods of quiescence. Deterministic prediction of events is impossible for both models. Non-linear increase of stresses before slip or variation of rupture velocity during slip is not included in either model.

nistic prediction of time, location, and magnitude of an imminent quake. The magnitude seems to be the most elusive parameter because of the additional non-linearity of the rupture process itself, which apparently can hardly be modeled by the slider block models.

The biggest advantages of these models are in the observed statistical self-similarity and in the visible transfer of stresses from one block to the other or – in nature – from a neighbouring fault or from nearby field of foreshock activity to a large or small location along a fault zone. In the following, some parameters of fault zones and rupture will be mentioned which are not involved in the slider block model parameters and might further complicate prediction efforts.

There is first the lateral variability of the dynamic friction along a fault segment. It is manifested in observations of structural heterogeneity (Blenkinsop 1991, Sibson 1984, Rutter et al. 1986) as well as in the spectral behaviour of earthquakes and the variations of rupture velocity (Schneider 1975, Kasahara 1981) which may reach from zero to nearly shear wave velocity V_S (Rikitake 1976). Irregularities of V_{Ru} have been observed for various large quakes and are found in many textbooks. Figure 4 shows a schematic example for

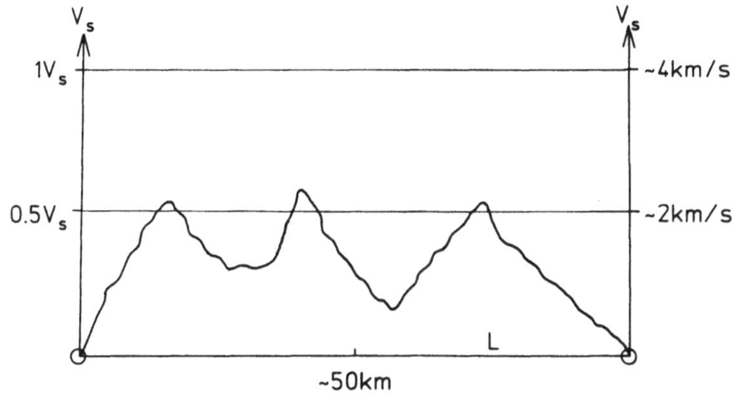

Fig. 4. Schematic view of the variation of rupture velocity V_{Ru} along a fault zone of about 100 km in length. (From observations of spectra and fault plane solutions). Seismic radiation is maximum at peaks of V_{Ru}.(Estimated after calculations of Schneider 1975 and Kasahara 1981).

the variation of the rupture velocity V_{Ru} along a long rupture zone. The average V_{Ru} is generally about half the shear-wave velocity V_S. But its peaks and its rugged appearance as well as the deformation of radiation patterns show that the rupture process is definitely non-linear. Another observation is the lateral change of parameters like grain size, grain orientation, or seismic velocities across a fault. Thickness variations of fault zone gouge or bends or curvatures of fault zones exist and might provide serious geometrical obstacles for the rupture process. Local and regional hydrolytic weakening processes in viscosity and velocity (Handy 1989) might further modify fault zone parameters.

The influence of the generally ductile lower crust below the "seismogenic" upper crust on rupture propagation is not known. The lower crust generally shows no seismicity (Meissner & Strehlau 1982) and is supposed to be responsible for a change in b-values (Pacheco et al. 1992) between small and large earthquakes, the latter having more two-dimensional focal configuration because of the depth limit of the seismogenic layer. Possibly differential creep in the ductile lower crust not only loads the fault zone in a non-linear manner; it might also provide complex interactions between upper and lower crustal levels during the rupture process, which might transfer sudden changes of stress and creep into the lower crust, which then might react in a brittle manner upon these short-time processes.

Conclusions

From the rather euphoric approach towards "real" earthquake prediction in the 1970s (Rikitake 1975, 1976) we have now reached – by trial and error – a level where the impacts of non-linearity are realized. Non-linearity is present in several stress accumulation processes and also in the rupture process. The introduction of chaos theory has certainly opened our eyes to the complexity of processes, and it is generally believed that many earthquakes will remain unpredictable. What is needed is a long- and short-time monitoring of those parameters which are directly influenced by stresses. This requires a dense network of stations (like the multiparameter stations in the Turkish German Earthquake Research Project) which can certainly not be installed in all endangered areas. In the future, continuous short- and long-time monitoring of small deformation by GPS methods or Radar Interferometry from satellites might develop into a rather inexpensive and powerful tool to detect smallest changes of deformations, caused by small changes in stress, studying pre-, co-, and post seismic deformations and monitoring locked, creeping, or slipped portions along fault zones. Other continuous measurements must join. In principle, weakly chaotic processes (which means that stretching in the phase space is small and Lyapunov exponents are only slightly larger than zero) are predictable (Schuster 1988) although only for a limited period of time as seen, for instance, in meteorology, where linearized versions of various non-linear processes and equations are used. Whether all the complex non-linear processes of stress accumulation and rupture will be fully understood, measured quantitatively or reduced to fractal connections is another question. Its answer has possibly to wait for future generations.

Acknowledgement. Thanks are due to my colleagues at the Institute for Geophysics, Kiel, and two anonymous reviewers for critical and helpful suggestions. Many ideas to this paper originated within the Turkish-German Earthquake Research Project, supported by the German Research Association (DFG).

References

Ambeh WB, Fairhead JD (1991) Regular, deep seismicity beneath Mt Cameroon volcano: lack of evidence for tidal triggering. Geophys J Intern 106: 287-291.

Bak P & Tang C (1989) Earthquakes as a self-organized critical phenomenon. J Geophys Res 94: 15635-15637.

Berckhemer H, Ergünay O, Zschau J (1991) The Turkish-German Project for Earthquake Prediction Research in NW Anatolia, a multidisciplinary approach to study the stress field, Proceedings of the International Conference on Earthquake Pred, State-of-the art, Council of Europe, Strasbourg, France, pp. 459-476.

Blenkinsop TG (1991) Cataclasis and processes of particle-size reduction. PAGEOPH 136: 59-86.

Burridge R, Knopoff L (1967) Model and theoretical seismicity Bull Seis Soc Am 57: 341-71.

Brown SR, Scholz CH, Rundle JB (1991) A simplified spring-block model of earthquakes. Geophys Res Lett 18: 215-218.

Gutenberg B, Richter CF (1954) Seismicity of the Earth and Associated Phenomenon. 2nd edition, Princeton University Press, Princeton.

Handy H (1989) Deformation regimes and the rheological evolution of fault zones in the lithosphere: the effects of pressure, temperature, grainsize, and time. Tectonophysics 163: 119-152.

Huang J, Turcotte DL (1990) Are earthquakes an example of deterministic chaos? Geophys Res Lett 17: 223-226.

Kagan,YY, Jackson DD (1991) "Seismic Gap Hypothesis: Ten Years After". J Geophys Res 96(21): 419-421, 431.

Kanamori H (1981) The nature of seismicity patterns before large earthquakes. In: Simpson DW & Richards PG (eds) Earthquake Prediction. American Geophysical Union, Washington DC, pp. 1-19.

Kasahara K (1981) Earthquake mechanism. Cambridge Earth Science Series, Cambridge.

Keilis-Borok VI, Rotwain IM (1990) Diagnosis of time of increased probability of strong earthquakes in different regions of the world: Algorithm CN. Phys Earth Planet Int 61: 57-72.

Kilston S, Knopoff L (1983) Lunar-solar periodicities of large earthquakes in southern California. Nature 303: 21-25.

Knopoff L (1964) Earth tides as a triggering mechanism for earthquakes. Seism Soc Am Bull 54: 1865-1870.

Korvin G (1991) Fractal modes in the earth's sciences. Elsevier, Amsterdam.

McCloskey J, Bean CJ (1992) Time and magnitude prediction in shocks due to chaotic fault interactions. Geophys Res Letter 19: 119-122.

Meissner R, Strehlau J (1982): Limits of stresses in continental crusts and their relation to the depth-frequency distribution of shallow earthquakes. Tectonics 1: 73-89.

Meissner R, Voss R, Kaestle HJ (1973): A_1-moonquakes, problems of determining their epicenters and mechanisms. THE MOON, 293-302.

Michael A, Langbein J (1993): Earthquake prediction lessons from Parkfield experiment. EOS: 145-155.

Nerzerov IL, Seminova AN, Simbirewa JG (1969) Physical basis of foreshocks. Nanka, Moskov (in Russian).

Nishenko SP, Sykes LR (1993) Comment on "Seismic Gap Hypothesis: Ten Years After". J Geophys Res 98(B6): 9909-9916.

Pacheco J-F, Scholz CH, Sykes LR (1992) Changes in frequency-size relationship from small to large earthquakes. Nature 355: 71-73.

Rikitake T (1975) Dilatancy model and empirical formulars for an earthquake area; PAGEOPH 113: 141-147.

Rikitake T (1976) Earthquake Prediction. Geophys Series 9, Elsevier, Amsterdam.

Rummel F, Frohn C (1982) Variation of ultrasonic velocity in granite and serpentinite during dilatant fracture under general triaxial compression. In: Schreyer W (ed) High pressure research in Geophysics, Schweizerbart, Stuttgart, pp 103-111.

Rutter EH, Maddock RH, Hall SH, White HS (1986) Comparative microstructures of natural and experimentally produced day-bearing fault gouges. PAGEOPH 124: 1-30.

Rydelek PA, Selwyn Sacks I, Scarpa R (1993) On tidal triggering of earthquakes at Campi Flegrei, Italy. Geophys J Int 109: 125-137.

Schneider G (1975) Erdbeben, Entstehung-Ausbreitung-Wirkung. Enke, Stuttgart.

Schuster HG (1988) Deterministic Chaos, VCH, Weinheim.

Shirley JH (1988) Lunar and solar periodicities of large earthquakes: southern California and the Alaska-Aleutian Islands seismic region. Geophys J 92: 403-420.

Shlien S (1972) Earthquake-Tide Correlation. Geophys JR astron Soc 28: 27-34.

Sibson RH (1984) Roughness at the base of the seismogenic zone: Contributing factors. J Geophys Res 89(B7): 5791-5800.

Simpson JF (1967) Earth tides as a triggering mechanism for earthquakes. Earth planet Sci Lett 2: 473-478.

Spetzler H, Mizutani H, Rummel F (1982) A model for time-dependent rock failure. In Schreyer W (ed) High pressure research in Geophysics, Schweizerbart, Stuttgart, pp 85-93.

Turcotte D (1992) Fractals and chaos in geology and geophysics. Cambridge University Press.

Wyss M, Habermann RR (1988) Precursory seismic quiescence. PAGEOPH 126: 319-332.

Young D, Zürn W (1979) Tidal triggering of earthquakes in the Swabian Jura? J Geophys 45: 171-182.

Zoback MD, Healy JH (1992) In situ stress measurements to 3.5 km depth in the Cajon Pass Scientific Research Borehole. J Geophys Res 97: 5039-5057.

Zoback MD, Zoback ML (1981) State of stress and intraplate earthquakes in the United States, Science 213: 96-104.

Zschau J, Ergünay O (eds.) (1989) The Turkish-German Earthquake Project, Institut für Geophysik, Kiel. Enke, Stuttgart.

Fractal Analysis as a Tool to Detect Seismic Cycle Phases

Giuliana Rossi
Istituto di Geodesia e Geofisica
Università di Trieste, I-34123 Trieste, Italy

Abstract. This chapter concerns the study of the temporal evolution of seismicity in a particular region, through a time repeated fractal analysis of the spatial distribution of the seismic events. The technique is applied to the NE-Italy Friuli seismic region, where a local seismic network has been active since 1977 (OGS 1977–1981, 1982-1990). Fractal dimensions have been calculated at fixed time intervals of thirty days. The spectrum of the time series thus obtained reveals that in this region the seismicity is characterized by a sort of periodicity, with periods that vary from one year to about four years, superimposed onto a longer period term.

The comparison of these results with the geodetic measurements in the same region suggests a 'periodic' variation of the stress field as the cause of both phenomena, and confirms the usefulness of this kind of fractal analysis in the study of the seismic process.

Introduction

The different phases of a seismic cycle may be detected through the analysis of the variations of different physical and geochemical parameters (Rikitake 1976). The most important information comes from the vertical and horizontal movements of rock masses, detectable through geodetic measurements, and overall, from the frequency and the spatial distribution of the seismic events in a region. The analysis of the time variations of the statistical properties of seismic activity allows one to monitor the microfracturing growth and coalescence of the anelastic phase that may precede an earthquake (Mogi 1985, Meredith 1990), or the particular seismicity pattern which Mogi (1985) calls "doughnut pattern", or the spatial migration, that normally characterizes the beginning of a new seismic cycle.

The fractal methods, with respect to the more conventional ones, have the great advantage of quantifying with a single figure, i.e. the fractal dimension, qualitative variations as described above (Mandelbrot 1977).

In the specific case treated here, a small fractal dimension indicates a spatial clustering of shocks, whereas a large fractal dimension suggests that the events tend to propagate in the whole region. A systematic scanning of the seismic catalogue with a temporal window allows the calculation of fractal dimensions in successive time intervals. A time series can then be obtained, representing the temporal evolution of the spatial distribution of the seismic activity, and therefore reflecting changes in the stress field. Fractal methods have been successfully applied to the study of different tectonic regions (Sadovskii et al. 1984, Rossi 1990, De Rubeis et al. 1993) and to the analysis of the different phases during a microfracturing experiment in laboratory (Hirata et al. 1987).

The region studied in this work is the Friuli seismic area, located in NE-Italy, and from a geological point of view, on the north-eastern border of the Adriatic plate. Two important structural systems, generated by the collision between the Adriatic plate with the European one, merge and interact with each other in this region (Fig. 1). Intense seismicity, the highest of the whole Alpine chain, indicates that the orogenic process is still active. Analysis of the fault plane solutions of the main earthquakes, which occurred in this century (Barbano et al. 1984), reveals the activation of Alpine and Dinaric mechanisms in turn, under the NS oriented compression. The seismic activity is limited to a narrow bent zone, and is highly clustered both in space and in time (Carulli et al. 1990). A migration of the hypocentres northwards was observed in the month following the M = 6.4 earthquake that partially destroyed Friuli on May 6th, 1976, causing the death of one thousand people (Finetti et al. 1979). The stress redistribution after its aftershock sequences caused a new migration of seismic events toward other seismic areas, at the boundaries of the principal one (Barbano et al. 1984).

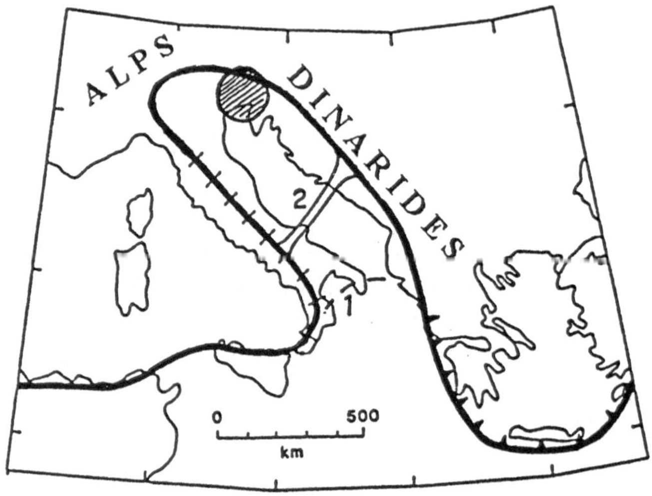

Fig. 1. Location of the region under study (dashed circle) in the frame of Adria plate. Triangles indicate active compression, while hatches indicate distensional mechanisms. *1* and *2* represent two possible locations of the southern boundary of the plate, according to Westaway (1990) and Anderson & Jackson (1987) respectively (from Westaway 1990, modified).

This work ideally continues a study of the seismicity properties, with particular attention to the Friuli case, through different methodologies. The Principal Parameters Method (Ebblin & Michelini 1986, Rossi & Ebblin 1990) gives the directions in which shocks tend to propagate, while fractal analysis (Rossi 1990) gives information about the clustering of the seismicity. The previous fractal analysis of the seismic catalogue of the local seismometric network managed by the Osservatorio Geofisico Sperimentale of Trieste, Italy, using a mid-term thirty days temporal window and a short-term thirty events one as well, evidenced some significant changes in earthquake spatial distribution before M > 4 events on both scales (Rossi 1990). In the same interval, an analysis of the strain field, using the data of the geodetic network managed by the Istituto di Geodesia e Geofisica of the University of Trieste, constituted by six tiltmeter stations, and one strainmeter (Zadro 1992), revealed the presence of recurrent strain field patterns, with a certain correlation with the occurrence of M > 4 seismic events (Zadro & Rossi 1991).

All these observations are at the basis of the present work, that continues and improves the analysis started by Rossi (1990), also comparing the results with the strain field properties.

Data Analysis

The catalogue used was obtained from the seismic local network, installed in 1977 in the Friuli seismic area and managed by the Osservatorio Geofisico Sperimentale of Trieste (OGS 1977–1981, 1982-1990). The hypocentral errors are of about 2 km, and the catalogue may be considered complete for events of magnitude greater than 1.8-2. For these reasons the events of magnitude smaller than 2 are disregarded, thus obtaining, from a previous one of about 5500 events, a new one of about 3500.

Among the various algorithms, the Correlation Integral Method (Grassberger 1983, Hirata et al. 1987) was adopted because of its ability to treat relatively small sets of data like the Friuli seismic catalogue. According to this method, the number of pairs of shocks at distances smaller than a certain length r, progressively reduced, is plotted on a bilogarithmical diagram versus the length itself. The slope of the straight line thus obtained gives the fractal dimension, or, following Grassberger (1983), its lower limit. To avoid the effect of the hypocentral errors, the space intervals r have been chosen, so that the smallest one is greater than 2 km, the hypocentral error of the catalogue.

As previously stated, a thirty day time window is used to follow any changes in seismic activity. For each interval, the fractal dimension quantifies the spatial distribution of hypocentres: the smaller the value, the greater the concentration. The choice of a fixed time window allows us to obtain a time series to which all

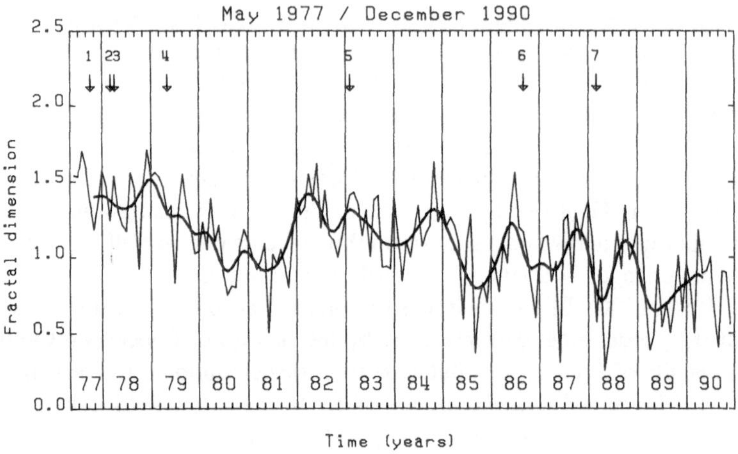

Fig. 2. Time variation of the fractal dimension relative to the spatial distribution of hypocentres, calculated for time intervals of thirty days (thin line). The same curve, filtered with a low band pass filter (cut off at 6 months) (heavy line). X-axis, time (months), y-axis fractal dimension. Arrows mark earthquake ($M > 4$) occurrence. *1*: M = 5.2; *2*: M = 4.1; *3*: M = 4.2; *4*: M = 4.8; *5*: M = 4.1; *6*: M = 4.1; 7: M = 4.2

the common time series analysis techniques, like filtering and spectral analysis, can be applied.

The analysis began on May 15th 1977, one year after the 1976 earthquake, and finished in December 1990. The results of the analysis are shown in Fig. 2: x-axis represents time in years, while y-axis represents fractal dimension. The time series obtained is represented with a thin line. Seismic events $M > 4$ occurred in the time interval considered are indicated with an arrow on the top of Fig. 2. The heavy line of Fig. 2 has been obtained applying a low band pass filter (cut off at 6 months) to the time series, so as to eliminate the contribution of the higher frequencies, i.e. the 2-3 months oscillations.

Previous analysis evidenced the presence of an annual term, and one of a longer period about 3.5 years (Rossi 1990). The present research adds more information about the frequency content. The main characteristic of the curves of Fig. 2 is their negative trend, to which oscillations of 1, 1.5 years` and about 4 years` period are superimposed.

For the same thirty day intervals, the number of seismic events has been calculated, obtaining the histogram of Fig. 3. A general decreasing trend may be observed, even if not as clearly as in the curves of Fig. 2. Some oscillations in the seismic activity level are evident, like the two minima of 1980–1981 and 1985-1986.

The February 1st, 1988 M = 4.2 earthquake, with a huge number of aftershocks was the most notable exception to this negative trend. It is noteworthy that at the same time, the fractal dimension (Fig. 2) shows on the contrary a minimum, due

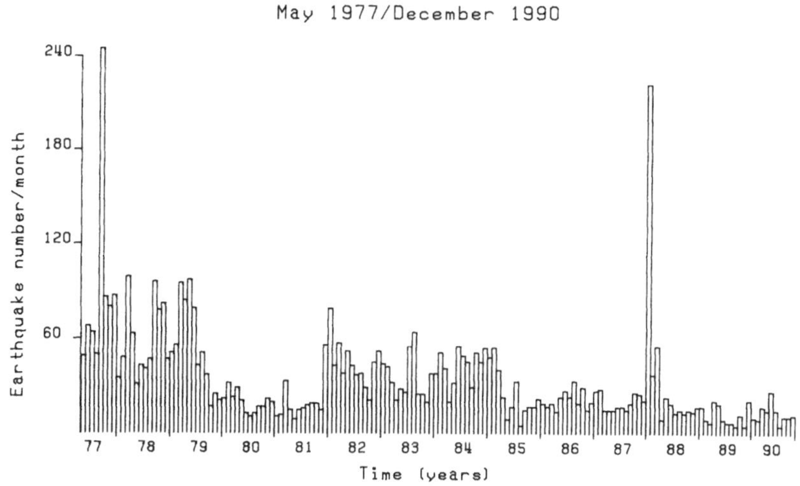

Fig. 3. Number of earthquakes per month for the time interval 1977–1990.

to the extreme clustering of shocks in a very limited area, which may therefore be assimilated to a point (Mandelbrot 1977). The distribution of the events in the region is shown in Fig. 4.

A completely different case is represented by 1984, characterized by an increase in the seismic activity (Fig. 3), accompanied by a fractal dimension increase (Fig. 2), with values between 1.5 and 2: the shocks tend therefore to distribute onto a plane (Mandelbrot 1977). From Fig. 5, it may be seen how the seismic events are diffused over the whole area, hence in agreement with the values obtained.

The presence of a sort of periodicity of 1 year and about 4 years is the most interesting feature of the curves of Fig. 2. The annual term is most probably due to local thermoelastic effects and overall to the variable underground water level, which exercises a pressure on the rock masses and may favour the mobilisation of them as well. It is noteworthy that an annual term has been found by an analysis of seismic activity in another Alpine region, in Switzerland (Roth et al. 1992), with a clear correlation with rain and snow occurrence.

Longer period components may be more directly related to tectonic and seismic processes. This appears to be the case of the 4 years' component. Since 1979, in fact, with the end of the aftershock activity following the 1976 event, $M > 4$ earthquakes succeeded with intervals that vary between 3 and 4 years: 1979–1983–1986–1988–1991. As already observed in other regions, the characteristics of the spatial distribution vary with the different phases of seismic cycle (von Seggern 1982, Ouchi & Uekawa 1986, Eneva & Pavlis 1991). A typical example is given by the 1983 event, preceded by the fractal dimension increase that began in 1982 and culminated in 1983. If, however, the high frequency oscillations are considered, a short-term fractal dimension decrease preceded the 1983 event,

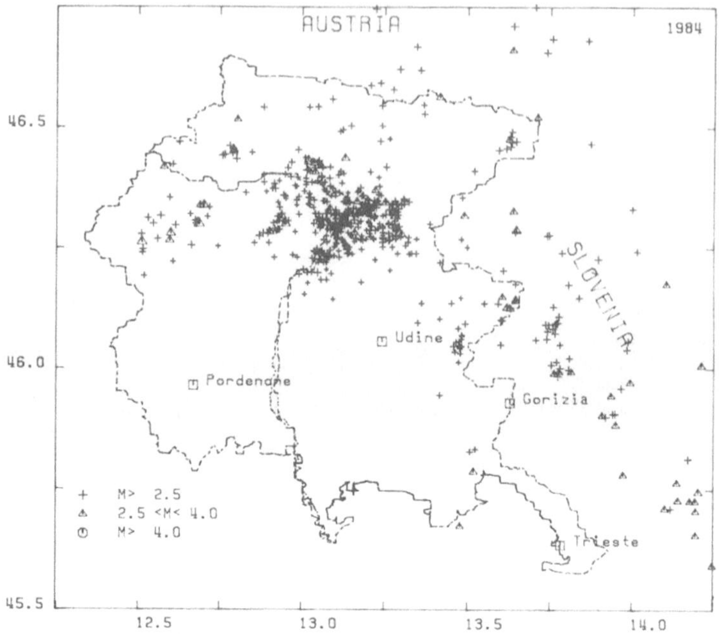

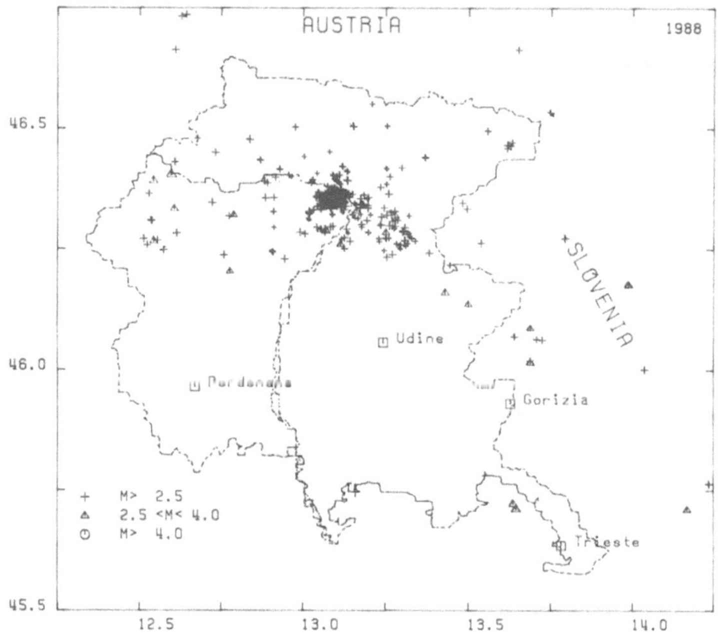

Fig. 4. (upper) Distribution of events in Friuli seismic area during 1988.

Fig. 5. (lower) Distribution of events in Friuli seismic area during 1984.

revealing the tendency of events to concentrate and delineate the future rupture plane. After this event, a new mid-term decrease begins, together with a new cycle, probably more related to the 1988 swarm than to the 1986 earthquake, located at the border of the area considered and covered by the network. It is noteworthy that a complete inversion of the stress field characterizes the phase that preceded and followed the 1983 event (Mao et al. 1989, 1990).

The apparent periodic character of the oscillation discussed till now is confirmed by the spectral analysis of the thin line time series of Fig. 2: the spectrum is shown in Fig. 6, with x-axis representing the periods involved, in years, and y-axis representing the relative spectral amplitude. The peaks appear centred on the annual band, at about 4 years, and on a longer period term. Because of the relative shortness of the time series considered, it is not possible to discriminate the longer time periods. They have therefore to be considered as only indicative.

These results become more significant when they are compared with the data of the geodetic network active in the same area since 1977. The single tiltmeter station of Grotta Gigante, near Trieste, indicated in the future as TS, has been active already since 1967. Fig. 7 shows the output of the two components (EW and NS) of this station. From the two components, it is possible to infer the deviation from the vertical of a site. The curves of Fig. 7 show a strong annual component, superimposed onto longer period ones (thin line). The heavy line

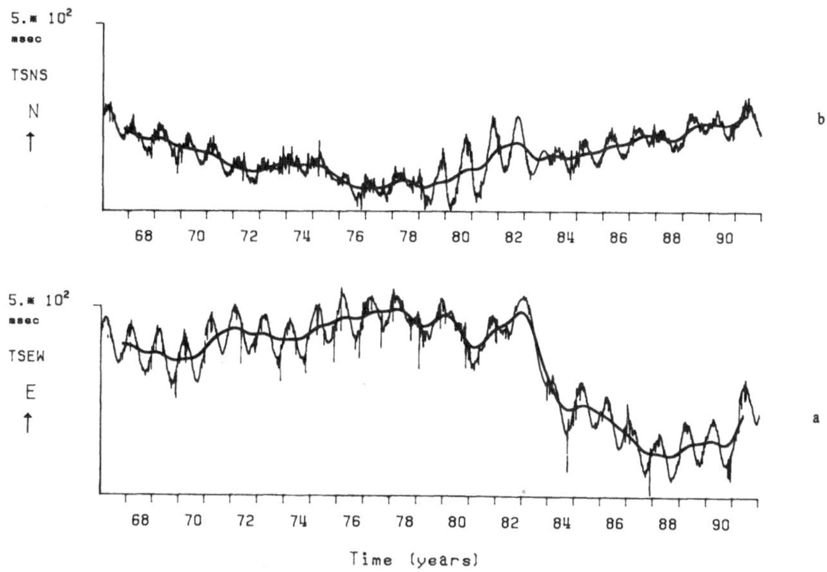

Fig. 6. Spectrum relative to the time series of Fig. 2 (thin line). X-axis: periods involved in years.

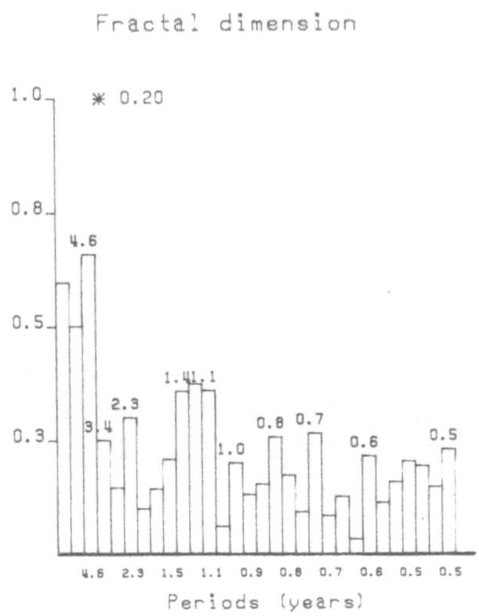

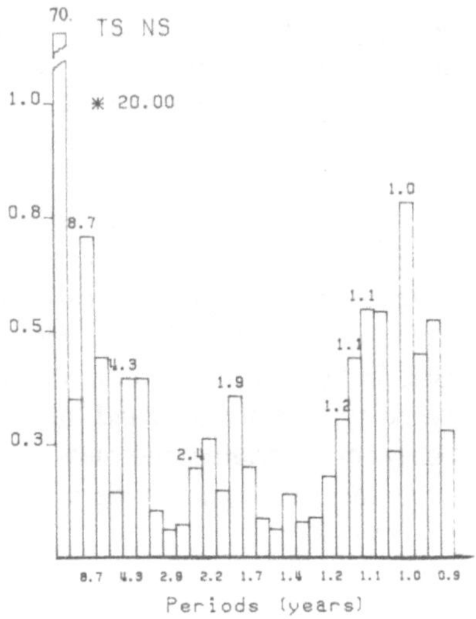

Fig. 7. (left) Tiltmeter components (a:NS; b:EW) of the tiltmeter station of Grotta Gigante nearby Trieste (thin line).The same curve, filtered with a low pass band filter (cut off at 2.5 years) (heavy line) X-axis: time in years, amplitude is in msec.
Fig. 8. (right) Spectrum relative to the curve of Fig. 7 (thin line). x axis: periods involved, in years.

represents the "secular term" obtained applying a low band pass filter (cut off at 2.5 years). The spectrum of the single NS component of TS is shown in Fig. 8, but the results are very similar for all the instruments. Again x-axis represents periods in years, while y-axis represents the relative spectral energies. The value of the first harmonic is indicated at the top of Fig. 8, being much greater than the others, and resulting so out of the scale.

In spite of the different length of the time series, and therefore of the frequency band definition, the spectrum of Fig. 8 confirms the observations made till now on the seismic activity. In fact, the spectral energy peaks mark both the annual term and the 4-5 years component. The longer time series allows us to discriminate a 8 years period term, while the longer period ones are still undefined.

Conclusions

The extraordinary agreement between the spectra of Fig. 6 and Fig. 8 demonstrates that, in both cases, the variation of the parameter observed (strain intensity in the last case, and earthquake spatial distribution in the first one) reflects the same phenomenon: the time variation of stress field in the region.

Besides, the most significant variations of the fractal dimension relative to spatial distribution of shocks are well correlated with the occurrence of mid energy earthquakes ($M > 4$). The post- and interseismic phases appear characterized by a fractal dimension decrease that reflects the activation of small clusters, due to local stress field inhomogeneities. The pre-seismic phase is characterized on the contrary by a diffused seismicity that covers the whole region, and by short term clustering.

Hence, in the 1977–1990 period the Friuli seismic zone appears to be dominated by an almost regular alternation of phases of accumulation and release of elastic energy, accompanied by recurrent $M > 4$ events. Recurrence phenomena are known in many seismic areas: one of the most famous cases is constituted by the Parkfield (California) zone (Bakun & McEvilly 1984). Periodic seismicity is still a controversial subject. The examination of a seismic catalogue often shows how seismic events are distributed in clusters separated by longer intervals of apparent quiescence (Scholtz 1990). In other cases, after an accurate statistical analysis of recurrence times, the process appears almost periodic (Nishenko & Bulland 1987). Periodicities (of 25, 12.5, 4 years) are evidenced by the analysis of the seismic energy release on a global scale (Xanthakis 1982) and local scales (Lindholm et al. 1991).

With regard to the periodicities found here, even taking into account the different methodologies and the different catalogues analysed, there is an amazing consistence in the results of this work. The time intervals considered in this work are, however, surely still too short to conclude that the seismic process in this region is dominated by these periodicities. Moreover, the analysis of strain data shows that, at least in this case, the phenomenon is not a stationary one.

This approach, however, with a time repeated fractal analysis, is very useful in analysing the statistical properties of seismicity in a region and its evolution in time. The use of a fixed time window allows the comparison of the results to those of other observations in seismic areas, and for the present case, the geodetic ones. This kind of approach constitutes an important contribution in the comprehension of the mechanisms with which seismic energy is accumulated and released during a seismic cycle.

Acknowledgments. I am indebted to the Osservatorio Geofisico Sperimentale of Trieste, and especially to Dr. Gianni Bressan for his kindness and generosity in giving me access to the data and answering my doubts. I am also indebted to the reviewers for the useful suggestions that contributed to improving the manuscript. This research was supported by MURST, 60 %, R 92, code 1417, contractor M.Zadro.

References

Anderson H, Jackson J (1987) Active tectonics of the Adriatic region. Geophys J R astron Soc 91: 937-983.

Bakun WH Mc Evilly TV (1984) Recurrence Models and Parkfield, California, Earthquakes. J Geophys Res 89: 3051-3058.

Barbano M S, Kind R, Zonno G (1984) Focal parameters of some Friuli earthquakes (1976 - 1979) using complete theoretical seismograms. J Geophys 58: 175-182.

Carulli GB, Nicolich R, Rebez A, Slejko D (1990) Seismotectonics of the Northwest external Dinarides. Tectonophysics 179: 11-25.

De Rubeis V, Dimitriu P, Papadimitriou E, Tosi P (1993) Recurrent patterns in the spatial behaviour of italian seismicity revealed by the fractal approach. Geophys Res lett 20: 1911-1914.

Ebblin C, Michelini A (1986) A principalparameteranalysis of aftershock sequences applied to the 1977 Friuli, Italy, sequence. Ann Geoph 4: 473-480.

Eneva M, Pavlis G (1991) Spatial Distribution of Aftershocks and Background Seismicity in Central California. PAGEOPH 137: 35-61

Finetti I, Russi M, Slejko D (1979) The Friuli earthquake (1976-1977). Tectonophysics 53: 261-272.

Grassberger P (1983) On the fractal dimension of the Henon attractor. Phys Lett 97A: 224-230

Hirata T, Satoh T, Ito K (1987) Fractal structure of spatial distribution of microfracturing in rock. Geophys. J R astron Soc 90: 369-374.

Lindholm C, Havskov DJ, Sellevoll MA (1991) Periodicity in seismicity:examination of four catalogs. Tectonophysics 191: 155-164.

Mandelbrot BB (1977) Fractals: form, chance and dimensions. Freeman, San Francisco, pp 364.

Mao WJ, Ebblin C, Zadro M (1989) Evidence for variations of mechanical properties in the Friuli area. Tectonophysics 170: 231-242.

Mao W J, Santero P, Zadro M (1990) Long - and middle-term behaviour of the tilt and strain variations in the decade following the 1976 earthquake in NE-Italy. PAGEOPH 132: 653-677.

Meredith PG, Main IG, Jones C, (1990) Temporal variations in seismicity during quasi-static and dynamic rock failure. Tectonophysics 175: 249-268.

Mogi K (1985) Earthquake prediction. Ac Press, Tokio.

Nishenko SP Bulland R (1987) Recurrence distribution for earthquakes forecasting. Bull Seism Soc Am 77: 1382-1399.

O.G.S. 1977-1981. Bollettino della rete sismologica del Friuli Venezia-Giulia O.G.S., Trieste.

O.G.S. 1982-1990. Bollettino della rete sismometrica dell'Italia Nord-Orientale, O.G.S., Trieste.

Ouchi T, Uekawa T (1986) Statistical analysis on the spatial distribution of earthquakes-variation of the spatial distribution of earthquakes before and after large earthquakes. Phys Earth Planet Int 44: 211-225.

Rikitake T (1976) Earthquake prediction. Elsevier, Amsterdam.

Rossi G (1990) Fractal dimension time variations in the Friuli (north-eastern Italy) seismic area. Boll Geof Teor Appl 127/128: 175-184.

Rossi G, Ebblin C (1990) Space (3-D) and space-time (4-D) analysis of aftershock sequences:the Friuli (NE-Italy) case. Boll Geof Teor Appl 125: 37-49.

Roth Ph, Pavoni N, Deichmann N (1992) Seismotectonics of the eastern Swiss Alps and evidence for precipitation-induced variations of seismic activity. Tectonophysics 207: 183-197.

Sadovski MA, Golubeva TV, Pisarenko VF, Snirman MG (1984) Characteristic dimensions of the rock and gerarchical properties of seismicity (on Russian). Fizika Zemli 2: 3-15.

Scholtz CH (1990) The mechanics of earthquakes and faulting. Cambridge Univ Press, Cambridge.

von Seggern DH (1982) Investigation of seismic precursors before major earthquakes and of the state of stress on fault planes. U.M.I. Dissertation Information, The Pennsylvania State University, Ph.D thesis.

Westaway R (1990) Present-day kinematics of the plate boundary zone between Africa and Europe, from the Azores to the Aegean. Earth Planet Sc Lett 96: 393-406.

Xanthakis J (1982) Possible periodicities of the annually released global seismic energy ($M > 7.9$) during 1898-1971. Tectonophysics 81: T7-T14.

Zadro M (1992) Tilt and strain variations in Friuli, NE-Italy after the 1976 earthquake. in: Boschi E & Dragoni M (eds) Earthquake prediction, Il Cigno Galileo Galilei, Roma, 295-315.

Zadro M, Rossi G (1991) Long-term strain variations in Friuli (NE-Italy) seismic area. Proc Intern Conf of earthquake prediction, state of the art, Strasbourg, pp 430-441.

Binary Descriptions of Stick-Slip Phenomena

Mario Markus[1], Heike Emmerich[1], Carsten Schäfer[1], Pedro Almeida[2] & António Ribeiro[2]
[1] Max-Planck-Institut für molekulare Physiologie, Postfach 10 26 64, D-44026 Dortmund, Germany
[2] Departamento de Geologia, Faculdade de Ciências, Universidade de Lisboa; 58, Rua da Escola Politécnica, 1200 Lisboa, Portugal

Abstract. A cellular automaton for seismological stick-slip processes is developed. Reducing complexity to a minimum, a binary description is obtained for a Poincaré section defined by the states of rest. This description allows us to compare the results with the binary number-theoretical automata by Wolfram. It is found that the Poincaré section can be approximated by stochastic modifications of Wolfram's automata, and a classification into the four automata classes proposed by Wolfram is possible. In aperiodic regimes, the spatial measure entropy reaches a constant minimum after transients. Gutenberg-Richter statistics yield large, resp. small, slopes for large, resp. small seismic moments, as well as a "characteristic earthquake", in agreement with seismological and geological data.

Introduction

Cellular automata have proven to be useful tools in an amazing variety of natural sciences including: fluid dynamics (Frisch et al. 1988), spin glasses (Farmer et al. 1984), ferromagnetism (Wolfram 1986), pattern recognition (Preston & Duff 1984), precipitation processes, immunological systems, brain patternization, ecology, and branching growth of fungi, bacterial colonies and vascular networks (Ermentrout & Edelstein-Keshet 1993), cytoskeleton dynamics (Dufort & Lumsden 1993), cellular calcium waves (Lechtleiner et al. 1991), coupling of mitotic cycles (Markus & Salvador 1992), excitable media (Markus & Hess 1990, Markus 1992), activator-inhibitor processes in morphogenesis (Schepers & Markus 1992), tectonics and seismology (Bak & Tang 1989, Ito & Matsuzaki 1990, Matsuzaki & Takayasu 1991, Nakanishi 1990, Feder & Feder 1991, Olami et al. 1992, Christensen & Olami 1992).

In the present work we will attempt to reduce a cellular automaton - aimed at describing some of the characteristics of earthquakes - to a minimum. To do so, we will describe it in a binary code with a goal of putting it in the general, number-theoretical framework resulting from stochastic modifications of Wolfram's (1984, 1986) automata.

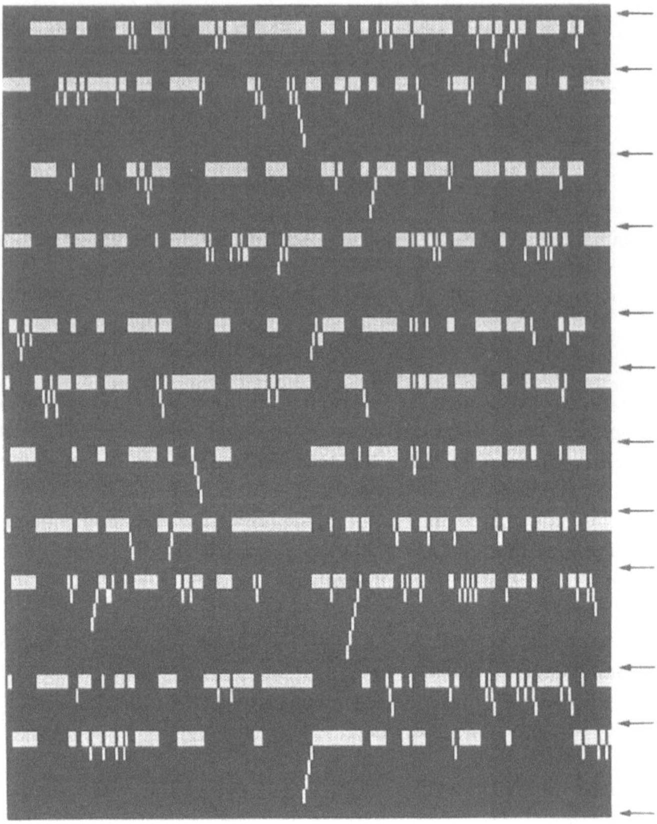

Fig. 1. Typical evolution of our automaton. The slipping cells are shown in white, the resting ones in black. The dissipation is p = 1.82. The arrows at the right indicate the state P_k where all cells are at rest.

Stick-Slip Models

A simple mechanical (one dimensional) model for earthquakes was proposed in the 1960s by Burridge & Knopoff (1967). It consists of a chain of blocks each of them connected to its nearest neighbours by springs as well as connected frictionally to a fixed rigid plate. The blocks are driven by the relative movement of the two rigid plates. When the total force on one of the blocks exeeds a certain threshold, the block slips, its energy being distributed among its neighbours. Later on, a number of 1D and 2D cellular automata using this slip-stick idea were proposed by the following authors:

i) Bak & Tang (1989), who assumed the driving force increases in each time step at a randomly chosen point.

ii) Ito & Matsuzaki (1990) who include plasticity, assuming that slipping occurs above threshold with a probability slightly smaller than 1.
iii) Matsuzaki & Takayasu (1991) who use a randomly chosen threshold and distribute the forces after a slip only to the neighbours that have not already slipped at that step.
iv) Nakanishi (1990), Feder & Feder (1991), as well as Olami et al. (1992), who introduce dissipation in different ways.
v) Christensen & Olami (1992), who introduce anisotropy.

All these publications have in common that they show that dynamics occur in a state of self-organized criticality (Bak et al. 1987), and that certain power laws observed in nature, such as the Gutenberg-Richter law (Gutenberg & Richter 1954) can be obtained from the cellular automaton approach.

Considering the diversity of automata just listed we learn that their essential predictions are quite independent of their particular assumptions. Therefore, we will not go the usual way of adding more sophisticated assumptions, but the opposite way, namely of reducing the automaton to a minimum, and of checking essential observations, such as Gutenberg-Richter's law.

Rules and Visualization

We considered a one-dimensional chain of n blocks. The force on block i at time t is $z_i(t)$. After each resting period the force is increased by 1. The block slips if, $[z_i(t)-p] \geq 2$, becoming zero after slipping and yielding $[z_i(t)-p]/2$ to each neighbour. The parameter p describes the dissipation. Note that we are implicitely incorporating two time scales: one (in nature of the order of seconds or minutes) for the slipping process and another one (in nature of the order of years or more) for the increase of the force by 1. For our calculations, however, the length of these two characteristic times is irrelevant: both can be considered as one calculation step. Thus:

$$z_i(t+1) = z_i^+(t+1) + z_i^-(t+1) + z_i^0(t+1)$$

where:

$$z_i^+(t+1) = (z_{i+1}(t) - p)/2 \quad \text{if} \quad z_{i+1}(t) \geq 2 \quad \text{and zero otherwise,}$$

$$z_i^-(t+1) = (z_{i-1}(t) - p)/2 \quad \text{if} \quad z_{i-1}(t) \geq 2 \quad \text{and zero otherwise,}$$

$$z_i^0(t+1) = z_i(t) \quad \text{if} \quad z_i(t) < 2 \quad \text{and zero otherwise.}$$

We define the set P_k (k=1,2,...) by the condition $z_i(t) < 2 \ \forall_i$, (states of rest between slip events). The set P_k is comparable to a Poincaré-section in continuous systems defined by a condition for the state variables. The state-of-rest-set P_k is a

b

c

d

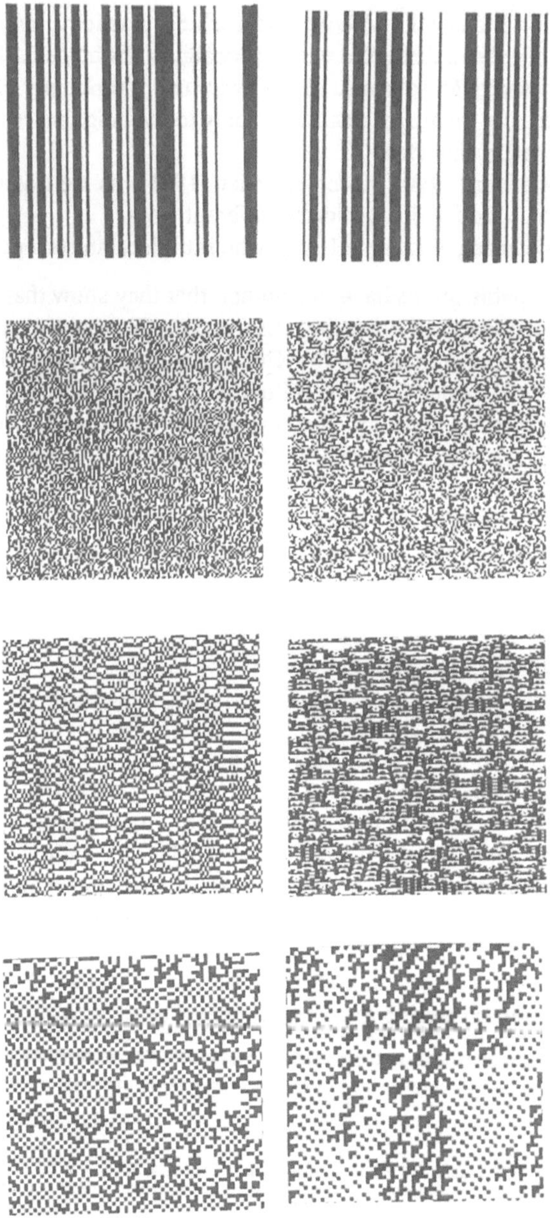

Fig. 2a-d. Binary automata (space: horizontal; time: vertical). Left: Results from our stick-slip simulations with p = 1 (**a**), p = 0.3 (**b**), p = 1.34 (**c**) and p = 0.725 (**d**). Right: Wolfram's binary automata (Wolfram 1986) with code 204, k = 2, r = 1 (**a**), code 10, k = 2, r = 2 (**b**), code 118, k = 2, r = 3 (**c**) and blocked pattern (showing each second time and second space steps) with code 110, k = 2, r = 1 (**d**).

represented as a binary sequence $\tilde{z}_i$ by setting $\tilde{z}_i = 0$ if $\tilde{z}_i < 1$ (shown in white) and $\tilde{z}_i = 1$ if $1 \leq \tilde{z} < 2$ (shown in black). In all figures, the spatial index $i=1,\ldots,n$ is represented horizontally and the temporal index k is representaI vertically. Our only control parameter in this model is the dissipation p. This model differs from the discrete cellular automaton models (Bak & Tang 1989, Ito & Matsuzaki 1990) and from the continuous versions (Matsuzaki & Takayasu 1991, Nakanishi 1990, Feder & Feder 1991, Olami et al. 1992) in that here the state variable is continuous, while the force increases discretely by 1 after each resting-state step.

Results from automata by Wolfram are represented by this author's code (Wolfram 1984) for totalistic automata with two states per cell and interactions within a range of radius r. "Totalistic" are automata in which the next state f of a cell depends only on the sum s of the states of the $2r + 1$ neighbours. The code is given by

$$C = \sum_{s=0}^{2r+1} 2^s f(s).$$

As an example for radius 2: if all sums s yield 1, except $s = 2$ and $s = 4$ then the code is $C = 2^0 \cdot 1 + 2^1 \cdot 1 + 2^2 \cdot 0 + 2^3 \cdot 1 + 2^4 \cdot 0 + 2^5 \cdot 1 = 43$.

Results

Comparison of Stick-Slip Simulations with Wolfram's Automata

Figure 1 shows a typical evolution of our automaton. The slipping blocks are represented in white, and the resting blocks are shown in black. The arrows on the right indicate the rows P_k consisting only of resting states between slipping events.

Figure 2 shows (on its left) the sets P_k of the states of rest for four different values of the dissipation p. At the right side of each picture, we show results from Wolfram's automata, which visually compare well with our sets P_k.

Wolfram classifies his automata into four classes, which he conjectures to be universal (Wolfram 1984). They are classified by their asymptotic (final) states:

Class I: Homogeneous states, comparable to point attractors in continuous systems.

Class II: Periodic structures, comparable to limit cycles.

Class III: Chaotic patterns.

Class IV: Complex localized structures, sometimes long-lived. These structures may die, become periodic, or propagate as predictable or unpredictable waves. No well defined chaotic attractors are obtained.

a

b

c

d

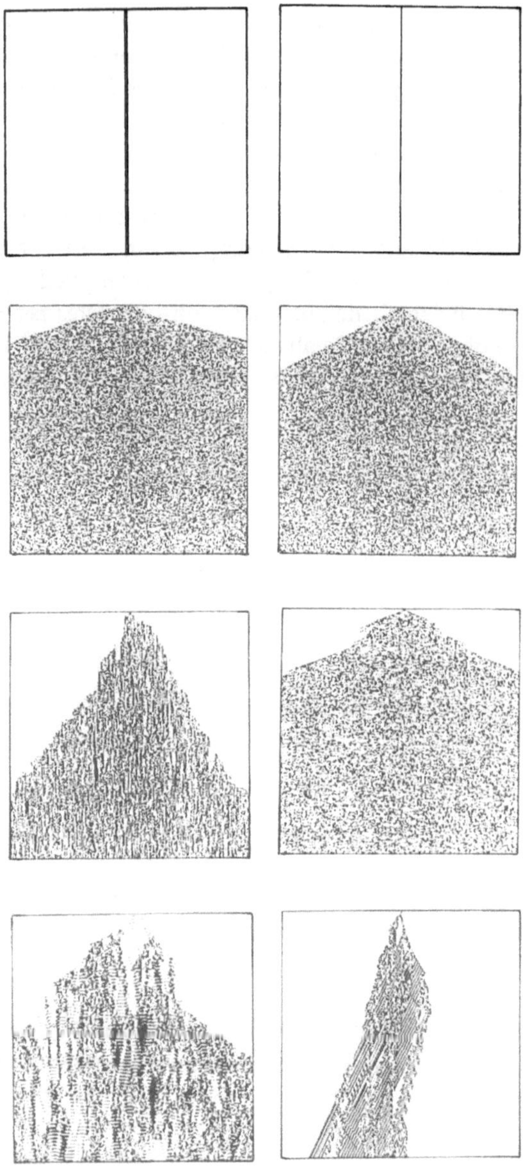

Fig. 3a-d. Difference patterns indicating the propagation of a perturbation in a single cell. The pictures correspond to those of Fig. 2.

While class III patterns are called unpredictable, class IV patterns are called undecidable. Class III patterns permit only limited prediction of future development but have some global invariants which are characteristic of the chaotic attractor, such as Lyapunov exponents and generalized dimensions. In contrast, class IV patterns do not even have such features in a global way.

In Fig. 2, case (a) corresponds to class II, cases (b) and (c) to class III, and case (d) to class IV.

We used a comfortable quantitative method for classifying the automata. This consisted in subtracting two patterns which are identical except for the initial conditions in one cell. Plotting the absolute values of the resulting values leads to a so-called "difference pattern" (Wolfram 1984).

Fig. 3 shows the diference patterns corresponding to Fig. 2 (for our stick-slip automata on the left, and for Wolfram automata on the right). The initial perturbation does not propagate for the (periodic) class II automaton (Fig. 3a). The propagation of the perturbation in the class III automata (Figs. 3b and 3c) occurs with more or less constant velocity, which is a well defined quantity (slope in the figures), just as the Lyapunov exponents for strange attractors. This is not the case in Fig. 3d (class IV): here, the global behaviour of the system changes unpredictably between no propagation, class III type propagation and even self-repair. Furthermore, the dynamics of the difference pattern for class IV depend on the initial random distribution of 0's and 1's, as shown in Fig. 4.

We posed the following question: does the union set of all difference patterns arising from different random initializations lead to some invariant quantity? The answer is certainly "yes" for class III automata since all the difference patterns have the same shape. But how about class IV? In Fig. 4 we show difference patterns for three different initial configurations for $p = 0.5$ (Figs. 4a to 4c) and the union set of 25 such configurations at the bottom (Fig. 4d). We can clearly see that although the single difference patterns show no well defined slopes, which is characteristic for class IV, the union set can be quantitatively described by a slope, just as for a single difference pattern in class III. However, we did find cases for which no well-defined slope could be determined in union sets.

Approximation of Stick-Slip Processes by Stochastically Modified Wolfram Rules

Return maps resulting from complex chaotic processes can be fitted by simple parabolas, which render realistic global features, although they ignore mechanistic details (Rössler et al. 1989, Markus 1990, 1992). It is with a similar intention that we ask if our states-of-rest-maps can be "fitted" by simple rules related to Wolfram's automata.

To answer this question, we determined (for a given r) the statistical expectation values E of each of the 2^{2r+1} possible configurations of 0's and 1's in our state-of-rest-map. Then we used as an approximation a statistical modification of Wolfram's rules: for a given binary configuration, 0's are set with probability 1-E and 1's with probability E (note that we do not restrict 'a priori' to totalistic rules). This procedure and its results are illustrated in Figs. 5 through 7.

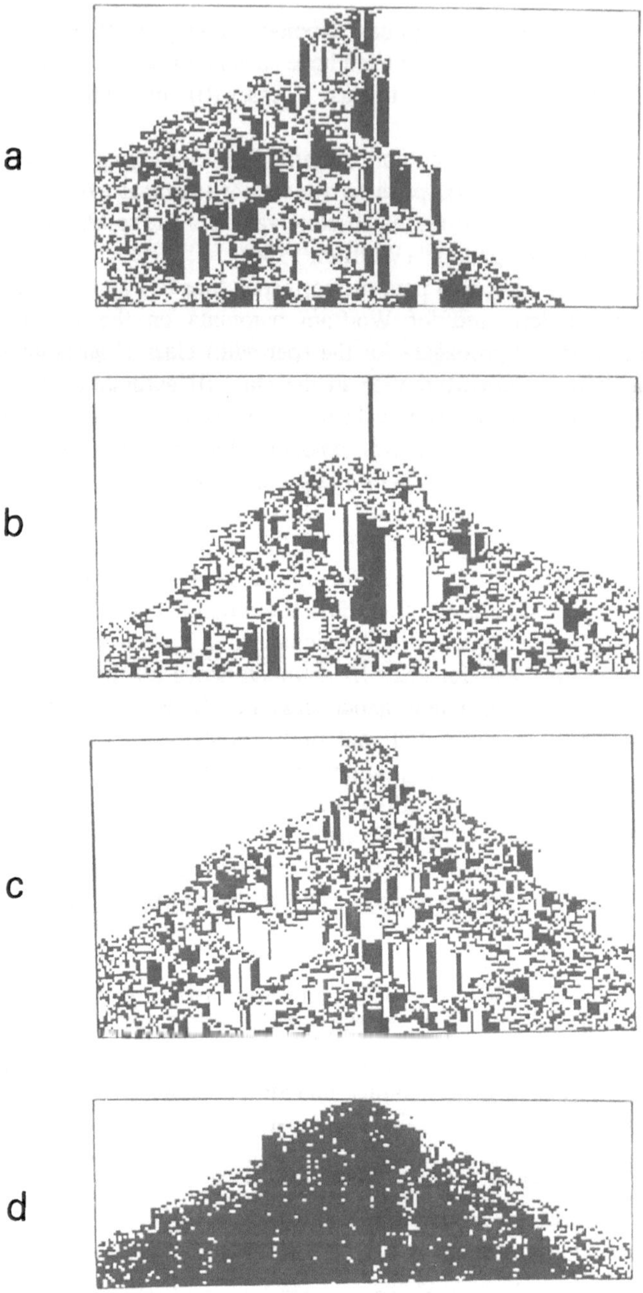

Fig. 4a-d. Difference patterns from our earthquake model for p = 0.5. **a**, **b** and **c** differ only in the initial random configurations. Note the undefined slopes, typical for class IV automata. **d**: the union set of 25 such patterns, having better defined slopes.

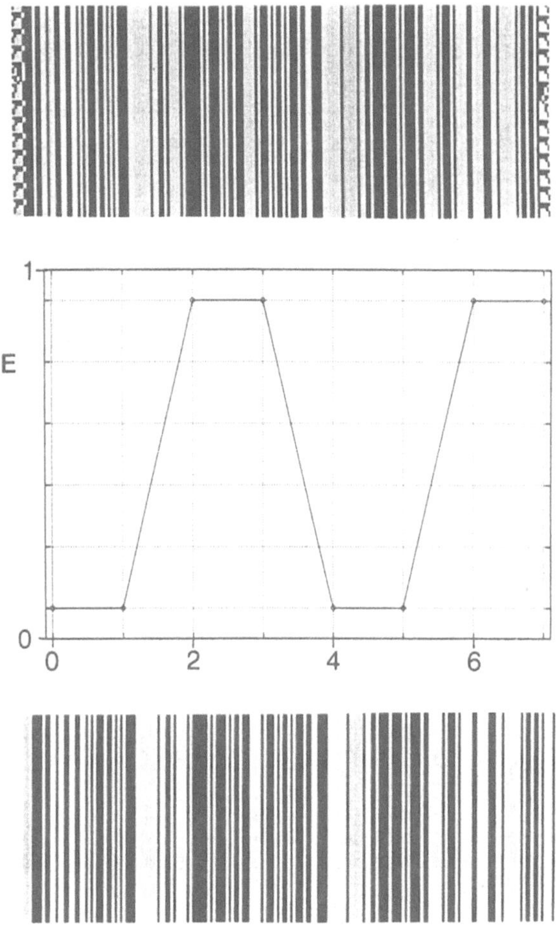

Fig. 5. Upper: stick-slip simulations with p = 1. **Middle**: expectation values E from the simulations above for all possible binary configurations with r = 1. **Lower**: Simulation with a stochastic modification of a Wolfram automaton, applying the rules resulting from the expectation values E.

In the upper part of these figures, we show our stick-slip maps (states of rest). In the middle part, we show the expectation values for all configurations of the chosen r (the abscissa is the decimal code of the configuration's binary sequence; examples: abscissa 3 in Fig. 5 corresponds to 011 since $r = 1$; abscissa 10 in Figs. 6 and 7 corresponds to 01010 since $r = 2$). In the lower part of Figs. 5 through 7 we show the dynamics resulting from the fitted approximation using stochastically modified Wolfram rules. The fit is perfect in Fig. 5 (class II) and only approximate in Fig. 6 (class III) and Fig. 7 (class IV). Note that if r is taken large enough, any binary pattern may in principle be fitted by a stochastic cellular automaton rule; therefore, the interesting feature of this result is the very small value of r (1 or 2) for which an approximation is obtained.

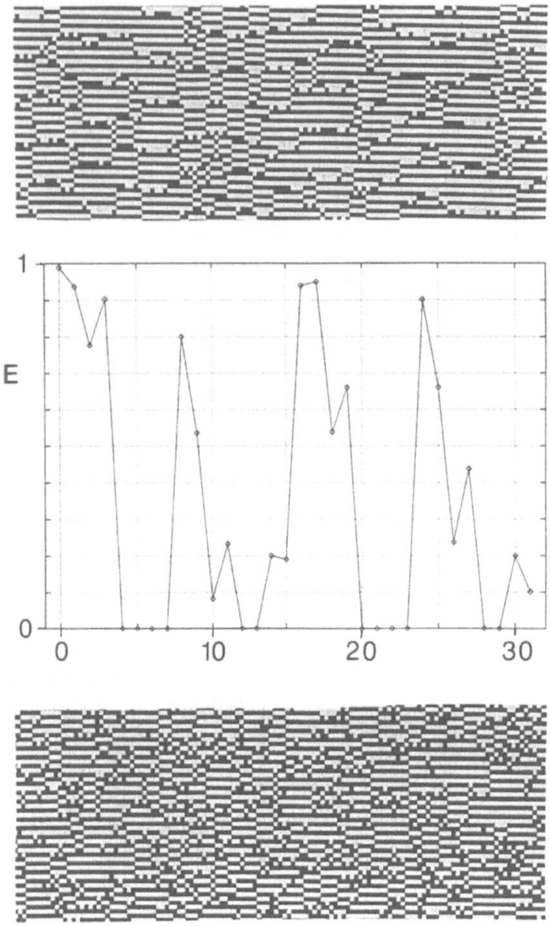

Fig. 6. As Fig. 5, but p = 1.82 and r = 2

Statistical Characterizations

The distribution of energy released during earthquakes has been found to obey the well-known Gutenberg-Richter law (1954). It is based on the empirical observations that the number N of earthquakes of magnitude greater than M is given by the relation

$$\log N = a - bM,$$

b being of the order of 1. M is proportional to log m, where m is the seismic moment and can be taken in our case to be propotional to $\Sigma_i\,[z_i(t+1)\text{-}z_i(t)]$, as in the model by Matsuzaki and Takayasu (1991). For this evaluation, we considered all automaton steps between the states of rest.

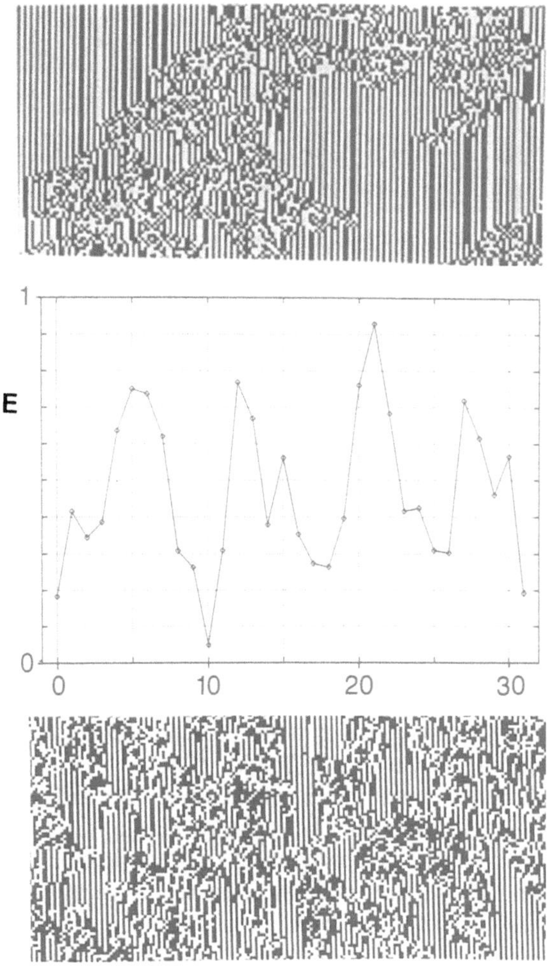

Fig. 7. As Fig. 5, but p = 0.5 and r = 2

Changing our free control parameter p in our one-dimensional model, we found best agreement with reported observations at $p = 1.82$. Fig. 8 (right part) shows a log-log-plot of N versus M. On the left part of Fig. 8, we show a curve obtained from seismological observations and from geological data. For the optimized $p = 1.82$ we obtained a maximum value of b (slope at lower M) of about half that of observations. However, our first results with a 2D-generalization of our automaton show that for 2D it is easy to adjust this quantity to observations.

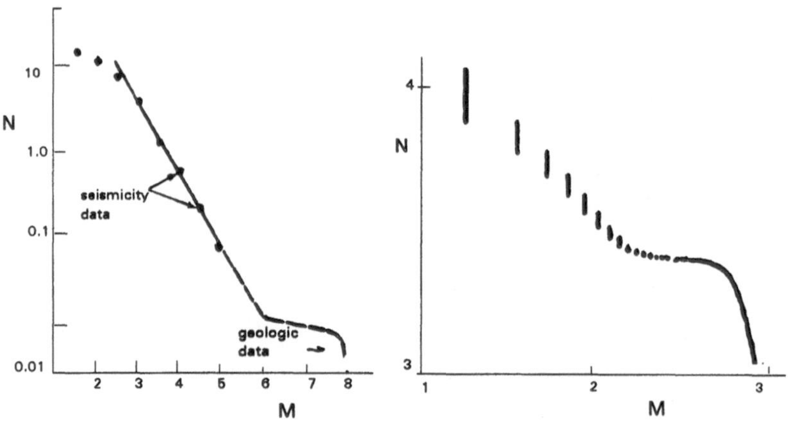

Fig. 8. Right: Gutenberg-Richter statistics for real data from the Wasatch fault zone (Schwartz & Coppersmith 1984). **Left**: Corresponding plot from our automaton simulations (p = 1.82). *N*: number of earthquakes, per unit time, larger than M. *M*: magnitude.

With our very simple 1D model, we have been able to reproduce a "characteristic earthquake". The "characteristic earthquake" is a consequence of segmentation of active faults with stable barriers, and has been proved in a large number of observation sites (Schwartz & Coppersmith 1984, 1986).

Our optimized value p = 1.82 is quite near the threshold value of 2 and thus contradicts the seismological models that assume perfectly elastic rheology. However, our result can be reconciled with a more general viscoelastic model of the seismogenic crust compatible with the increasing evidence that in many plate boundaries 50% or more of plate motion is aseismic (e.g. Ekstrom & England 1989).

As a statistical characterization of the states of rest, we determined the spatial measure entropy defined by

$$S = -\frac{1}{x}\sum_{j=1}^{2^x} p_j^{(x)} \log_2 p_j^{(x)},$$

where $p_j^{(x)}$ is the probability of finding the jth binary configuration of lenght x (Wolfram 1984). Fig. 9 (left) shows how well S saturates with respect to x and to the number of resting states k. For this computation, we used the optimized value $p = 1.82$. The corresponding states-of-rest-map is given in the upper part of Fig. 6. Although we have here a class III (chaotic) regime we see on the right side of Fig. 9 that S tends to a minimum. Transients thus have an entropy which

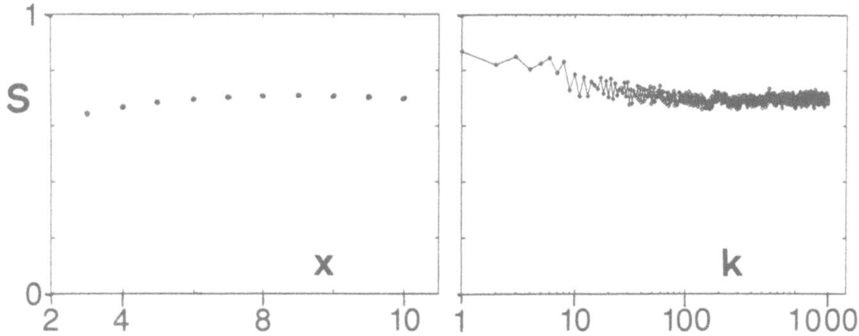

Fig. 9. Left: spatial measure entropy S, as obtained from our simulated resting states, as a function of configuration length x. **Right**: Saturation value of the left plot, as a function of the number of resting states *k*.

decreases as the system self-organizes into a chaotic attractor. It is certainly remarkable that the concept of entropy can be generalized to such a complex dynamic state.

Discussion

To our knowledge, this is the first work in which Wolfram's rules, subjected to probabilities, have found an application in modelling processes in nature. In fact, our stick-slip model dynamics can be fitted (Figs. 5 through 7) by Wolfram's simple arithmetics rules (modified with appropriate statistics as described in section "Approximation of Stick-Slip..."). Furthermore, the classification into four automaton classes, as introduced for Wolfram's automata, is applicable to stick-slip processes. However, attempting to display the dependence of automaton classes on *p*, we found such complicated relationships that we left the study of this dependence to future research. In any case it is interesting that we found a convergence between a simulation of a physical process and phenomena that had been considered in a pure number-theoretic context.
A propagation of perturbations (the so-called "butterfly effect") that is not exponential in time but $\sim t^c$ for class III dynamics (see difference patterns in Fig. 3b and 3c) has also been shown by us in the turbulent Belousov-Zabotinskii reaction (Markus & Schäfer 1991). While here c ≈ 1, there a diffusion-type of propagation (c ≈ ½) is obtained. In any case, this type of propagation $\sim t^c$ is characteristic for self-organized criticality.

Gutenberg-Richter plots with a small, resp. large, slope at small, resp. large, magnitudes (as in Fig. 8) have already been obtained in the stick-slip automaton

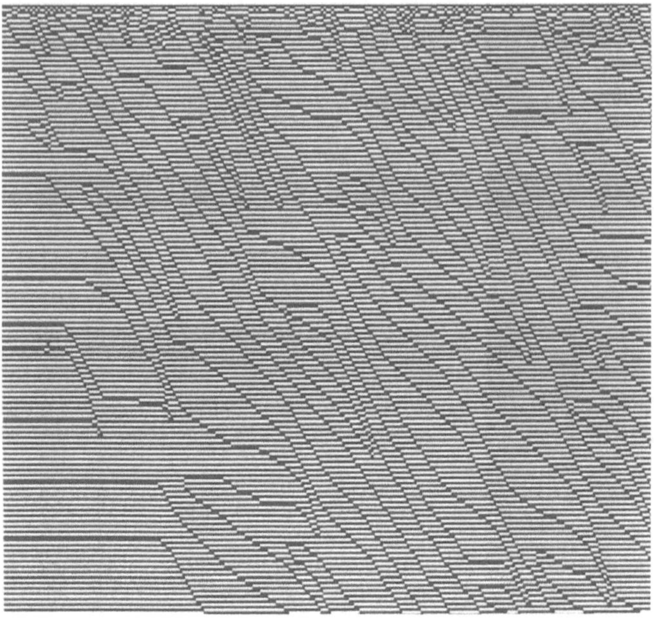

Fig. 10. Evolution of a 2D stick-slip automaton assuming a von Neumann neighbourhood and p = 1.4. By means of a crosscut of the three-dimensional system defined by the fault plane and time, we represent here one coordinate of the fault plane (horizontal) as a function of time (vertical).

proposed by Nakanishi (1990). However, he required a more complex model involving a rather artificial functional dependence of the state after slipping from the amount of force exceeding the threshold.

Our states-of-rest-maps have the following advantage: they make a complex spatiotemporal process easy to grasp with one glance. Of course, things become more difficult for 2D automata. The 2D model produces a 3D output (x, y, t), where (x, y) is the fault plane, and t is time. In order to get an easy reading of these results, one can display one row of blocks (constant x or y) against t, so that t is represented vertically and the spatial distribution horizontally. We have obtained first results with this technique, as exemplified here in Fig. 10 for the case of a 2D, class IV calculation. We would like to point out that fault's states of rest, as displayed throughout this work, can nowadays be monitored by geodetic measurements.

We think that the most remarkable result of this work is that, in spite of the long and complex processes between the states of rest, the latter can be approximated by simple rules and described in a quantitative way by a perturbation propagation ~t and by the spatial measure entropy.

Acknowledgements. We thank the Commission of the European Communities for financial support. P.A. acknowledges a Master of Science Scholarship from Junta Nacional de Investigaçäo Científicã e Tecnológica (Programa CIENCIA). This research was funded in part by project TESIS (J.N.I.C.T.) and Seismotectonics of Portugal (G.P.S.N., J.N.I.C.T.)

References

Bak P, Tang, C (1989) Earthquakes as a self-organized critical phenomenon. J Geophys Res 94: 15.635-15.637.

Bak P, Tang C, Wiesenfeld K (1987) Universality and complexity in cellular automata. Phys Rev Lett 59: 381-384.

Burridge R, Knopoff L (1967) Model and theoretical seismicity. Bull Sei Soc Am 57: 341-371.

Christensen K, Olami Z, (1992) Scaling, phase transitions, and nonuniversality in a self-organized critical cellular-automaton model. Phys Rev A 46: 1829-1838.

Dufort PA, Lumsden CJ (1993) Cellular automaton model for the actin cytoskeleton. Cell Motility and the Cytoskeleton 25: 87-104.

Ekstrom G, England P (1989) Seismic strain rates in regions of distributed continental deformation. J Geoph Res 94: 10231-10257.

Ermentrout GB, Edelstein-Keshet L (1993) Cellular automata approaches to biological modelling. J Theor Biol 160: 97-133.

Farmer D, Toffoli T, Wolfram S (1984), (ed) Physica 10D, all volume.

Feder HJS, Feder J (1991) Self-organized criticality in a stick-slip process. Phys Rev Lett 66: 2669-2674.

Frisch U, Hasslacher R, Pomeau Y (1988) Lattice-gas automata for the Navier-Stokes equations. Phys Rev Lett 56: 1505-1508.

Gutenberg B, Richter CF (1954) Seismicity of the Earth. Princeton Univ Press, Princeton.

Ito K, Matsuzaki M (1990) Earthquakes as self-organized critical phenomena. J Geophys Res 95: 6853-6860.

Lechtleitner J, Girard S, Peralta E, Clapham D (1991) Spiral calcium wave propagation and annihilation in Xenopus laevis oocytes. Science 252: 123-126.

Markus M (1990) Chaos in maps with continuous and discontinuous maxima. Computers in Physics, Sept/Oct: 481-493.

Markus M (1992) Are 1D maps of any use in ecology? Ecological Modelling 63: 243-259.

Markus M, Hess B (1990) Isotropic cellular automaton for modelling excitable media. Nature 347: 56-58.

Markus M, Salvador A (1992) Cellular automata modelling of cell cycle heterogeneity and synchronization. In: Hildebrandt G et al. (ed) Chronobiology and Chronomedicine, Peter Lang, Frankfurt, pp. 589-602.

Markus M, Schäfer C (1991) Spatially periodic forcing of spatially periodic oscillators. In: Seydel R et al. (ed) Bifurcations and Chaos: Analysis, Algorithms and Applications. Birkhäuser, Boston, pp. 263-275.

Markus M, Nagy-Ungvarai Z, Hess B (1992) Photoaxis of spiral waves. Science 257: 225-227.
Matsuzaki M, Takayasu H (1991) Fractal features of the earthquakes phenomenon and a simple mechanical model. J Geophys Res 96: 19925-19931.
Nakanishi H (1990) A cellular automaton model of earthquake with deterministic dynamics. Phys Rev A 41: 7086-7089.
Olami Z, Jacob A, Feder HJS, Cristensen K (1992) Self-organized criticality in a contiuous, nonconservative cellular automaton modeling earthquakes. Phys Rev Lett 68: 1244-1247.
Preston K, Duff MJD (1984) Modern Cellular Automata. Theory and Applications. MIT Press, Boston.
Rössler J, Kiwi M, Hess B, Markus M (1989) Modulated nonlinear processes and a novel mechanism to induce chaos. Phys Rev A 39: 5954-5960.
Schepers H, Markus M (1992) Two types of performance of an isotropic cellular automaton: stationary (Turing) patterns and spiral waves. Physica A 188: 337-343.
Schwartz DP, Coppersmith KJ (1984) Fault behavior and characteristic earthquakes: examples from the Wasatch and San Andreas fault zones. J Geophys Res 89: 5681-5698.
Schwartz DP, Coppersmith KJ (1986) Seismic hazards: new trends in analysis using geologic data. in: Active Tectonics, Nat Ac Press, Washington DC, 215-230.
Wolfram S (1984) Universality and complexity in cellular automata. Physica 10D: 1-35.
Wolfram S (1986) (ed.) Theory and Applications of Cellular Automata. Advanced series on complex systems, Vol. 1. World Sci Publ Co, Singapore.

Main Topics of Fractal Research into Earthquakes in China, a Review

Shouzhong Diao & Hongtai Chao
Seismological Bureau of Shandong Province
Jinan, 250021, The People's Republic of China

Abstract. In the mid 1980's, fractal research into earthquakes developed in China and became an active branch. This chapter gives a review of the general situation of fractal research into earthquakes in China, on the important topics which have been investigated and on the problems which still exist and should be solved.

The main topics of fractal research into earthquakes in China are as follows: (a) It is testified by much evidence that the temporal, spatial and magnitude distributions of the earthquakes are self-similar on some levels and scales, and are fractal in the statistical sense. (b) Before and after a strong earthquake, the fractal structures of the medium-small earthquakes around the epicenter region are complex: some show deepen dimension while others show rising dimension. (c) The fault systems on the Chinese continent have fractal structures. In general, the thrust structure has a lower fractal dimension than the tensional. (d) It is also found that the tectonic topography, which is related to neotectonic and earthquake activity, also has fractal structures, such as the channels, ridges, and so on. (e) Fractal research into earthquakes has led to the development of a new discipline in seismology and new ideas in geology.

Some problems should still be solved: (a) There is still a lack of scientific demonstration as to how many statistical samples are enough for fractal analysis of earthquakes. (b) There is no unified quantitative standard to determine the upper and lower limits of the scaling range. (c) The parameters Dq and $f(\alpha)$ should be studied intensively.

Outline

The fractal geometry presented by Mandelbrot in 1970's has become a new tool to study natural phenomena and provided a new approach to describe the complex systems. As the geoscience studies progress, it is found that many complex phenomena are difficult to explain by traditional theory. Thus seismology and other branches of geoscience naturally seek support from a new theory, such as fractal geometry.

In the early 1980's, the term "fractal dimension" appeared in the seismological documents in China. Since the mid 1980's, especially in the 1990's, the fractal theory has developed greatly in seismological research in China and a series of important achievements have been obtained, which are shown in the following.

1. The programs of fractal research into earthquakes have been supported enthusiastically, and an active research organization has formed. Since the mid 1980's, there are more than 50 programs that have been supported by the Office of Joint Seismology Fund of China, the Foundation of Academia Sinica and so on. The programs and funds have grown year after year. Moreover, fractal research into earthquakes has been integrated into the major plans in the Eighth Five-years Plan (1991-1995) by the State Seismological Bureau of China. Thus fractal research into earthquakes has from the beginning been done by organizations instead of individuals. Now an active team has been formed, which consists of several research groups.

2. Academic exchanges have been launched enthusiastically. In the past few years, the State Seismological Bureau of China, jointly with the nonlinear center of the Science and Technology University of China, the nonlinear center of Institute of Geophysics of Academia Sinica and the nonlinear center of Beijing University, have organized a series of nonlinear scientific pursuits, including the fractal character of earthquakes, such as the 1986 (Hangzhou), 1987 (Hangzhou), 1988 (Beijing), 1990 (Changli of Hebei), 1991 (Beijing), 1992 (Baoding) conferences. At the same time, many fractal researchers actively contributed to the first (Chengdu, 1989) and second (Wuhan, 1991) conferences on fractal theory and its application, the annual meetings of the Geophysical Society of China and the scientific conferences of the Seismological Society of China. Moreover, they also contributed to the international conferences on fractal research into earthquakes.

3. Many research achievements have been obtained. From 1987 to 1992, there were more than 130 papers on the fractal character of earthquakes published in scientific journals in China, and the quantity every year showed a growing tendency. Five new monographs or special issues on the fractal character of earthquakes have also been published. This shows the exciting progress and the broad prospects in fractal research into earthquakes in China.

In brief, fractal research into earthquakes in China is past its embryonic stage and is becoming an active field in seismology.

Main Topics

1. Many modeling methods have been established and introduced to study a series of fractal dimensions of the temporal and spatial distribution of earthquakes and other fractal dimensions in geoscience related to earthquakes.

To give the methods to calculate the fractal dimension of earthquakes correctly is the first problem to be solved. Kagan et al. (1980), Aki (1981), Kagan (1981), King (1983), Mandelbrot et al. (1984), Smalley et al (1987), Aviles & Scholz (1987) and Hirabayashi et al. (1991) etc. have made great efforts. Chinese researchers have also done much on the study and introduction of the methods.

The relationship between the Hausdorf dimension (D_0) of the energy of earthquakes and the b-value: $D_0 = b / 1.5$, has been deduced by the relations of magnitude-frequency and magnitude-energy ($\log N = a + bM$ and $\log E = 11.8 + 1.5\, M_s$) presented by Gutenberg and Richter in 1954 (Hong Shizhong et al. 1987). The relationship between the fractal dimension of length system of the seismic faults and the b-value: D=2b, has also been deduced by the relation between the energy of earthquakes and the scale of seismic faults. The statistical function of the scale of ruptures in experimental rocks has testified to the relationship between the b-value of acoustic emission and the fractal dimension of ruptures in rocks (Niu Zhiren & Shi Xigjue 1992). This also suggests that the energy structures of earthquakes are fractal and the relationship between the fractal dimension (D) and the b-value is simple. All the results mentioned above are in keeping with that achieved from other approaches (Aki 1981, King 1983). In the past, the b-value of earthquakes has been studied, but it has not been taken as a parameter related to the fractal dimension.

Now, the methods to calculate the temporal and spatial fractal dimension are an active field studied by Chinese researchers. The basic mathematical models are that the temporal distribution of earthquakes is generally regarded as a stochastic Cantor set, while the spatial distribution of earthquakes is regarded as a stochastic Sierpinski pad. Methods to calculate fractal dimension have been introduced or established; for example, the D_0 can be calculated by the box-counting method which is deduced directly from the definition of the fractal dimension (Mandelbrot et al. 1984; Hong Shizhong & Hong Shiming 1987; Chen Yong 1988); the informational dimension (D_1) of earthquakes was calculated by introducing the Renyi information and using the distributing ratio (i.e. $n_i / \Sigma n_i$) instead of the distributing probability (Chen Yong 1988; Zhu Lingren et al. 1990); the second order informational dimension (D_2) was calculated by introducing the Grassberger and Procaccia method (Luo Jiuli 1989, Hu Ping et al. 1990) and then the calculating method was improved by Wang Binghong et al. (1992); the multifractal dimension (D_q) or $f(\alpha)$ spectrum was calculated by the fixed-radius method or the fixed-mass method (Hong Shizhong & Huang Dengshi 1991). Moreover, some methods were developed to calculate the fractal dimension of active faults (Ma Jing et al. 1990, Shi Telin et al. 1990); the fractal dimension of ruptures in experimental rocks (Xie Heping & Chen Zhida 1988, Shi Xingjue et al. 1992); the fractal dimension of drainage systems (Hong Shizhong & Hong Shiming 1988; Huangfu Gang et al. 1990) and other fractal dimensions related to crustal deformation, electrical resistivity and groundwater, whose variations are concerned with earthquakes. All these

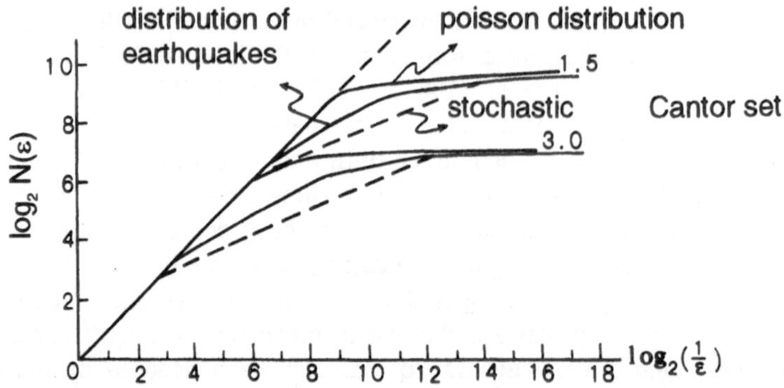

Fig. 1. Several kinds of distributions of log N(ε) - log ε^{-1}(from Ma Wenjing & Zhang Xiadong 1992). The ε is the scaling range and here $\varepsilon = 2^{-n}$. The N(ε) is the numbers of lattices for different ε. This figure gives three distributions, and we can see that the earthquake distribution is a statistical fractal, which is between the Poisson distribution and the Cantor set.

applications show that fractal research into earthquakes in China has made convincing progress.

2. A large number of examples have suggested that on some scales the temporal and spatial structures of earthquakes are fractal.

In the past few years, the temporal and spatial fractal dimensions of earthquakes that occurred in ten seismic areas or zones within Chinese territory, including big earthquakes of M > 6 and the small seismic sequences, have been studied intensively. The conclusions show that the temporal and spatial distribution of earthquakes is fractal in the statistical sense in some time and space domains, and is different from the completely stochastic distribution and also from the theoretical model, being between the two (Fig. 1). It must be pointed out that the earthquakes are not always fractal. Fig. 2 shows that the earthquakes are stochastic in certain time, space and magnitude ranges (Diao Shouzhong et al. 1990, Han Weibin et al. 1991). Thus the temporal and spatial distributions of earthquakes are very complex: some are fractal, some are stochastic and some are quasi-periodic. Sometimes the fractal structures dominate and sometimes the stochastic.

3. It has been found that the temporal and spatial fractal dimensions of regional medium-small earthquakes have a characteristic variation before strong earthquakes.

Many researchers have taken the temporal and spatial distribution of earthquakes as a dynamic system and then studied its evolution, trying to look for forecasting information on strong earthquakes. Chen Yong (1989a, 1990) proved

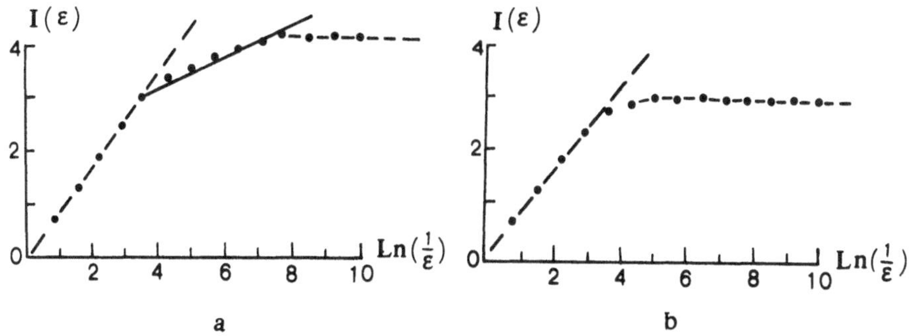

Fig. 2. Fractal and stochastic structures of earthquakes (from Diao et al. 1992). The temporal and spatial structures of earthquakes are complex. Sometimes the fractal structures dominate (*a*) and sometimes the Poisson distribution (*b*). This figure gives two examples in northern China. The ε is the scaling range and the I (ε) is the seismic entropy of information.

that the dimension and entropy of the small earthquakes in epicentral areas decrease before strong earthquakes. Many phenomena of temporal and spatial fractal dimension reduction before strong earthquakes in the Chinese continent have been reported. Fig. 3 gives some examples of the fractal dimension reduction before some earthquakes of M > 6. All these examples show that in the 1-5 years (often 1-3 years) before some strong earthquakes in the Chinese continent, the small regional earthquakes have a tendency of temporal and spatial fractal dimension to reduce, as observed by b-value reductions before strong earthquakes. Zheng Jie et al. (1983) and Chen Yong (1989b) also found that local deformations in rocks during experiments appeared before the ruptures occurred. This is considered a phenomenon of fractal dimension reduction of the distribution of micro-ruptures in rocks. Moreover, an example of fractal dimension reduction of a strong rupture in a mine was given by Zhang Shaoquan (1990). Thus the rupture processes on different scales including natural earthquakes, mine breaks and rock ruptures all have fractal dimension reduction, which should be given great attention.

However, we should not omit the fact that some researchers have shown other variations in fractal dimension before strong earthquakes, i.e. rising then dropping,dropping then rising or rising monotonously, and so on (Han Weibin et al. 1991). Some strong regional earthquakes are strongly active in the fractal dimension reduction period, but some are different, which shows the divergence among the regions. Thus we can see that the variations of the temporal and spatial fractal dimension are complex. Some phenomena should be studied by accumulating more examples. Thus it is still difficult to forecast earthquakes using the fractal dimension.

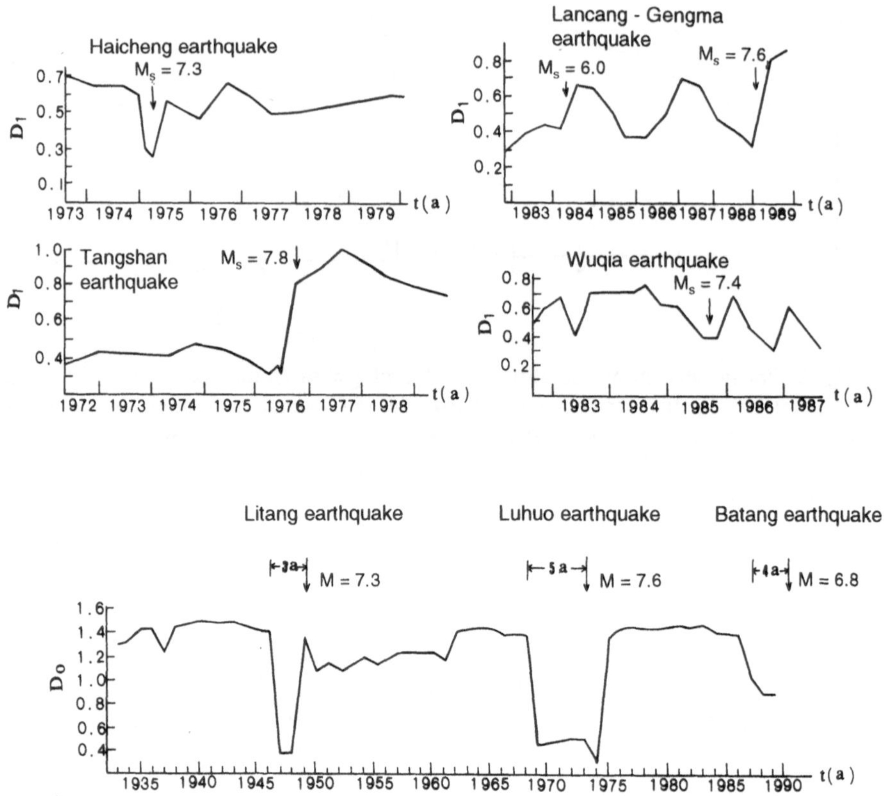

Fig. 3. **The variations of the fractal dimension value of the regional medium-small earthquakes in parts of the Chinese continent show a phenomenon of fractal reduction before strong earthquakes (from Feng et al. 1992). The D_0 is the Hausdorf dimension and the D_1 is the informational dimension.**

4 The fractal dimension of a seismic sequence should probably be related to the type of seismic sequence.

The seismic sequence is the result of the rupture of many locked segments with different scales along the seismic faults or the expansion of the seismic faults. To study the type of seismic sequence is helpful for understanding the physical process of the rupture in the epicenter region. It is also helpful for organizing disaster reduction after strong earthquakes. Thus this research has great theoretical and practical significance. By analysing the results of fractal analysis of many seismic sequences, we gain the following understanding: (1) The fractal dimension value of most pre-shock sequences is lower than that of the aftershock sequences (Diao Shouzhong et al. 1990, Jiang Haikun et al. 1990, Xu Yebang et al. 1990, Dai Weile et al. 1991, Wang Haitao et al. 1991). Sometimes, the fractal dimension value of the pre-shock sequence is larger than that of the aftershock sequence (Fan Zengjie et al. 1990; Zhang Xiaodong et al. 1991). (2) The fractal dimension of the earthquakes before the second main shock in the sequence of two-main-shocks type or the fractal dimension of the earthquakes before the strong aftershocks in sequence shows a reduction tendency (Diao Shouzhong et al. 1989, 1990, Jiang Haikun & Diao Shouzhong 1990, Zhang Xiaodong 1991). (3) The fractal dimension value of the micro-earthquake group is generally large, and it is difficult to distinguish the forewarning earthquake group.

5. The fractal dimension of earthquakes in a large active fault zone is different along different segments and shows some variation with time.

A large active fault zone is generally a seismic zone in which large earthquakes are concentrated. The preliminary study suggests that the fractal dimension is again different along different segments and also shows a temporal variation. For example, the fractal dimension of earthquakes along the southern segment of the Anninghe fault zone has been lower than that along the northern segment in the past 20 years, which probably corresponds with the fact that the seismicity on the southern segment is stronger than that on the northern segment. In the Tancheng-Lujiang fault zone, a large active fault zone in eastern China, the temporal and spatial distributions of the earthquakes of $Ms \geq 4.7$ within two active periods (total about 600 years) have a similar evolution process, which is: $\rightarrow$ stochastic distribution (in a quiet period) $\rightarrow$ fractal distribution (in an active period) $\rightarrow$ fractal dimension reduction (high tide period of strong earthquakes) $\rightarrow$ tochastic distribution (new quiet period) (Diao Shouzhong & Wang Hongwei 1992). As these phenomena are very useful, they should be studied intensively. These investigations are, however, not enough.

6. The fractal dimension of an active fault system is probably related to the stress character.

The fractal dimension of some of the active fault zones on the Chinese continent, which have clear surface traces and have been geologically mapped on a large scale, have been calculated by the ruler-counting method (Table 1). The conclusions suggest that: (1) The fractal dimension values of some active fault systems on the Chinese continent are between 1.01 and 1.6. The upper limit of 1.6 coincides with that achieved from theoretical modeling ($D = 1.6$). (2) The strike-slip faults with thrust components or thrust faults have a low fractal dimension value, such as the Xianshuihe fault zone, the Longshoushan fault zone, the Qilianshan fault zone, the Changma fault zone, and so on. The normal active fault zones have large fractal dimension values, such as the Honghe fault zone. Generally, the thrust fault zones are in great danger of strong earthquakes.

7. The fractal dimension of rock ruptures is related to the brittleness of rocks. The fractal dimension of rock ruptures is helpful for understanding the fractal dimension of active faults and earthquakes. The fractal dimensions of the ruptures in marble, limestone, granite and sandstone have been calculated by the ruler-counting method according to the length of th micro-ruptures; the values

are between 1.01 and 1.03 (Shi Xingjue et al. 1992), which are similar to those of some main faults (such as the Qilianshan fault zone). The fractal dimension of the marble is surveyed by the slit island method. This method is introduced from the fractal surveying of the two-dimensional surface of the fractured metals (Mandelbrot et al. 1984, Pande et al. 1987). The result shows that the fractal dimension of rock ruptures is related to the brittleness of rocks and the fractal dimension value ($D = 1.18$) of along-crystal ruptures is lower than that ($D = 1.31$) of cross-crystal ruptures, which agrees with the theoretical value that is 1.26 of along-crystal ruptures and 1.37 of cross-crystal ruptures (Xie Heping & Chen Zhida 1988). Based on the new fractal theory of rock ruptures, Niu Zhiren & Shi Xingjue (1992) further testisfied that the fractal dimension of rock ruptures is an important factor for determining the property of the brittleness of rocks.

Table 1. The fractal dimension of some active faults in the Chinese continent.

Active faults	Segment or part	D_0	Features of activity	References
Xianshuihe fault zone	all of the zone	1.15	strike-slipping	MaJing 1988
Northern segment of Anninghe fault zone	northern part	1.15±0.02	strike-slipping	Bai Lanxiang 1991
	central part	1.42±0.02		
	southern part	1.39±0.02		
Honghe fault zone	northern segment	1.62	strike-slipping with trusting	Jing Denghui 1989
	southern segment	1.32		
Active faults in Western Yunnan Province	Dali area	1.45	strike-slipping with normal component	Hangfu Gang 1991
	Jianchuan area	1.39		
	Lijiang area	1.39		
	Yongsheng area	1.41		
Longohouchan fault zone	all of the zone	1.03	strike-slipping with trusting	Shi Teling et al. 1990
Qilianshan fault zone	all of the zone	1.01	strike-slipping with trusting	Shi Teling et al. 1990
	western segment	1.01		
	middle segment	1.02		
	eastern segment	1.01		
Changma fault zone	all of the zone	1.09	strike-slipping with trusting	Hou Kangming 1990
Yishu fault zone	all of the zone	1.34	strike-slipping with trusting	Xue Ge et al. 1992
	northern segment	1.41		
	southern segment	1.29		

They proved that before one micro-crack grows, the inelastic volumetric strain (ε_n) in rock is: $\varepsilon_n = \left[K\gamma d_m^2(D-1)\right] / \left[(4-D)\mu V\right]$. Here γ is the surface energy of rock, d_m is the scale of the maximum dislocation pile-up group, μ is the shearing modulus, V is the volume of rock, D is the fractal dimension of rock cracks and K is a constant. In fact this formula is a test for determining the brittleness of the rock, and the smaller the ε_n is, the more brittle is the rock. Thus it is easy to see that the smaller D is, the smaller is ε_n, e.g. the more brittle is the rock. But the minimum strength of rock (σ_n) can be described in the following formula: $\sigma_n = (4\mu\gamma / \pi d_m)^{1/2}$ and it is found that the strength of the rock has no relations to the fractal dimension. This is an important discovery!

8. The fractal dimensions of drainage systems and linear structures are related to neotectonic activity and seismicity.

Drainage systems are controlled by neotectonic activity. Thus their fractal dimensions must include information on neotectonic activity. The methods to calculate the fractal dimension of drainage systems have been deduced from the Horton law. The fractal dimensions of drainage systems in parts of the Chinese continent have been calculated. Because the basic data and calculating methods of different investigations are different, the results cannot be compared now, but we can see from these results that the fractal dimension values are relatively large in strongly active areas, where the seismicity is relatively strong, and where many $M \geq 7$ earthquakes have occurred.

The linear or approximately linear topography or geological structures on the earth's surface are called linear structures. The different linear structures have different fractal dimension values, which show the complexity and active features of the structures. The preliminary results suggest that the larger the fractal dimension value is, the stronger are the neotectonic activity and seismicity (Kong Fanchen et al. 1991).

New Ideas on Seismology

As the seismological phenomena are studied intensively by the new mathematic tool of fractal dimension, some traditional viewpoints on seismology have been changed, which has led to the development of a new discipline in seismology and new ideas in geology.

1. The essence of the b-value of earthquakes is the description of the geometric structure of the rupture systems.

The famous Gutenberg-Richter formula: $\lg N(M) = a - bM$, to show the relationship between the earthquake frequency N(M) and the magnitude (M), is the most universal formula in seismology; the coefficient a and b are two statistical constants with no relations to M. Based the experiments on rock rupture, Mogi (1962) considered that the b-value is mainly controlled by the inhomogeneity of the matter, but Scholz et al. (1967) considered that the stress is the main factor to control the b-value. Because variation in b-value can be observed before some strong earthquakes, Scholz's idea is often taken as the physical basement of earthquake prediction. As mentioned above, Aki (1981), King (1983) and Niu Zhiren & Shi Xingjue (1992) have deduced the formula of rock rupture system from different approaches: $D = 2b$, the essence of the b-value of earthquakes is the description of the geometric structure of rock rupture systems. This has strengthened Mogi's viewpoint. However, the geometric structure of rupture systems can change with a rise in differential stress, and then lead to a variation of the b-value indirectly, so Scholz's idea is only one aspect of truth. Diao Shouzhong et al. (1990) studied a sequence of induced earthquake generated by water injection and found that the b-value of pre-shocks is larger than that of aftershocks. This is a good example.

2. Study on the chaos of earthquakes

The new topics of nonlinear dynamics including fractal geometry have naturally induced seismologists to present the question: are earthquakes chaotic? Chen Yong (1989a) has given a systematic review. Because we cannot establish an equation group to describe the dynamics of earthquakes now, some researchers try to calculate the second order informational dimension, the Kolmogorov entropy, and the Lyapunov index and use these three parameters of generalized systems to describe and determine the dynamic behaviors of earthquake systems. Taking the temperature data of groundwater in Maoya Spring, Sichuan Province, from Feb. 1975 to Mar. 1976 as the temporal sequence data studying the earthquake-generated system, Luo Jiuli et al. (1989) studied the pregnant system of the 1976 Songpan earthquake ($M = 7.2$) in Sichuan Province and got the chaotic attractor of $D_2 = 2.5$ before the earthquake. Based on the earthquake catalogue data, An Zhenwen et al. (1992) obtained the chaotic attractor in temporal sequences of the regional earthquakes before the 1975 Haicheng earthquake ($M = 7.3$), the 1976 Tangshan earthquake ($M = 7.8$) and the 1976 Longling earthquake ($M = 7.5$); the D_2-values are 1.7, 1.8 and 2.1 respectively, and considered that the earthquakes have a chaotic distribution with low fractal dimension before the main shock and a stochastic Poisson distribution after the main shock (Fig. 4). Gu Haoding & Sun Wenfu (1992) gave the chaotic attractors $D_2 = 2.8$ of the aftershock sequence of the Haicheng earthquake ($M = 7.3$) and the Tangshan earthquake ($M = 7.8$), and considered that the aftershock sequence system had the features of chaos. Guo Zengjian & Qin Baoyan (1991) and Hu Ping et al. (1990) presented a question on the chaos and the time limit of the earthquake prediction. If chaos in earthquake systems is really applicable, the idea of a definite earthquake prediction will form an additional challenge.

It is a promising approach to use the second order informational dimension to predict earthquakes. But there is a strong requirement for the data samples and sampling to calculate the D_2 care should also be taken to study whether the

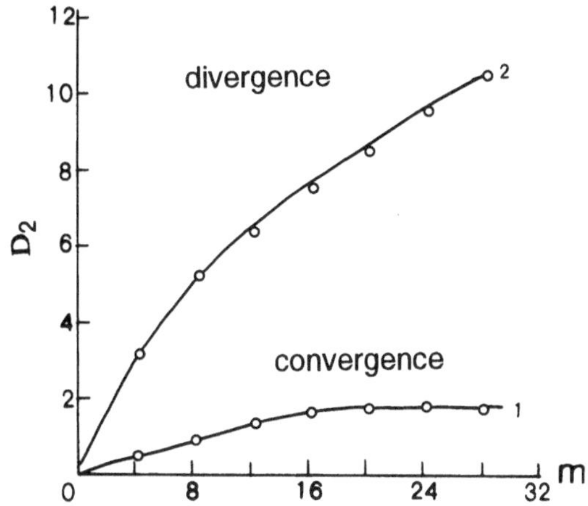

Fig. 4. The D_2 variations before and after the Tangshan M = 7.8 earthquake on July 28, 1976, show that the regional earthquake activities have a behavior of low fractal dimension and weak chaos before the main shock (curve 1) and are stochastic after the main shock (curve 2) (from An et al. 1992). The D_2 is the correlation dimension and the m is dimension of the embedding space.

seismological catalogue data and geophysical data samples are satisfactory to calculate the D_2. So far, there have not been many examples, so we are not sure whether earthquakes are chaotic.

Problems and a Promising Future

1. Statistical samples

Earthquakes are fractal in a statistical sense, so it is necessary to consider the scale of the statistical samples for fractal calculation. So far, there is no strict theory to testify what prerequisites and conditions are needed for fractal calculation. Generally, it is determined by experiments. For example, if we want to calculate D_q, we must have $N_{min} \geq 500$ samples. Smith (1988) suggested that the N_{min} should be equal to or greater than 42^M samples to calculate the D_2 by the Grassberger-Procaccia method (M is the integer part of the D_2). If M = 3, then $N_{min} = 74088$. This is a strict requirement. Thus a high price has to be paid to achieve the fractal dimension of the seismic dynamic attractor using the observation data.

2. Scaling range

There is another problem in calculating the fractal dimension: the scaling range is generally determined by the naked eye from observation curves, such as

$I(\varepsilon)$-$\log\varepsilon^{-1}$ curve. Here $I(\varepsilon)$ is the informational entropy of earthquakes and ε is the scale) after giving some constraints. Scientific standards to select the range are still lacking. Zhu Lingren et al. (1992a) found that when the fractal dimension is calculated by the range-changing method, the formula: $N(\varepsilon) = \varepsilon^{-D}$ is not right for all surveying ranges ε, even for the ideal self-similar structures. Only when ε is the characteristic scale of α^n, is this formula right. Thus the surveying range is important for selecting the scaling range. All these studies are continuing.

The two problems mentioned above are the main problems studied in China. Moreover, the physical meanings of the temporal and spatial fractal dimension of earthquakes and the upper and lower limits of the scaling range are also problems which have been solved.

Fractal research into earthquakes has made promising progress in China and shows a promising future for research and applications. In the future, the Chinese researchers will make more progress on calculating methods and models, provide more data and multifractal examples, and discuss the chaotic aspect of earthquakes intensively. It is expected that the continuous fractal research into earthquakes will provide a deeper understanding for the complex phenomena of earthquakes and provide a new impulse in geoscience.

References

Aki K (1981) A probabilistic synthesis of precursory phenomena. In: Simpson DW and Richards PG (ed) Earthquake prediction-an international review. Maurice Ewing Series 4, American Geophisical Union, Washington, pp 566-574.

An Zhenwen, Wang Linying, Yao Donghua, Zhu Chuanzhen (1992) Chaotic characteristics of the seismogenic process. Acta Seismolagica Sinica 14(4): 393-399 (in Chinese with English abstract).

Aviles CA, Scholz CH (1987) Fractal analysis applied to characteristic segments of the San Andreas fault. J G R 92(B1): 331-344.

Cai Jianguan, Tang Jin, Xu Zhaoyong (1992) The distribution and fractal characteristics of stress drop in different seismic regions. Acta seismologica Sinica 14(3): 289-295 (in Chinese with English abstract).

Chen Yong (1989a) Applications of the fractal and chaotic geoscience. The Publishing House for Scientific Reference, Beijing (in Chinese).

Chen Yong (1989b) The phenomena of dimension reduction in the process of rock failure. Acta Geophysica Sinica 32(Supp.1): 132-143 (in Chinese with English abstract).

Diao Shouzhong, Jang Haikun (1990) Fractal dimension features of earthquakes resulting from water flooding in Jiao 7 oil well, Shengli oil field, Shandong province. North China Earthquake Science 8(1): 76-82(in Chinese with English abstract).

Diao Shouzhong, Wang Hongwei (1992) Fractal structure in large scale distribuition of earthquake of Tancheng-Lujiang seismic belt. In: Chen Yong et al. (ed) Special issue on the application of nonlinear science methods to seismology. Seimological Press, Beijing, pp 82-83 (in Chinese).

Diao Shouzhong, Jiang Haikun, Ren Qingwei, Li Mengluan (1990) The time fractal characteristics of four earthquake sequences occurred in Lingwu-Wuzhong area of Ningxia. Northwestern Seismological Journal 12(2): 42-48 (in Chinese with English abstract).

Fan Zengjie, Niu Zhiren (1990) Preliminary study of the fractal characteristics of the aftershock sequence. Northwestern Seismological Journal 12(4): 26-29 (in Chinese with English abstract).

Feng Deyi, Liu Xilan, Jiang Chun, Zheng Xingming, Li Minzhou (1992) Fuzzy fractal dimension and its some applications in earthquake researches. Acta Geophysica Sinica 35(4): 459-468 (in Chinese with English abstract).

Gu Haoding, Sun Wenfu (1992) Self-organization and evolution of seismicity. Acta Geophysica Sinica 35(1): 25-36 (in Chinese with English abstract).

Guo Zengjian, Qin Baoyan (1991) Chaos and time limit of earthquake prediction. Recent Developments in World Seismology (3): 9-11 (in Chinese with English abstract).

Grassberger P, Procacci I (1983) Characterization of strange attractors. Phys Res Lett 50(5): 346-349.

Han Weibin, Xu Yebang, Luo Jiuli (1991) Seismic fractal and earthquake prediction. Earthquake Research in Sichuan (3): 9-15 (in Chinese with English abstract).

Hirabayashi T, Tto K, Yoshii T (1990) Multifractal analysis of earthquakes. Mathematical Seismology V: 102-146.

Hong Shizhong, Hong Shiming (1987) The fractal dimension and its application in seismology. Earthquake Research in Sichuan (1): 39-46 (in Chinese with English abstract).

Hong Shizhong, Hong Shiming (1988) A study of fractal in geoscience: drainages,earthquakes and others. Exploration of Nature 7(2): 33-40 (in Chinese with English abstract).

Hong Shizhong, Huang Dengshi (1991) Multifractals and earthquake. Earthquake Research in Sichuan (1): 8-18 (in Chinese with English abstract).

Hu Ping, Yang Peicai, Li Wei, Zhao Meng (1990) Research on seismic dynamical behavior and predictability. Acta Geophysica Sinica 33(6): 647-656 (in Chinese with English abstract).

Huangfu Gang, Han Ming, Wang Jinnan (1991) Study on fractal geometry of faults in northwestern Yunnan Province. Seismology and Geology 13(1): 61-66 (in Chinese with English abstract).

Jiang Haikun, Diao Shouzhong (1990) Characteristics of the temporal fractal dimension of the Haicheng and Tangshan earthquake sequences. Earthquake Research in China 6(2): 53-56 (in Chinese with English abstract).

Jiang Haikun, Diao Shouzhong (1991) Characteristics of information dimension D_1 of spatial distribution of shocks around Haicheng and Tangshan large earthquakes. Earthquake 2: 37-46 (in Chinese with English abstract).

Jing Denghui, Ma Jing (1990) Fractal investigation of fault geometry and earthquake activity-Honghe river as an example. In:The proceedings of the 2nd national conference on tectonophysics. Seismological Press, Beijing, pp 75-84 (in Chinese).

Kagan YY (1981) Spatial distribution of earthquakes: the three-point moment function. Geophys J R Astr Soc 67: 697-717.

Kagan YY, Knopoff L (1980) Spatial distribution of earthquakes: the two-point correlation function. Geophys J R Astr Soc 62: 303-320.

King G (1983) The accommodation of large strains in the upper Lithosphere of the earth and other solids by self-similar fault systems: the geometrical origin of b-value. PAGEOPH 121(5/6): 761-815.

Kong Fenchen, Ding Guoyu (1991a) Fractal geometric analysis of drainage and loess gully systems in Shanxi and its adjacent areas. Seismology and Geology 13(3): 221-229 (in Chinese with English abstract).

Kong Fanchen, Ding Guoyu (1991b) The implication of the fractal dimension values of lineaments. Earthquake 5: 33-37 (in Chinese with English abstract).

Li Haihua, Zhang Wenxiao, Zhang Yongli, Ma Wenjing, Guo Yaping (1987) The temporal fractal structure of small seismic activity before the Menyuan earthquake (M = 6.4). Northwestern Seismological Journal 9(4): 15-20 (in Chinese with English abstract).

Luo Jiuli, Liu Donyan (1990) Earthquake activity: chaotic or deterministic?--An exploration to seismic recurrence attractor. Exploration of Nature 9(1): 26-31 (in Chinese with English abstract).

Luo Shuoli, Luo Wei, Zhu Hang (1991) The self organization and similarity of seismic activity. Earthquake Research in Sichuan 3: 1-8 (in Chinese with English abstract).

Ma Jing, Ma Shingli, Lei Xingling (1990) The geometry of the faults in Xianshuihe fault zone and the seismicity. In:The proceedings of the 2nd national tectonic physics conference. Seismological Press, Beijing, pp 58-67 (in Chinese).

Ma Wenjing, Zhang Xiaodong (1992) The preliminary study on generalized fractal dimension for temporal distribution of earthquake sequence. Inland Earthquake 6(1): 1-6 (in Chinese with English abstract).

Mandelbrot BB (1977) The fractal geometry of nature. Freeman, San Francisco.

Mandelbrot BB, Passoja DE, Paullay AJ (1984) Fractal character of fracture surface of metals. Nature 308: 721-722.

Mogi K (1962) The influence of the dimensions of specimens on the fracture strength of rocks: comparison between the strength of rock specimens and that of the Earth's crust. Bull Earthq Res Inst 40: 175-185.

Niu Zhiren, Shi Xingjue (1992) Statistical theory of rock fracture. Acta Geophysica Sinica 35(5): 594-603 (in Chinese with English abstract).

Pande CS et al. (1987) Characterization of fractured surface. Acta Metal 35: 1633-1638.

Peng Chengbin, Chen Yong (1989) On the fractal structure of earthquakes. Earthquake Research in China 5(2): 19-26 (in Chinese with English abstract).

Shi Telin, Guo Daqing, Yang Yuheng, Guo Jiankang, Xiao Lizhu, Sun Jingfang (1990) The fractures of fractal dimension in Qilian Mountain fault and Longshou Mountain fault. Northwestern Seismological Journal 12(3): 23-32 (in Chinese with English abstract).

Shi Xingjue, Niu Zhiren, Xu Heming, Fan Zengjie, Zhou Taixi (1992) The fractal dimension of the fractured surface of rocks. Acta Geophysica Sinica 35(2): 154-159 (in Chinese with English abstract).

Smalley RF Jr., Chatelain JL, Turcotte DL, Prevot R (1987) A fractal aproach to the clustering of earthquakes: application to seismicity of the New Hebrides. Bull Seis Soc Am 77: 1368-1381.

Wang Binghong, Li Dongsheng (1992) Calculation of the multifractal of earthquakes in Tangshan area using the popularized Grassberger-Procaccia method. In: Chen Yong et al. (ed) Special issue on the application of nonlinear science methods to seismology. Seimological Press, Beijing, pp 105-111 (in Chinese).

Xie Heping, Chen Zhida (1988) The fractal geometry and the rock fractures. Acta Mechanica Sinica 20(3): 264-274 (in Chinese with English abstract).

Xu Yebang (1991) A study of the characteristics of the information dimension D_1 of the temporal and spatial distributions of earthquakes in an active fault zone. Acta Seismologica Sinica 13(3): 372-379 (in Chinese with English abstract).

Zhang Shaoquan (1990) The natural ruptures and the concept of viewing the earthquake as a whole. Earthquake (6): 41-52 (in Chinese with English abstract).

Zhang Xiaodong (1991) Information content and fractal dimension of earthquake sequences. Earthquake (3): 59-68 (in Chinese with English abstract).

Zheng jie, Yao Xiaoxin, Chen Yong (1983) A experimental study on localization of deformation of rock. Acta Geophysica Sinica 26(6): 554-563 (in Chinese with English abstract).

Zhou Chuanzhen (1991) Some understanding of the characteristics of seismic fractals. Recent Developments in World Seismology(3): 4-6 (in Chinese with English abstract).

Zhu Chuanzhen, An Zhenwen, Wang Linying, Yao Donghua (1991) Fractal feature of earthquakes and its significance in earthquake prediction. Journal of Seismological Research 14(1): 73-86 (in Chinese with English abstract).

Zhu Lingren, Zhou Shiyong (1992a) Calculating fractal dimension by scale transformation - one of the basic considerations in simple calculation of fractal dimension. Inland Earthquake 6(3): 217-222 (in Chinese with English abstract).

Zhu Lingren, Zhou Shiyong, Wang Haitao, Bai Chaoying, Gong Yuqing (1992b) Research on the information dimension of seismic activity. Acta Seismologica Sinica 14(4): 385-392 (in Chinese with English abstract).

Evidence for Self-Similarities in the Harmonic Development of Earth Tides

Hans-J. Kümpel
Geologisches Institut, Rheinische Friedrich-Wilhelms-Universität Bonn,
Nußalle 8, 53115 Bonn, Germany

Abstract. Recent models of the earth's tidal potential include more than 10^3 individual harmonic constituents (Tamura 1987, Xi Qinwen 1987, 1989), enough to screen these data for the existence of self-similarities. It is found that the number of constituents surpassing a certain amplitude level shows a fractal distribution over at least four orders of magnitude. Its fractal dimension is close to 0.5. Accordingly, a refined tidal model which is complete should include about 10 times as much harmonic constituents as a coarser model, when the smallest amplitude of the refined model is 100 times smaller than that of the coarse one. Whether this feature masks an inherent structure in the spatial distribution or in the dynamics of celestial bodies governing the tidal forces is yet unknown.

Introduction

Tides are a well known phenomenon on earth, resulting from differential forces of mainly lunar and solar origin. Tidal forces not only lead to marine tides but also to variations in the gravity field, in rock strain and tilt, and to the occasionally observed diurnal and semidiurnal changes in the height of water levels in wells (e.g. Melchior 1978). The forces are often expressed through the tidal potential for a rigid earth, usually normalized to 0.75 $GM_L R_E^2 / r_L^3 = 2.6277\ m^2/s^2$ (Doodson's constant), where G is the gravitational constant, M_L the mass of the moon, R_E the earth's radius, and r_L the mean distance between moon and earth.

Since the time varying part of the earth tidal potential depends on the relative position of a few celestial bodies only, it is known to a rather high precision (Merriam 1992; Dahlen 1993). For tidal analyses it is useful to develop the potential, ϕ, into a series of individual harmonic constituents, i.e. as

$$\phi = c \sum_{i=1}^{n} H_i \cos(\omega_i t + \theta_i) \tag{1}$$

(Wilhelm & Zürn 1984). c is a coefficient, H_i, ω_i and θ_i are the normalized amplitude, the frequency, and the phase of the i-th harmonic constituent of a given development, and t is time. A set of n such constituents is called a tidal model. The largest amplitude is that of the semidiurnal lunar wave M_2, which in the model of Tamura (1987) is 0.908184.

Various tidal models have been suggested in the past. According to the growing needs to analyse time series of increasing resolution, especially gravity recordings, the models became more and more detailed (Table 1). An attempt to describe a so far hypothetical self-similar structure in tidal models is the subject of this paper. If such a structure exists, the completeness of more extensive models to be elaborated in the future can be grossly checked by extrapolation. The presence - or absence - of fractal features in tidal models might also be reasoned theoretically, but this will not be done here.

Table 1. Parameters of compared tidal models. N = number of individual harmonic constituents; A_{min} = lowest normalized amplitude included; R = sum of amplitudes of hypothetically omitted constituents; left: normalized to Doodson constant; right: in percentage of the normalized M_2-amplitude [eq.(5) and assumptions in the text].

Author	N	A_{min}	R_N^*	R_N^* (%M_2)
Doodson (1921)	379	10^{-4}	0.04216	4.6
Cartwright & Tayler (1971)	505	5×10^{-5}	0.03165	3.5
Cartwright & Edden (1973)				
Büllesfeld (1985)	649	3×10^{-5}	0.02463	2.7
Tamura (1987)	1200	3×10^{-6}	0.01333	1.5
Xi Qinwen (1987, 1989)	2845	10^{-6}	0.00562	0.61

Fractal Analysis

If plotted as a times series, the tidal potential appears to have a rather regular shape (Fig. 1). The series is dominated by the superposition of a few diurnal and semidiurnal waves, leading to the fortnightly return of stronger and weaker amplitudes known as spring and neap tides at sea.

The line spectrum of the potential reveals a more complex structure (Fig.2). Individual constituents are grouped in tidal bands; within each band the constituents are arranged in subgroups; the subgroups are divided in sub-subgroups, and so on. A few constituents have amplitudes bigger than 10^{-1}, quite many have amplitudes bigger than 10^{-3}, many more bigger than 10^{-5}. We may

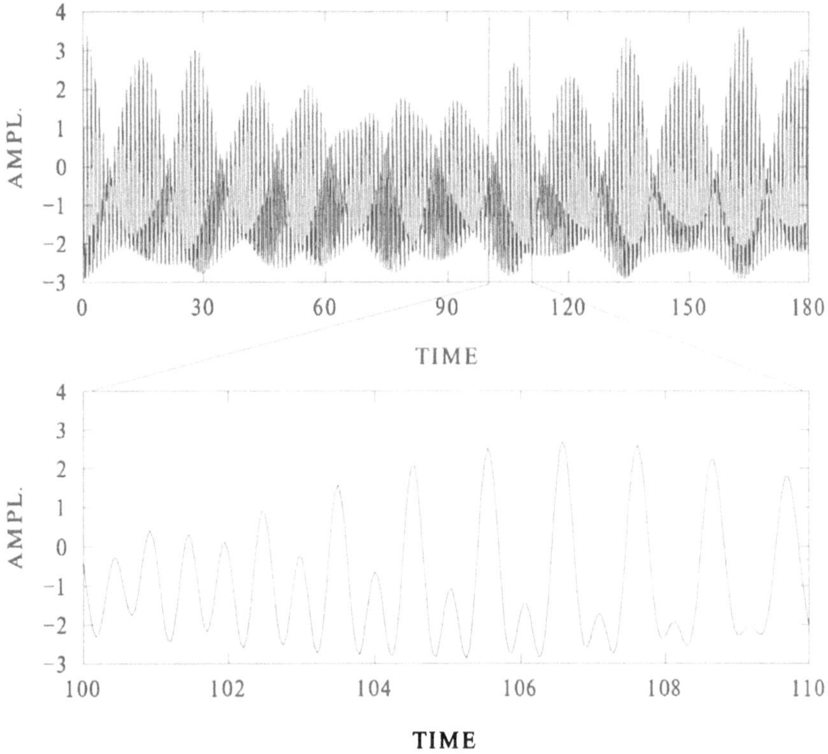

Fig. 1. Time variation of the tidal potential based on the model of Tamura (1987), computed with ETGTAB (Wenzel 1992) for a site in the city of Bonn, Germany, 50 m above sea level. **Top**: Days from January 1, 1995, 0 UT. **Bottom**: Zoom for days 100 to 110. Amplitudes are in m^2/s^2.

ask for the relation between the amplitude of some constituent and the number of constituents having an amplitude bigger than this one. The question is similar to asking for the frequency-magnitude relation of tidal harmonic constituents (instead of that of earthquakes in a specified region within a representative period of time).

Let A denote the normalized amplitude of an arbitrary harmonic constituent, and N(A) the number of constituents of amplitude bigger than or equal to A. The log-log-plot of N(A) over A for the tidal models listed in Table 1 is given in Fig.3. Obviously, the overall relation is roughly linear over at least four orders of amplitudes. Data from more recent models seem to resume the trend of earlier, less complete models. There is not much difference in this behavior whether the amplitudes of all constituents are considered or only those of the longperiodic, diurnal, semidiurnal, or terdiurnal bands (as demonstrated for the Tamura model in Fig. 3).

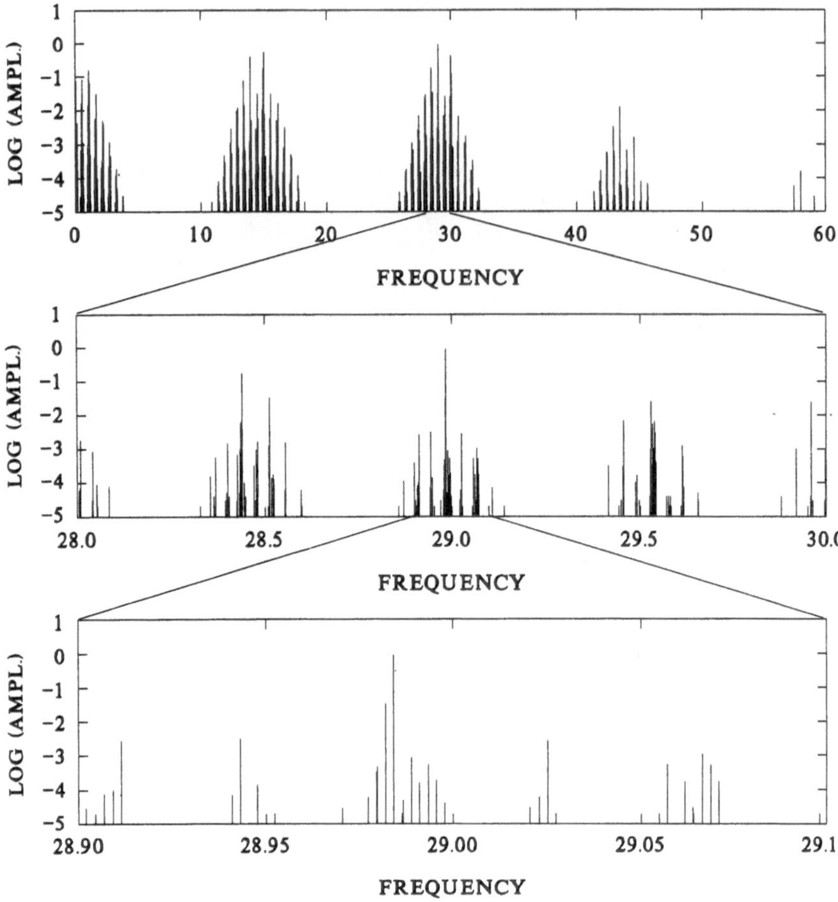

Fig. 2. Line spectrum of tidal potential (Tamura 1987) used for computation of the time series in Fig.1. Frequencies in deg/h; amplitudes normalized. **Top:** Longperiodic, diurnal, semi-, ter-, and quarterdiurnal tidal bands (from left to right). **Middle:** Zoom of part of the semidiurnal band, from 28.0 to 30.0 deg/h. **Bottom:** Zoom of part of the tidal band around 29.0 deg/h. The highest peak in all graphs is that of the M_2 wave (= 0.908184).

The overall trend of the log-log-relationship closely follows the simple equation

$$N = 4 * A^{-0.5} \qquad (2)$$

It is easily recognized that a relation with 3 or 5 instead of 4 as the coefficient in eq.(2), or with -0.4 or -0.6 instead of -0.5 as the exponent, yields a much worse fit. The rather poor fit within the high amplitude range can be attributed to the low number of constituents having such big amplitudes. It does not oppose the validity of a fractal law. Given that the amplitude of a harmonic constituent is its characteristic linear dimension, the set of all constituents may be stated to follow a fractal distribution (Turcotte 1989), characterized by a fractal dimension close to 0.5.

Note that within the amplitude range 10^{-6} to 10^{-3} the tidal models differ somewhat from each other: Büllesfeld's model includes more constituents within the range $3 \cdot 10^{-5}$ to 10^{-3} than any other model, and the relative decay of the number of constituents of Tamura's model below 10^{-5} is not confirmed by Xi Qinwen's model. But this one too deviates from eq.(2) for amplitudes below the 10^{-4} level: about 1150 constituents are missing on the 10^{-6} level if eq.(2) were be valid. One is tempted to speculate that even this most extensive tidal model is incomplete in the low amplitude range; not only because constituents with amplitudes below the 10^{-6} level have been neglected for the present lack of practical use, but also because of systematic omissions. Candidates could be the effect of the earth's ellipticity as pointed out by Dahlen (1993), or the omission of the lunar inequality, nutation, and planetary effects (Merriam 1993). Why, on the other hand, should the trend of eq.(2) continue beyond the 10^{-5} level (except for the beauty of nature)?

It is sometimes argued that the knowledge of the fractal dimension of a pattern is of no benefit. Here are some examples opposing this opinion. Let us assume that all the tidal models in Table 1 are more or less correct for amplitudes above the 10^{-4} level. They barely differ from each other and reasonably follow eq.(2) for this range. If then, for any reason, eq.(2) were correct for amplitudes below the 10^{-4} level, several conclusions may be drawn:

- When for the refinement of a tidal model the lowest amplitude to be included is 100 times smaller than in the original one, the number of constituents making up the refined model will be about ten times higher.
- For a complete model, down to infinitely small amplitudes, the number of constituents is infinite.
- Since, on average, the normalized amplitude A_i of the i-th biggest constituent equals $16/i^2$ [inversion of eq.(2)], the sum of all amplitudes for i from 1 to infinite can be obtained by computing the sum of the corresponding infinite series. This sum is $8 \cdot \pi^2/3$ or 26.3189451... (e.g. Bronstein & Semendjajew 1970). However, according to the poor fit of eq.(2) in the amplitude range above the 10^{-2} level (altogether 38 constituents, not counting the zero frequency constituent), a better estimate for the total sum of all tidal amplitudes is

$$S_{tot} = \sum_{i=1}^{38} A_i + 16 * \sum_{i=39}^{\infty} \frac{1}{i^2} = 3.81216 + 16\left(\frac{\pi^2}{6} - \sum_{i=1}^{38} \frac{1}{i^2}\right) = 4.22831 \qquad (3)$$

- Knowing the number N of constituents included in a model, the expected sum of the corresponding amplitudes is then

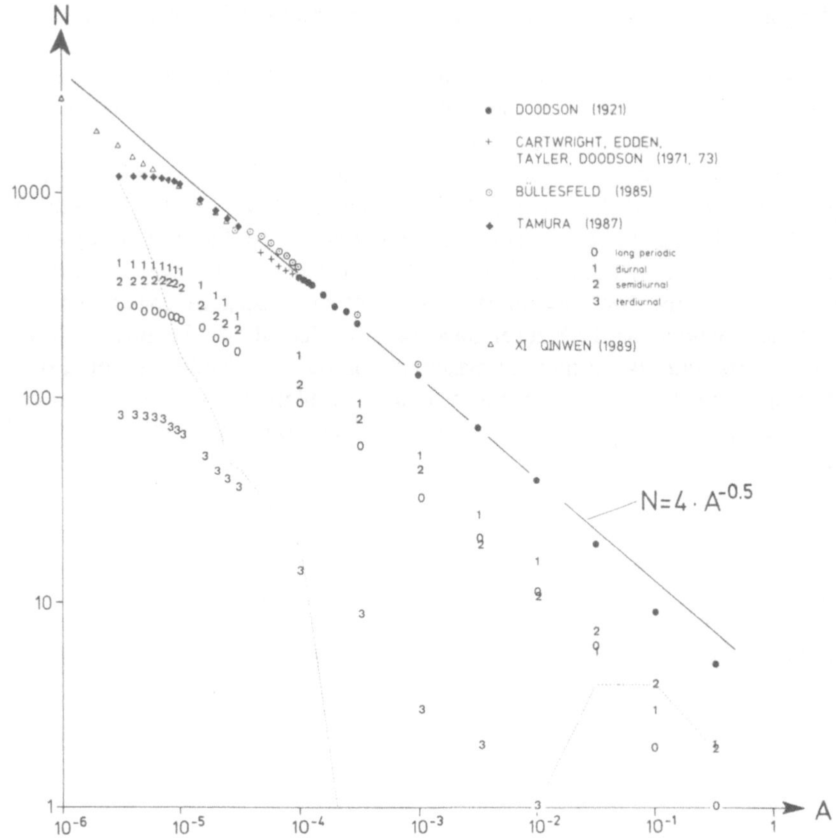

Fig. 3. Log-log-plot of number N of harmonic constituents of amplitude equal or greater than some normalized amplitude A of the tidal potential, over A, displaying a 'frequency-magnitude relation' close to $N = 4A^{-0.5}$. Tidal potential according to models of various authors, as indicated. Separately plotted are constituents of longperiodic, diurnal, semidiurnal, and terdiurnal tidal bands for Tamura's model. For bigger amplitudes two points per decade are displayed, but up to eight points per decade in the low amplitude range. In the range above the 10^{-4} level all models nearly coincide, except for that of Büllesfeld. The broken line traces the difference in the number of constituents between the $N = 4A^{-0.5}$ relation and Tamura's model.

$$S_N^* = 3.81216 + 16 * \sum_{i=39}^{N} \frac{1}{i^2} \tag{4}$$

– The expected residue with respect to a putatively complete model (the sum of all the neglected amplitudes) can be estimated to be

$$R_N^* = S_{tot} - S_N^* \tag{5}$$

Again, it has been assumed in the above that all models are complete down to the lowest amplitude included, or that no constituent with an amplitude bigger than the smallest one has been omitted. If this type of completeness is attributed to all the models cited here, their residues may be obtained from eq.(5). They are listed in the last rows of Table 1. Extrapolating this scheme below the 10^{-6} level one finds that 16,000 constituents have to be incorporated into a model to lower its residue to 0.11% of the M_2-amplitude, or 160,000 to lower it to 0.011%, and so on.

Discussion

The findings presented above are purely empirical. So far no argument has been given why the distribution of the tidal harmonic constituents could or should show a fractal dimension close to 0.5 . Note that a fractal dimension of 0.5 can be visualized through a simple form of 'Cantor dust', namely, by infinitely applying an algorithm that divides a finite line into four parts and removes two of them (e.g. Turcotte 1989). In the search for a theoretical reasoning it may be worthwhile to investigate whether *any* form of the frequency-magnitude relation of the tidal harmonic development could be justified, or if a fractal distribution reflects some intrinsic, natural aspect in the phenomenon of tides. One regulating aspect is the tendency for locking of a body's rotation period into fractional parts of the body's period of revolution around a bigger celestial, as e.g. effective in the systems Moon-Earth and Mercury-Sun (Schroeder 1991, pp. 338, for a broader discussion of this matter in the context of chaos). The fractal distribution in the tidal frequency-magnitude relation, as just another display of a self-organized structure in the universe, might be easier to accept than some unexplained deviation from a fractal distribution.

Acknowledgements. I should like to thank W. Zürn and H.-G.Wenzel for fruitful comments, and the latter for allowing me to use his program ETGTAB and for sending the datafile Xi-1989 (Xi Qinwen 1989). Reviews and comments from R. Meissner and R. Wang are gratefully acknowledged.

References

Bronstein IN, Semendjajew KA (1970) Taschenbuch der Mathematik. Harri Deutsch, Zürich.

Büllesfeld F-J (1985) Ein Beitrag zur harmonischen Darstellung des gezeitenerzeugenden Potentials. Dt Geodät Kommission, Bayer Akad der Wiss, Reihe C, vol 314.

Cartwright DE, Edden AC (1973) Corrected tables of tidal harmonics. Geophys J Roy Astron Soc 33: 253-264.

Cartwright DE, Tayler RJ (1971) New computations of the tide-generating potential. Geophys J Roy Astron Soc 23: 45-74.

Dahlen FA (1993) Effect of the earth's ellipticity on the lunar tidal potential. Geophys J Int 113: 250-251.

Doodson AT (1921) The harmonic development of the tide-generating potential. Proc Royal Soc London Ser A100: 305-329.

Melchior P (1978) Tides of the planet earth. Pergamon Press, Oxford.

Merriam JB (1992) An ephemeris for gravity tide predictions at the nanogal level. Geophys J Int 108: 415-422.

Merriam JB (1993) A comparison of recent tide catalogues and the consequences of catalogue error for tidal analysis. In: Melchior P (ed) Bull d'Inform Marrees Terrestres. Brussels, pp 8515-8535 (vol 115).

Schroeder M (1991) Fractals, chaos, power laws. Freeman, New York.

Tamura Y (1987) A harmonic development of the tide-generating potential. In: Melchior P (ed) Bull d'Inform Marrees Terrestres 99, Brussels, pp 6813-6855.

Turcotte DL (1989) Fractals in geology and geophysics. PAGEOPH 131: 171-196.

Wenzel H-G (1992) Program ETGTAB, Version 920107. Geodätisches Institut, Universität Karlsruhe.

Wilhelm H, Zürn W (1984) Tidal forcing field. In: Landolt-Börnstein, Zahlenwerte und Funktionen aus Naturwissenschaft und Technik, Neue Serie, Group V (vol 2). Springer, Berlin.

Xi Qinwen (1987) A new complete development of the tidegenerating potential for the epoch J2000.0. In: Melchior P (ed) Bull d'Inform Marrees Terrestres 99. Brussels, pp 6786-6812.

Xi Qinwen (1989) Datafile of tidal model Xi_1989 (unpublished).

Global Relief: Evidence of Fractal Geometry

Sergey S. Ivanov
P.P. Shirshov Institute of Oceanology,
23 Krasikova Str, 117218 Moscow, Russia

Abstract. On the base of a one-degree grid of elevation values lengths of different topographic isolines were computed for seven grades of resolution. It was ascertained that elevation contours have fractal dimensions significantly greater than unity at the scales from 100 to 2000 km. Some additional considerations extend these limits to 5 km and 20000 km respectively. Average value of fractal dimension of global relief as a whole is stated to be 1.37. To test this estimation the topography of the equatorial belt was examined with the help of the method of variance which provided the value of 1.38. An original method of dispersion counter-scaling was also applied and proved the existence of spatial variations in self-similar properties of regional topography.

Introduction

This chapter treats Mandelbrot's (1967, 1975) concept of fractality of the Earth's surface relief to a global test.

The length of a line, L, measured with different step L_o is different; one can incorporate the concept of fractal dimension D in the following way

$$D = \log L / \log L_o \tag{1}$$

The theoretical basis of this expression is described in Mandelbrot (1967). Dimensional values under the logarithm in (1) are expressed in certaini length units l, hence $D = [\log(L/l)] / [\log(L_o/l)]$ and

$$L = L_o^D \, l^{D-1} \tag{2}$$

where L_o^D is Hausdorff's measure. In this way fractal dimension is defined by Barenblatt et al. (1984).

These expressions show that the length of a one-dimensional rectifiable curve may be expressed as $L = (L/l)l$ or $L = L^1 l^0$, which coincides with (2) if $D = 1$. Hence the fractal dimension of a rectifiable curve is a unity, i.e. fractal and geometrical dimensions are equal. Thus fractality is a quality of objects with non-integer dimensions. In this sense fractal objects may be opposed to Euclidean ones, whose dimension is necessarily integer (Krohn 1988).

The position of any point relative to a given origin on a Euclidean one-dimensional curve may be exactly defined by one value (coordinate). For a fractal curve one coordinate does not suffice because its value depends on scale. This is one explanation of why a fractal line has a dimension greater than a unity (Zeldovich & Sokolov 1985).

Global Topography Analysis

On this basis calculations imitating measurements of contour lines lengths with various steps were done. The data by Gates & Nelson (1975) representing a bidimensional array of topography elevation values averaged in 1° x 1° quadrangles offer an exellent baseline for this study. This permits us to easily calculate the length of any isolevel line with regard to latitudinal position of every segment of it. The grid also provides a good possibility to recalculate values to other regular grids with lesser density.

There is no need to describe computer procedures since they are routine, the procedure used seven arrays of average topography values for rectangles with the sides of 2, 3, 5, 10, 15, and 20 degrees respectively . We may regard these arrays as images of the same pattern viewed from various distances, in other words with various resolution. Descending from scale to scale, from 20° to 1° grid, one can see more and more details while the whole image becomes progressively complicated. Naturally the length of every isoline increases with decreasing of scale and the goal was to reveal whether this change is governed by any self-similarity rules.

For each scale total lengths of topography isolevel lines for different levels with 0.5 km step were estimated. As far as calculated, the contour consists of rectangular links oriented along parallels and meridians the computed length value is exagerated. Thus the computed value L' of a straight line inclined at an angle φ to latitude would be L' = L (sinφ + cosφ), where L is the real line length. For a closed line if all directions are equally probable (i.e. for a circle) the calculated length is

$$L' = L\int_0^{2\pi}\left(\sin\phi + \cos\phi\right)d\phi - \frac{4L}{\pi}$$

so it is 1.27 times greater than the real length. This value of exageration may also be obtained from geometric considerations if we correspond the perimeter of a square with the length of enclosed circle. So in order to obtain realistic values of isoline lengths we must reduce the values calculated over a rectangular net with a factor of 0.785.

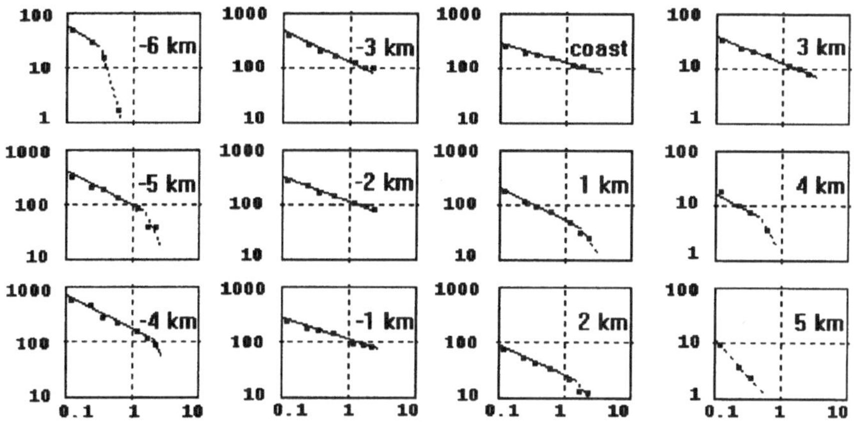

Fig. 1. Computed length-resolution dependencies for various contour lines. Both values are given in thousands of km. The level of contour line is labelled on each graph. Note the closeness of experimental points to straight lines. It is also clear that the contour lines at extreme levels are not long enough to reveal their self-similarity, and we observe descending tails of dependencies tending to zero at less detailed scales.

Changes of Scaling Exponents with Height

The results of these calculations show that the length of each contour line falls with increasing grid element without any exception, as is presented on Fig. 1 where length of isolines is plotted as a function of step (that is side of grid element in km) in bilogarithmic scale. Striking closeness of the calculated points to appropriate straight lines (the regression coefficients are mostly greater than 0.99) strongly suggests that the Earth's surface contour lines are really self-similar at these scales with fractal dimensions varying from 1.3 to 2.4.

The changes in fractal dimension seem to be regular, as seen on Fig. 2, where a graph of its dependence on topography is displayed. A well defined minimum at -0.5 km divides the whole graph into two different parts, say oceanic and continental. The left, oceanic part may be readily approximated with a straight regression line

$$D = 1.260 - 0.055\ H\,(\text{km})$$

with regression coefficient of -0.97. The right, continental one is rather more complicated and contains a minimum and a maximum within it. If we try to approximate it with a straight line as shown on Fig. 2 ignoring the points from 2 to 4 km, we shall obtain regression line $D = 1.375 + 0.171\ H$ (km) with regression coefficient as high as 0.999.

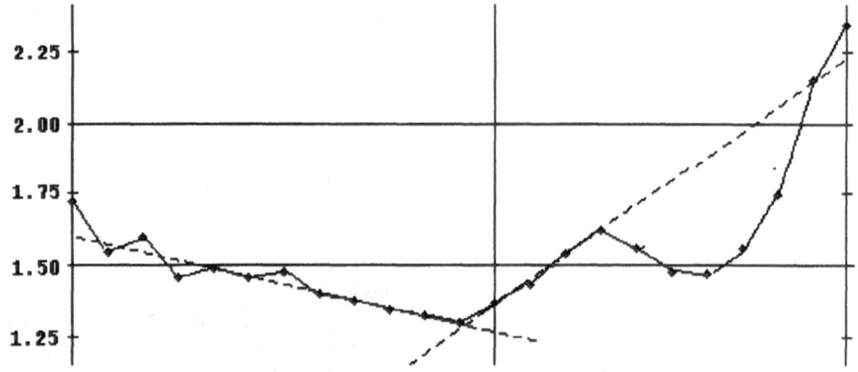

Fig. 2. Fractal dimension of global relief contour lines as a function of contour level. Dashed are regression lines for oceanic and continental realms obtained as explained in text.

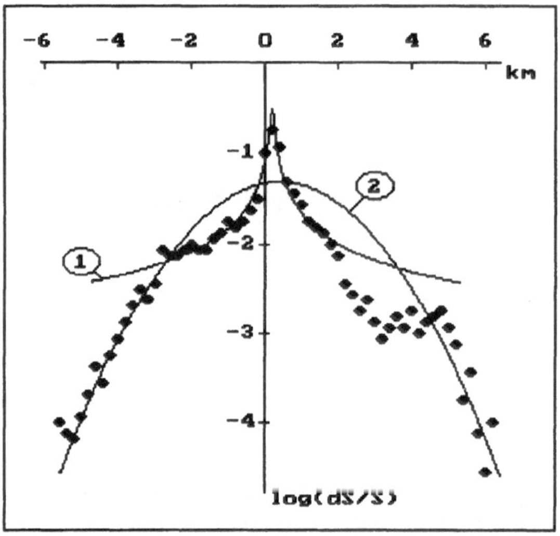

Fig. 3. Interpretation of continental heights distribution based on the analysis of relief potential energy (Ivanov 1989a, b). *1* - curve of hyperbolic sine distribution law; *2* - curve of normal distribution law; *3* - experimental points after (Cogley 1985) representing relative areas occupied by heights values within each 200 m height interval.

The grounds for such approximation are derived from our previous study of global continental heights distribution (Ivanov 1989a, b). It was stated from physical considerations that this distribution is governed by two principal laws – normal law and the law of hyperbolic sine. Experimental data sufficiently fitted this theoretical forecast except for the altitude range from 2 to 4.5-5 km (Fig. 3).

As far as these heights are anomalous in the sense of their energetic stipulation it is logical to suppose that this very range may be anomalous in the sense of self-similar properties as well. In this way the approximation of Fig. 2 appeared. If it is correct, a new and interesting problem emerges of explaining interrelations between fractal and energetic properties of global relief.

Scale Range of Self-Similarity

As is clear from the description of the procedure the change of scale was imitated by averaging the data to a grid with lesser density. A logical outgrowth of this approach is to imagine the extreme situation, when the grid consists of two points each representing one of the hemispheres. Since the mean level of the solid Earth's surface is about -2.5 km, one of the two average values for hemispheres should be slightly less and the other slightly greater than -2.5 km regardless of how the whole sphere is divided into two parts (northern and southern, eastern and western, or any other way). Thus in this case we obtain the length of the -2.5 km contour equal to the length of the Earth's great circle, that is 40000 km, and the side of this "grid" will be 180°, so the step value is 20000 km. Plotting this point together with the data obtained for the -2.5 km isobath (Fig. 4) we find them in a perfect agreement fitting to a straight line representing the fractal dimension of 1.39.

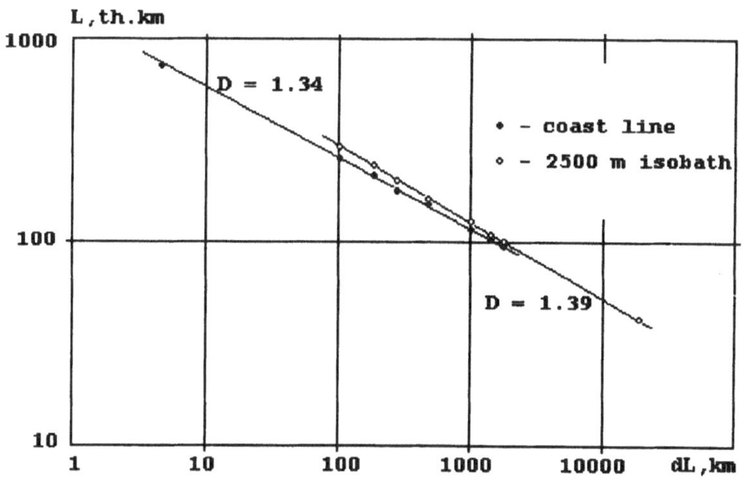

Fig. 4. Length-step dependence for coastline (solid squares) and -2500 m isobath (open squares). The point for -2500 m with step 20000 km is plotted according to considerations in text; the point for coastline with step 5 km is after (Lukyanova & Kholodilin 1975).

Another possibility of extending the scale limits of this test is provided by Lukianova & Kholodilin (1975), who measured the length of the world coastline by walking a pair of dividers set to an opening of 2 mm along the world map in the scale of 1:2500000. Thus the step is estimated as 5 km, and the resulting coastal length is reported to be 7770000 km (Lukyanova & Kholodilin 1975). This gives us one more point that is plotted on the graph together with the data on sea-level contour computed above (Fig. 4). As in the previous case the fit is very good, and the fractal dimension of the coastline is 1.34.

From these considerations we are led to believe that self-similar properties of global relief are manifested in the scale range from 5 km to 20000 km (the latter value means infinity for spherical Earth) and fractal dimension of contour lines is approximately uniform within these limits, having an average value of 1.37. One may take this result to be really dramatic, especially taking into account that the extreme points of the graph of Fig. 4 were obtained by other methods relative to internal ones. This phenomenon may be regarded as a case history of fractal geometry in nature.

It is interesting that the same value of fractal dimension (1.37) was obtained by Nakano (1984) for the coastline and some selected relief contours of Northwestern Japan in the step range from 1 to 10 km, that provides additional suupport to the conclusions presented above. The temptation of comparing the global average with regional values is very strong but it is clear that it has no reasonable base. Nevertheless it is difficult to find more than one successful attempt at estimating the global value, that is the estimate of Turcotte (1989), who showed on the base of the analysis of global topography spectrum that fractal dimension of global relief is approximately equal to 1.5. Now we may regard this value as a first rough approximation, where the influence of extremal elevations was slightly exagerated.

Scaling of Regional Topography

In order to rectify the conclusions that were drawn above, some other apprroaches to the problem were applied. The Earth's surface in three-dimensional space may be regarded rather as a self-affine than a self-similar object due to the gravity that emphasizes one of the three spatial dimensions. So the methods of analysis of self-affine sets are adequate to be deployed to the surface as a whole (and not to its horizontal cross-sections, which are topography isolines).

First the method of variation scaling suggested by Mark & Aronson (1984) for examining self-affine objects was used. To avoid difficulties emerging because of the Earth's sphericity, this method was applied only to the equatorial belt of the Earth's surface 40° width, where curvature distortions may be neglected. The squared differences between elevations of all possible pairs of points (for the total number of 14400 points) were analyzed with respect to the distance between points. The corresponding graph is presented on Fig. 5. We see a striking linear dependence of variance with space step in the scale range from 100 to 1000 km with the slope of 0.62 that gives the fractal dimension value of 1.38. For larger scales one can find manifestations of latent periodicity with basic wavelength of approximately 4000 km, which completely shadows self-similar properties revealed by perimetric method at these scales. Probably the mentioned periodicity is an artefact and is caused by the limited width of the band under study which is equal to 4400 km. Still the coincidence of fractal dimension values obtained by both methods for lesser scales is very good.

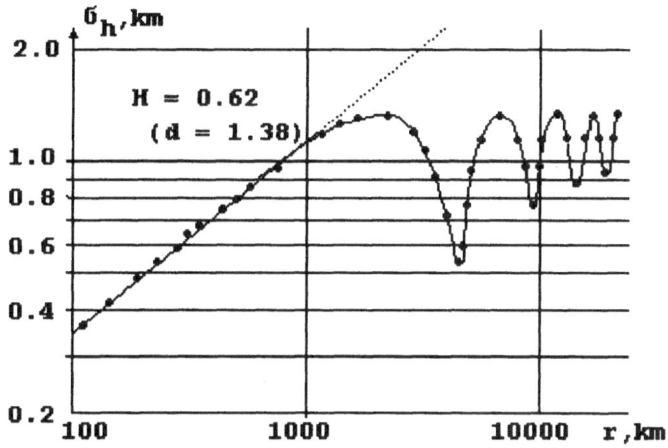

Fig. 5. Variance-step plot of the Earth's topography within the 40° equatorial belt.

Another method to estimate fractal dimension of a self-affine set was elaborated during this study and was called the "dispersion counter-scaling" method. More details on its fundamentals may be obtained from Ivanov (1994). The object under study was the same equatorial belt, subdivided into 9 squares with the side of 40°. For each square the scaling of internal and external dispersions was traced in the space scale range from 100 to 4000 km (Fig. 6). As was supposed, corresponding values of fractal dimension may be regarded as local and global dimensions of self-affine set.

The conclusions that may be drawn from these plots are not quite clear yet. Nevertheless it is obvious that local fractality of regional relief (as estimated by internal dispersion scaling) manifests itself rather definitely and fractal dimension so determined is in agreement with the global value obtained earlier. External dispersion seems to be less scale dependent and gives greater values of global fractal dimension. This result is rather unexpexted and worth further investigation.

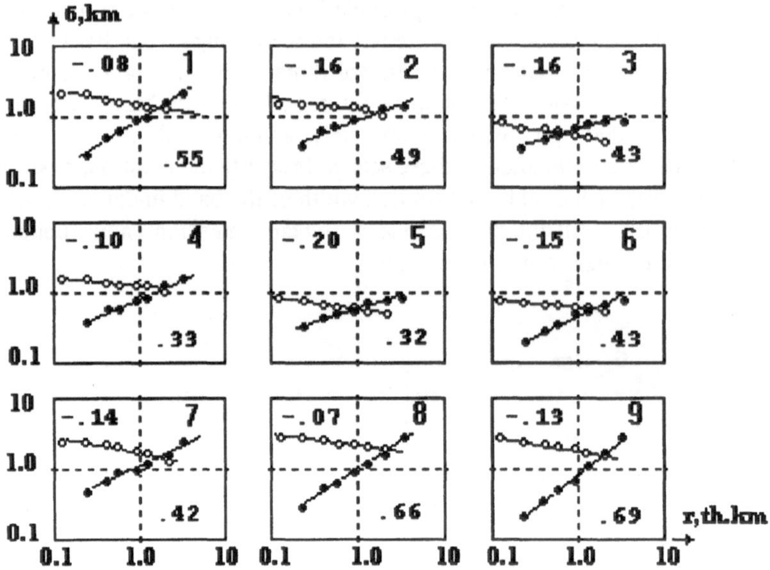

Fig. 6. Plots of dispersion "counter-scaling" of the Earth's topography for nine consequent square areas within the 40° equatorial belt.

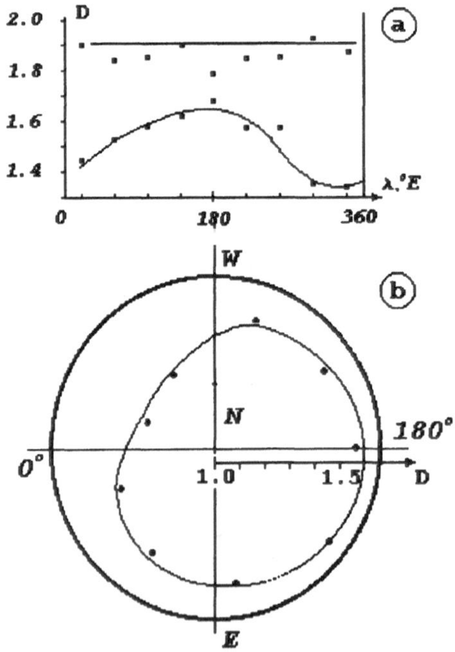

Fig. 7a, b. Variations of fractal dimensions of the equatorial belt topography with longitude. **a** plot of dimension values as determined from internal (circles) and external (dots) dispersions; **b** polar diagram of the internal (local) dimension as seen from the North.

While external dispersion (and corresponding fractal dimension) of regional relief seem to be uniform along the equatorial belt, the internal one reveals a strong dependence on longitude (Fig. 7) assuming certain asymmetry in fractal properties of the Earth's surface.

Conclusions

Fractality of the Earth's topography is a manifestation of stochastic properties of relief. One may suppose that a challenge of long standing will be the conjugation of these results with deterministic regularities of global relief such as energetic stipulation of continental heights distribution (Ivanov 1989a, b), or thermally induced subsidence of oceanic lithosphere (Parker & Oldenburg 1973, Sorokhtin 1973). So these results pose some special problems of explaining of such a striking self-similarity of global relief in a wide range of scales – ver four orders of magnitude.

As could be easily seen, the method of variance gives a unique value of fractal dimension for the topography as a whole and the counter-scaling method gives two different values. On the other hand perimetric scaling provides a spectrum of dimensions dependent on height which is very close to the concept of multifractal fields with threshold-controlled dimension values in the sense of Schertzer & Lovejoy (1983). We have tried to avoid any comments on this topic, leaving room for further discussion.

Acknowledgements. This research was supported by Russian Foundation of Fundamental Studies, grant 93-05-8943. I would like to thank Prof. G.I. Barenblatt for numerous helpful discussions.

References

Barenblatt GI, Zhivago AV, Neprochnov YuP, Ostrovsky AA (1984) Fractal dimension: a quantitative characteristic of ocean bottom relief. Oceanology 24: 695-697.

Cogley G (1985) Hypsometry of the continents. Z Geomorhp, Supplementbd 53, VIII: 1-48.

Gates WL, Nelson AB (1975) A new (revised) tabulation of the Scripps topography on a 1° global grid. Repts 1276, 1277, Rand Corp., Sta. Monica.

Ivanov SS (1989a) Energetic interpretation of the hypsographical curve of the continents. Dokl Akad Nauk SSSR 306 (5): 1087-1091 (in Russian).

Ivanov SS (1989b) Continental heights distribution: an energetic approach. In: 28th Int Geol Congress, Washington, Book of Abstracts 2: 102-103.

Ivanov SS (1994) "Counter-scaling" Method for Estimation of Fractal Properties of Self-Affine Objects. In: Kruhl JH (ed) Fractals and Dynamic Systems in Geoscience. Springer, Berlin Heidelberg New York, pp 391-397 (this volume).

Krohn C (1988) Sandstone fractal and Euclidean pore volume distributions. J Geoph Res 93 (B4): 3286-3296.

Lukianova SA, Kholodilin NA (1975) The length of the coastline of the World Ocean and different types of shores and coasts. Vestn Mosk Univ Ser geogr 1: 48-54 (in Russian).

Mandelbrot BB (1967) How long is the coast of Britain? Statistical self-similarity and fractional dimension. Science 156: 636-638.

Mandelbrot BB (1975) Stochastic models for the Earth's relief, the shape and the fractal dimension of the coastlines, and the number-area rule for islands. Proc Natl Acad Sci USA 72, 3825-3828

Mark D, Aronson P (1984) Scale-dependent fractal dimensions of topographic surfaces: an empirical investigation, with applications in geomorphology and computer mapping. Math Geol 16(7): 671-683.

Nakano T (1984) A systematics of "transient fractals" of rias coastline; an example of rias coast from Kamaishi to Shizugawa, Northeastern Japan. Ann Rep Inst Geosci Univ Tsukuba 10: 66-68.

Parker RL, Oldenburg DW (1973) Thermal models of oceanic ridges. Nature 242: 137-139.

Schertzer D, Lovejoy S (1983) The dimension and intermittency of atmospheric dynamics. In: Bradbury LJS, Durst F, Launder BE, Schmidt FW, Whitelaw JH (eds) Turbulent Shear Flows 4, Springer, Berlin Heidelberg New York Tokyo, pp 7-33.

Sorokhtin OG (1973) On the dependence of mid-ocean ridges topography on spreading rate. Dokl Acad Nauk SSSR 208: 1338-1341 (in Russian).

Turcotte D (1989) Fractals in Geology and Geophysics. PAGEOPH 131: 171-195.

Zeldovich YaB, Sokolov DD (1985) Fractals, similarity, and intermediate asympthotics. Usp Fiz Nauk 146(3): 493-506 (in Russian).

Part III
Formation, Structure and Distribution of Minerals and Matter

Fractal Geometry and the Mining Industry, a Review

Brian H. Kaye
Physics Department, Laurentian University
Sudbury, Ontario P3E 2C6, Canada

Introduction

Fractal geometry is the creation of Mandelbrot, who published his first book on the subject in English in 1977 (Mandelbrot 1977). Fractal geometry is the geometry of rugged systems, that is objects having "non-smooth" boundaries and non-Euclidean shapes. From a study of rugged systems fractal geometers have derived many descriptive parameters for describing fractured systems such as broken rock and the random space filling structures of sedimentary systems such as sandstone and other porous systems. The term fractal geometry was coined by Mandelbrot from a Latin word meaning fractured. In the 15 years since the publication of Mandelbrot's book, fractal geometry has found many applications in the mining industry.

It is the purpose of this chapter to indicate the various areas where fractal geometry theorems are proving to be useful concepts for tackling difficult problems in the mining industry (Kaye 1989, 1993). When discussing the applications of fractal geometry to the mining industry it is useful to differentiate between theoretical fractal geometry and applied fractal geometry. In Figure 1 the famous Mandelbrot set is shown. A Mandelbrot set is a mathematical figure which has been reproduced in many articles on fractal geometry. It has the interesting property that the boundary is infinite. If one looks at magnified portions of the boundary it looks likes the larger image, a property which is described as self-similarity. Because of self-similarities, if one looks at a portion of the boundary of the Mandelbrot set it is impossible to know the magnification at which one is inspecting the boundary. The complexity of the mathematical theory associated with the Mandelbrot set has tended to delay implementation of fractal geometry in the applied sciences. However the structure of the Mandelbrot set is virtually irrelevant to the problems of the applied industries. The properties of items such as fractured rock are fractal because if one inspects the surface of a fracture at higher and higher magnification, one has a range of inspections in which there is a statistical self similarity of the surface structure. The ruggedness of the boundary can be described using the concepts of fractal geometry. One of

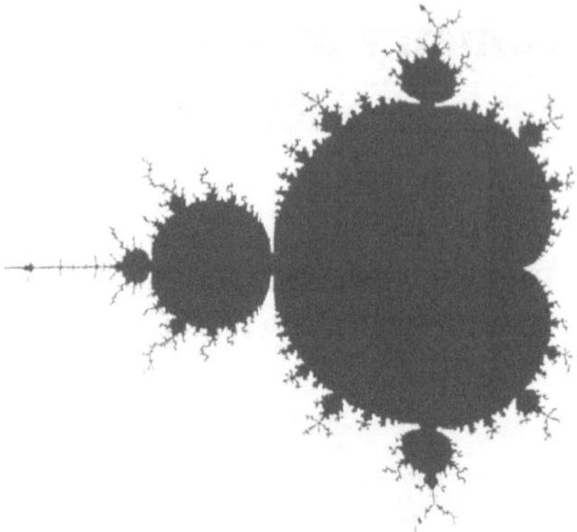

Fig. 1. The Mandelbrot set is a theoretical or ideal fractal system which has created great interest in fractal geometry but which is not relevant to the technology of the mining industry.

the major differences between ideal fractal systems such as the Mandelbrot set and real systems is that natural fractals such as rock fractures only exhibit fractal structure over a limited range of magnifications. The range over which fractal behaviour is manifest in natural systems provides information on the formation dynamics and the functional behaviour of the systems.

Fractal Dimensions: a Term with Several Meanings

The ruggedness or space filling ability of a system can be described using the concept of fractal dimensions. As so often happens in the early days of a new technology, the term fractal dimension has been used loosely in various industries and one should carefully examine the context of a published report to determine what any specific author means by the term fractal dimension. To illustrate the various ways in which the term fractal dimension can be used consider the carbonblack profile of Figure 2a. The ruggedness of the projection of the boundary of the profile can be described by a boundary fractal dimension. The boundary fractal dimension is measured by estimating the perimeter of the projected profile at a series of inspection resolutions. These polygons are created by various techniques. Perhaps the simplest technique to explain is what is known as the structured walk technique. In the structured walk technique one strides around the profile using an exploration step of size λ. The polygon perimeter becomes the estimate of the perimeter of the profile at the inspection resolution λ as shown in Figure 2b. It is usual to normalize the perimeter estimate and the step size using the maximum projected length of the profile known as the

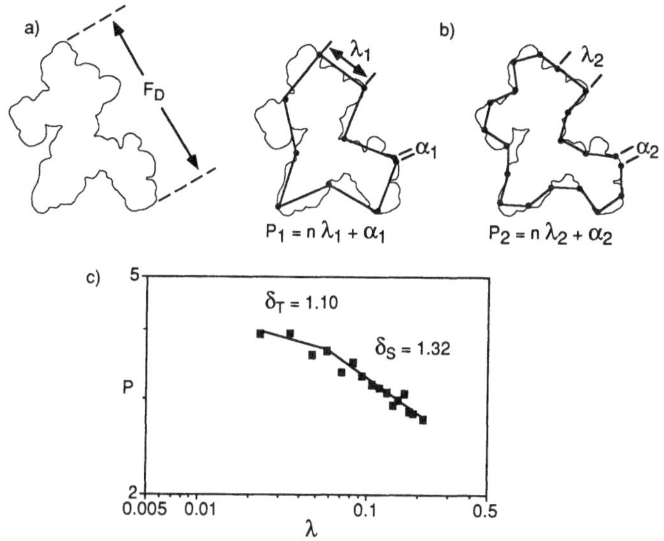

Fig. 2. The structure of a carbonblack profile can be described by means of several fractal dimensions. **a** Magnified profile of a carbonblack pigment. **b** Polygons stepped out on the profile at two different resolutions. **c** Richardson plot of a structured walk exploration of the profile of **a**. λ : Stride length, resolution, of inspection. α : short final step required to complete the polygon. P : Perimeter estimate of the profile at resolution λ. F_D : maximum projected length of the profile.

Feret diameter, F_D. The data for a series of explorations of the profile of the carbonblack shown in Figure 2a is shown in Figure 2c. The data for the exploration measurements plotted in log-log format is known as a Richardson plot in honor of one of the pioneers of boundary evaluation (Mandelbrot 1977). It can be shown that the slope of the data line on a Richardson plot is a fractional addendum to the topological dimension of a boundary which describes the ruggedness of a boundary. It can be seen from the data plotted in Figure 2c that for this particular carbonblack profile there are two linear regions in the data plot. The fractal dimension deduced from the linear relationship at coarse resolution can usefully be described as a structural fractal dimension, δ_S. This fractal dimension describes the gross physical behaviour of the fineparticle such as the aerodynamics of the particle moving in the lung and the capture probabilities if it passes through a filter. It also appears to contain information on the formation dynamics of the agglomerate (Kaye et al. 1991). The fractal dimension manifest in the data for high resolution exploration can be described as the textural fractal dimension of the profile, δ_T This fractal dimension is useful for describing the chemical reactivity and surface phenomena of the agglomerate. In other studies of the systems such as the carbonblack of Figure 2a, the structure of the system is examined by means of light scattering or neutron scattering experiments. When this type of experiment is carried out the resultant structural information is described as the mass, or space filling ability, fractal dimension. In this term the adjective mass is used if one is looking at an agglomerate and the term space filling ability is used if one is looking at the space occupied by a spreading crack system.

A different type of physical fractal structure of interest to the mining engineer is the Sierpinski carpet shown in Figure 3a. The Sierpinski carpet is a two dimensional version of a structure which in three dimensions is known as a Menger sponge. The Menger sponge can be used as a model of a catalyst structure, the porous structure in a tailings pond or the probable pathways through an oil bearing sandstone. A discussion of the three dimensional structure is beyond the scope of this chapter and the interested reader is directed to Kaye (1989, 1993b). Again the Sierpinski carpet is an ideal system which if the construction algorithm is continued ad nauseam results in an invisible structure which contains an infinite number of infinitely thin threads! In a real system the fractal structure is manifest over a range of inspection resolutions and the range over which fractal structure is manifest is itself information on the way in which the system is formed. Natural systems have a random structure which is described as being statistically self-similar rather than exactly self-similar.

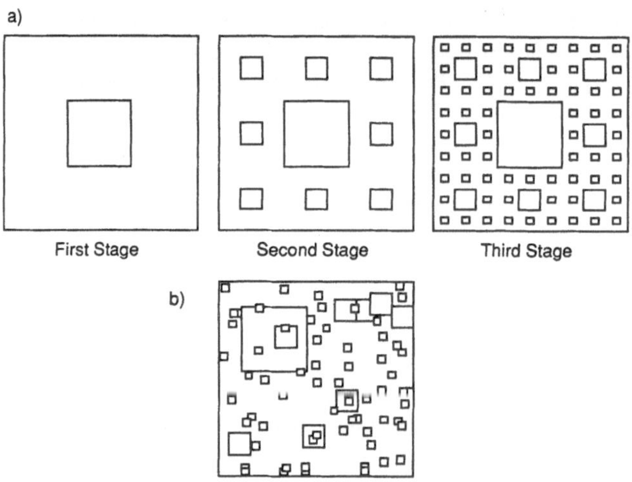

Fig. 3. The structural dimension of an equivalent Sierpinski carpet is another concept drawn from fractal geometry which is useful for describing systems of interest to the mining industry. **a** Three successive stages in the construction of an ideal Sierpinski Carpet having a fractal dimension of 1.89. **b** A randomized, statistically self-similar version of the third stage of part (a) which can serve as a model for porous bodies such as sandstone and ground water percolation in tailings ponds.

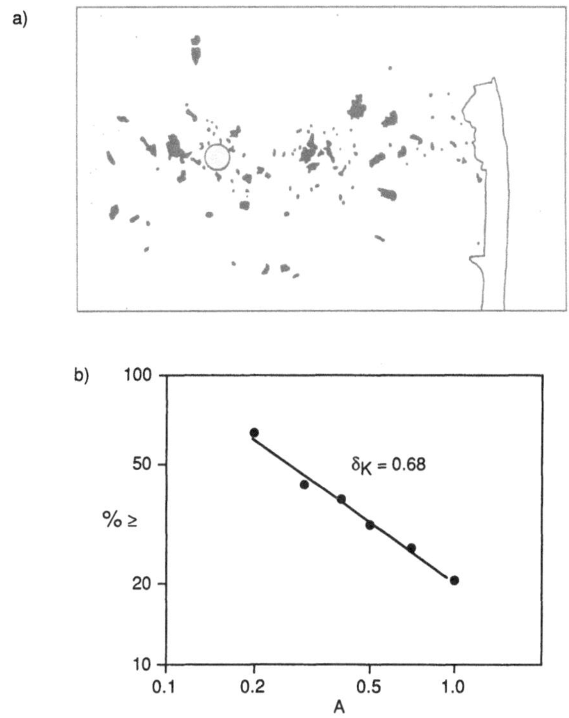

Fig. 4. Fractal dimension in data space can be used to describe the size distribution of fragments produced in a ballistic fragmentation process. **a** Tracing of a high speed photograph of a stell ball breaking through a plastic resin board. **b** Cumulative size distribution of the fragments produced in (a). A ; projected area of a fragment. $\% \geq$: percent of profiles with the same or greater area than stated.

The various fractal dimensions discussed so far are based upon the physical structure of systems. A fractal dimension in data space is a different type of fractal dimension which is not a physical identity but a parameter, descriptive of a distribution associated with a system, which is based on the concepts of fractal geometry. To understand what is meant by this term consider the system shown in Figure 4a. A high speed ballistic impact on the piece of phenolic resin board creates a set of fragments. The system was photographed microseconds after impact. If one measures the size distribution of these fragments and plots the number of fragments greater than or equal to a given size (in this case the area of the fragment) one obtains a linear relationship as shown in Figure 4b. For reasons that have been discussed in detail elsewhere it is useful to regard the slope of such a linear relationship as the fractal dimension in data space of the fragments produced by the ballistic fragmentation. The fragmentation process in ballistic fragmentation is essentially different from that in stress fragmentation

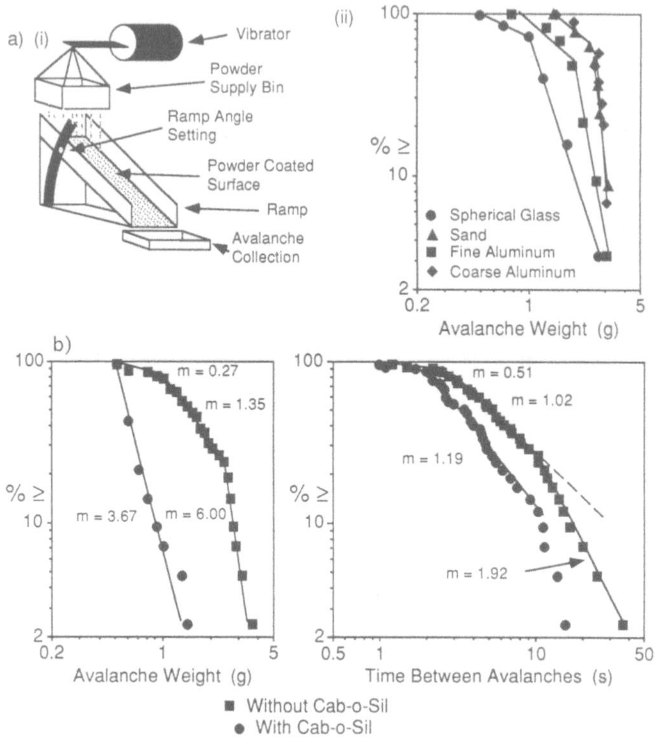

Fig. 5. **The avalanching behaviour of powder heaps can be described by means of fractal dimensions in data space. a A summary of the dynamic avalanching behaviour of several powders i Equipment used to study avalanching behaviour. ii Avalanching behaviour of various powders. b Avalanching behaviour of rock tailings powder with and without flow agent.**

which operates by applying pressure in a system such as crusher. In stress fragmentation cracks move through the body, exploiting the flaw structure present in the body whereas in ballistic fragmentation failure takes place in regions of tensions created by the interaction of reflected shock waves in a piece of material subjected to high speed collisions with other objects. A hypothesis which is causing great excitement in the scientific community is that the boundary fractal dimension of fragments produced by ballistic fragmentation is related to the data space fractal dimension of the fragment distribution. Should this hypothesis be proved, one will be able to predict the size distribution of fragments from a study of the boundary fractal dimension of the fragments (Kaye 1993, Brown et al. 1993).

Other fractal dimensions in data space are used to describe events generating a time series. Thus in catastrophe theory, and a study of what is known as critically self-organized systems, the behaviour of an avalanching powder heap is describable by fractal dimensions in data space when the frequencies of avalanches of a given size are plotted against the size of the avalanche on log-log graph paper (Bak et al. 1988, Bak & Chen 1991, Held et al. 1990, Kaye et al. 1992). The original work on the study of powder behaviour in avalanching heaps studied pseudo-static heaps in which grains of sand were added one at a time to a growing heap. Kaye and co-workers have extended this work to study dynamic avalanching in which powder is poured onto a slope and the avalanche is measured at the end of the slope using the equipment shown in Figure 5a(i). Of particular interest to the mining industry is the avalanching behaviour of rock tailing powders which are some times poured into heaps to create backfill or into tailing ponds. In Figure 5b the behaviour of rock tailings from a nickel mine with and without flow agent is shown. The higher the fractal dimension of such data lines the more free flowing is the powder. Flow agents are often added to dry powders to facilitate their movement in process equipment. The three data lines of the untreated powder illustrate that basically avalanches of three kinds are produced in the dynamic behaviour of the untreated powder. For the main bulk of the avalanching behaviour a fractal dimension in data space of 1.35 is descriptive of the behaviour. However once the build up of powder in this equipment reaches approximately 3 grams of powder the system becomes more unstable and the avalanches have a fractal dimension of 6. Below an avalanche weight of 1.2 grams some mechanism suppresses the formation of small avalanches.

The behaviour of the powder is totally changed by the addition of a small amount of Cab-o-sil. flowagent as illustrated in the diagram. Later in this chapter we will show how such data space fractal dimensions are of use in the description of rockburst frequencies.

Now that we have described the various ways in which the term fractal dimension is used in applied fractal geometry, we can explore direct applications of fractal geometry to various topics of interest to the mining engineer.

Applying the Concepts of Fractal Geometry to Real Problems in the Mining Industry

One of the main activities of the mineral processing engineer is to fragment lumps of an ore to release the valuable mineral. Vast quantities of energy are consumed by the mining industry in this process. As mentioned in the previous section, ballistic fragmentation generates fragments of ore which have boundary fractals related to the surface activity of the fragments and there is also the possibility that the size distribution of the fragments can be described by fractal dimensions in data space. Several workers are pursuing the applications of fractal geometry to the fragmentation process. Often the first step in the fragmentation process is a blasting of the rock and Crum has applied fractal concepts to a study of bench blast fragmentation (Crum 1990). Heping Xie and co-workers have made several studies of the way in which the concepts of fractal geometry can be applied to the fragmentation process (Xie 1989, 1991, 1992). As discussed earlier, the possible relationship between the ballistic fragmentation of a rock and the size distribution of the fragments is being studied by Miles and co-workers at the University of Nottingham (Brown et al. 1993). If an object is made to crack by the application of forces to a point, a crack system can be modelled by specifying a growth algorithm of the crack from the tips of the crack spreading out in space. (Such stress cracks differ from the way in which a body fails under ballistic failure. See earlier comments.) Bradt and co-workers are studying the way in which fracture cracks are generated in laminated tempered safety glass panels and other ceramics. Their work will obviously have applications in the mining industry (Sakai et al. 1990, 1991).

Fractal Structure of Porous Bodies of Interest to the Mining Engineer

When a fluid is injected into a porous material the material fingers into the material to produce a structure such as that shown in Figure 6. This system was generated by Feder and co-workers, who injected an epoxy fluid into an array of glass beads simulating a synthetic sandstone (Feder et al. 1986). The space filling fractal dimension of such an array, which is a measure of the efficiency with which a fluid can drive out oil in a sandstone, can be measured by superimposing a set of rings on the system and counting the intersection frequency as one moves out from the center (Kaye 1993). Other workers who have explored the fractal structure of porous bodies include Schlueter and co-workers, J. Muller and co-workers, Po-Zen Wong, Thompson and co-workers (Schlueter et al. 1991, Muller

Fig. 6. When a fluid is injected into a porous body the liquid often moves out from the injection point as a fractal finger. As illustrated by the Epoxy fingers penetrating a synthetic sandstone layer made with glass beads.

et al. 1990, Wong 1988, Thompson et al. 1987, Krohn 1988). Lenormand and co-workers have made extensive studies of the fractal structure of porous bodies (Kaye 1989). Others workers have studied the fractal structure of porous coal (Johnston et al. 1991). Here at Laurentian University we are applying fractal geometry to a study of the percolation path through piles of crushed ore used in bio-hydrometallurgy techniques. We are also modelling the randomwalk of radon gas through porous beds of gravel. The percolation of water through tailing beds is also a potential area for applying fractal geometry to the problems of the mining industry.

Fractal Geometry and Rockbursts

For the purposes of this chapter a rockburst can be described as a mining-induced low level earthquake. During the mining of an ore body stresses may build up in the rock until there is a sudden release of energy similar to that which occurs when an earthquake releases stress energy in the rocks of the earth's crust. Rockbursts can range in energy from slightly felt tremors to disastrous failings of the area being mined with the subsequent danger to miners. Unfortunately some rockbursts lead to loss of life in the mining industry. Fractal geometry is being applied to the study of rockbursts in two ways. First of all the distribution of the occurrence of rockbursts of different energies can be described by a fractal dimension in data space in the same way that scientists have been using fractal geometry concepts to describe the occurrence of earthquakes of given energy. Rockburst data for a mine known as the Lucky Friday mine is shown in Figure 7. This data is from a personal communication from Jean-Marc Noel of Laurentian University who has been studying rockburst phenomena under the auspices of a grant from the Canadian Mining Research Directorate. An interesting aspect of this graph is that the frequencies of low level rockbursts cannot be used to predict the frequency of major rockbursts. This fact is counter to some of the widely held opinions in the mining industry at the time of writing. Another way in which fractal dimensions are being used to study the occurrence of rockbursts is shown in Figure 8. If we plotted the location of rockbursts in a mining space we could choose an arbitry center for the cluster of points and measure the density of occupation of the dataspace using a set of search circles. The slope of the line when we plot the number of events within a given search area defines a space occupancy fractal dimension as illustrated in Figure 8b. The set of events in Figure 8a is actually a random distribution simulated using a random number table. One of the ways in which research is proceeding is that one can simulate the density of occupation data such as that of Figure 8a and compare it to a real set of data to determine if the rockburst pattern is random. There is some indication that the clustering of rockburst events in space changes when a major rockburst within the system is imminent. The possibility of using such data

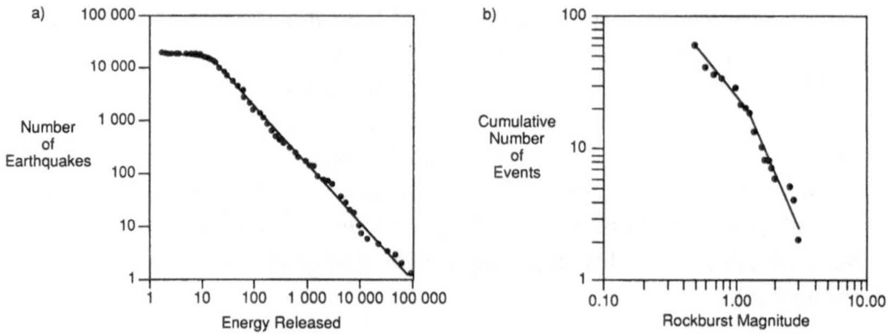

Fig. 7 The frequency of rockbursts in a mine are describable by fractal dimension in data space, presented above is rockburst data for the Lucky Friday mine (graph supplied by J.M. Noel, Laurentian University).

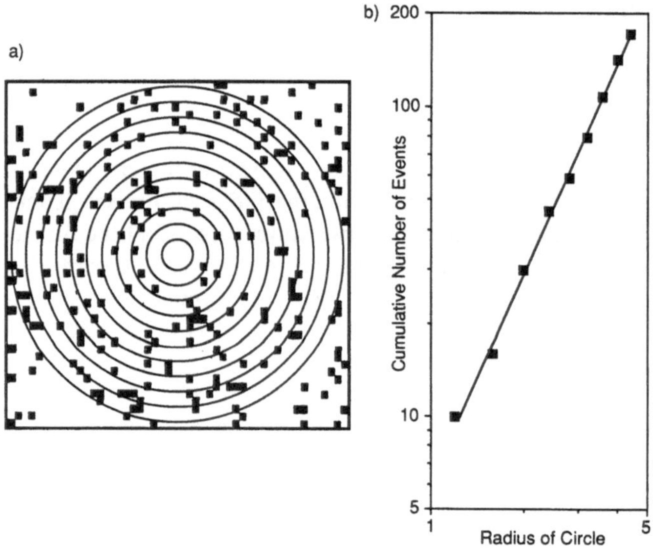

Fig. 8. Space occupancy fractal dimensions are also being used to study the possible patterns of spatial clustering of rockburst events in mining space. There is some indication that significant changes in spatial occupation fractal dimension are related to the occurrence of major rockburst events. The system shown above is a simulated random distribution of rockbursts.

processing to discover patterns in rockburst occurrence is proceeding and is one possible route in which one may be able to predict the occurrence of major rockbursts. (This possibility is a subject of a study proceeding at Laurentian

University.) A similar type of fractal dimension of space occupancy is being used by others to study the possibility of describing the occurrence of ore deposits over a given space by means of a fractal dimension (Carlson 1991).

Conclusions

In this chapter the various areas of the mining industry in which the new ideas of fractal geometry are proving to be fruitful areas of research have been explored. Because of space limitations it has not been possible to discuss the health problems posed to mining personnel from fractally structured dust such as diesel exhausts and other fumes. The interested reader is referred for a discussion of these topics to Kaye (1989, 1993a, b). Other areas of fractal geometry of potential interest to the mining engineer are that typical records from a bore hole log are fractal in structure. The physical significance of such bore-hole fractal structures has yet to be established. We can expect to see many areas of development of fractal geometry and the mining industry (Turcotte 1991, Scholz 1990, Kubi 1985).

References

Bak P, Chen K (1991) Self-Organized Criticality. Sci Am, January: 46-53.

Bak P, Tang C, Weisenfeld K (1988) Self-Organized Criticality. Phys Rev A 38(1): 364-374.

Barton CC, Scholz CH (1994) The Fractal Size and Spation Distribution of Hydrocarbon Accumulations: Implications for Resource Assessment and Exploration Strategy. In: Baron CC and La Pointe PR (eds.), Fractal Geometry and its Application in Petroleum Geology, AAPG Memoir, in press.

Blenkinsop T (1994) The Fractal Distribution of Gold Deposits: Two Examples from the Zimbabwe Archaean Craton. In: Kruhl JH (ed) Fractals and Dynamic Systems in Geosciences. Springer, berlin heidelberg New York, pp 247-258 (this volume).

Brown GJ, Miles NJ, Hall ST (1993) Fractal Characterization of Pulverized Materials. Part Part Syst Charact 10(1): 1-6.

Cab-o-sil. is a trade name of the Cabot Corporation. See trade literature of Cabot Corp., Tuscola, Il., U.S.A., 61953.

Carlson, CA (1991) Spatial Distribution of Ore Deposits. Geology 19: 111-114.

Crum SV (1990) Fractal Concepts Applied to Bench-Blast Fragmentation. In: Hustrulid and Johnson (eds) Rock Mechanics Contributions and Challenges. Balkema, Rotterdam, pp. 913-919.

Feder J, Jossang T, Maloy KJ, Oxaal U (1986) Models of Viscous Fingering. In: Engleman R and Jaeger Z (eds), Ann Isr Phys Soc, 8. (Fragmentation, Form and Flow in Fractured Media,. Proc of conference held Neve Ilan, Israel, 6-9th, January 1986)

Hedley DGF A Five Year Review of the Canada-Ontario Industry Rockburst Project. Canmet Special Report, SP90-4E., Canadian Energy Mines, 55 Booth Street, Ottawa, Ontario.

Held GA, Solina DH, Keane DT, Haig WJ, Horn PM, Grinstein G (1990) Experimental Study of Critical Mass Fluctuations in an Evolving Sandpile. Phys Rev Lett 65(9): 1120-1123.

Johnston P, McMahon P, Reich MH, Snook IK, Wagenfeld HK (1991) The Effect of Processing on the Fractal Pore Structure of Victorian Brown Coal. J of Colloid and Interface Sci, September.

Kaye BH (1989) A Randomwalk through Fractal Dimensions. VCH, Weinheim.

Kaye BH (1993a) Applied Fractal Geometry and the Fineparticle Specialist. Part I: Rugged Boundaries and Rough Surfaces. Part Part Sys Charact 10 (3): 99-110.

Kaye BH (1993b) Chaos and Complexity: Discovering the Surprising Patterns of Science and Technology. VCH, Weinheim.

Kaye BH (1993c) Fractal Dimensions in Data Space: New Descriptors for Fineparticle Systems. Part Syst Charact 10(4): 191-200.

Kaye BH, Clark GG (1991) Formation Dynamics Information: Can it be Determined from the Fractal Structure of Fumed Fineparticles? Chapter 24 in: Provder,T. (ed) Particle Size Distribution II, Assessment and Characterization. Am Chem Soc, Washington, D.C.

Kaye BH, Liu Y, Gratton J, Clark GG (1992) Dynamic Shape Factors from Catastrophic Tumbling Studies and the Catastrophic Collapse of Powder Heaps. Proc. Fifth European Symposium on Particle Characterization, Nurnberg, Germany, 24-26 March (1992).

Krohn CE (1988) Fractal Measures of Sandstones, Shales and Carbonates. J Geophys Res 93: 3297-3305.

Kubi KK (1985) Fractal behaviour in Ore Deposits. Extended abstract from Fall meeting of the Materials Research Society, (Symposium), Laibowitz, Mandelbrot and Passoja (eds), M.R.S., Pittsburgh.

Lung CW (1989) Fractal Dimension of the Fractured Surface of Materials. In: Harony A and Feder J (eds) Fractals in Physics. Physica D. 38: 242-245.

Mandelbrot BB (1977) Fractals, Form, Chance and Dimension. Freeman, San Francisco. Mandelbrot first described his theories of fractal geometry in "Fractals, Form, Chance and Dimension". In 1983 Mandelbrot published an updated and expanded version of the book, "The Fractal Geometry of Nature", this book is considered by Dr. Mandelbrot to be the definitive book on the subject. (Personal Communication).

Muller J, Hansen JP, Skjeltorp AT, McCauley J (1990) Multifractal Phenomena in Porous Rocks. Proc Mat Res Soc Symp 176: 719-723.

Sakai T, Ramulu M, Ghosh A, Bradt RC (1990) A Fractal Approach to Crack Branching (Bifurcation) in Glass. Fractography of Glasses & Ceramics 17: 131-146.

Sakai T, Ramulu M, Ghosh A, Bradt RC (1991) Cascading Fracture in a Laminated Tempered Safety Glass Panel. International Journal of Fracture 48: 49-69.

Schaefer DW (1988) Fractal Models and the Structure of Materials. MRS Bulletin 13(2).

Schlueter EM, Cook NGW, Witherspoon PA (1991) Fractal Dimensions of Pores in Sedimentary Rocks and Relationship to Permeability. Preprint provided by the author from Materials Science and Mineral Engineering, University of California, Berkeley.

Scholz CH (1990) The Mechanics of Earthquakes and Faulting. Cambridge Univ Press, Cambridge.

Thompson AH, Katz AJ, Krohn CE (1987) The Microgeometry and Transport Properties of Sedimentary Rock. Advances in Physics 36(5): 625-694.
Turcotte DL (1991) Fractals in Geology: What Are They and What Are They Good for? GSA Today 1(1): 2-4.
Velde B, Dubois J, Moore D, Touchard G (1991) Fractal Patterns of Fractures in Granites. Earth Planet Sci Lett 104: 25-35.
Wong P (1988) The Statistical Physics of Sedimentary Rock. Physics Today, Dec: 24-32.
Xie H (1989) The Fractal Effect of Irregularity of Crack Branching on the Fracture Toughness of Brittle Materials. International J Fracture 41: 267-274.
Xie H (1991) Fractal Nature on Damage Evaluation of Rock Materials. Chinese Journal of Rock Mechanics and Engineering 10: 697-704.
Xie H (1992) Fractals in Rock Mechanics. Balkema, Rotterdam.
Xie H, Bhaskar R, Li J (1991) Generation of Fractal Models for Fine Particle Characterization. Submitted to The Society of Mining Engineers.

The Fractal Distribution of Gold Deposits: Two Examples from the Zimbabwe Archaean Craton

Tom Blenkinsop
Department of Geology, University of Zimbabwe
PO Box MP 167, Mount Pleasant, Harare, Zimbabwe

Abstract. Gold deposits in two study areas in the Zimbabwe Archaean craton have fractal spatial distributions over a length scale from 2.5 to 25 km. The number of squares of side d necessary to cover every deposit is proportional to a power of the length of the square, and the number of deposits within a circle of radius r is proportional to a power of the radius. The fractal dimension, given by the exponent of the length scale in each method, is approximately 1. The fractal relations for both study areas are very similar. The distributions of deposits are interpreted as the result of hydrothermal mineralization by fractal fluid systems focussed in deformation zones.

Introduction

Description of the spatial distribution of gold deposits is one of the most basic requirements for both successful exploration and ore reserve evaluation. The more geologically interesting problems of ore genesis can be approached by combining a knowledge of deposit distributions with geological information. Accurate description of the distribution of ore deposits is therefore essential to both applied and theoretical aspects of ore geology.

Mandelbrot (1983) first suggested that mineral distribution in the earth could be considered in fractal terms. If ore deposits have a fractal distribution, they are clustered to some degree, and they can be treated as fractal dusts. Dimensionality analysis can discriminate clearly between a random (Poisson) distribution, which is not clustered, and fractal dusts. Furthermore, the distribution of fractal dusts can be described accurately and precisely by standard techniques of dimensionality analysis.

A comprehensive fractal analysis of ore deposit distributions was made by Carlson (1991), who studied 4775 precious metal deposits in the Basin and Range, western U.S.A., and demonstrated that clustering of deposits could be described very well by two fractal relationships over different length scales. This study pointed towards a possible explanation of these relationships by hydrothermal and plutonic processes of mineral transport and deposition.

The objectives of this chapter are to investigate whether the spatial distribution of gold deposits can be described by fractal relationships in a different geological setting, and, if so, to interpret these relationships by integrating geological knowledge about the deposits with the fractal relationships. Carlson's approach has been applied to two study areas in the Zimbabwe Archaean craton (Fig. 1), where most mineralization is late Archaean in age, compared to the Mesozoic age of mineralization in Carlson's study. A subsidiary objective is to compare two techniques of dimensionality analysis. The distributions of the gold deposits in the two study areas shown in Figures 2 and 3 clearly suggest non-random, clustered distributions, that might be treated as fractal dusts.

Sources of Data

Locations and gold production statistics for both study areas are found in the bulletins and accompanying 1:100,000 maps of the Geological Survey of Zimbabwe (Wilson 1964, 1968, Stidolph 1977). These bulletins contain a comprehensive list of all production recorded from the last years of last century up to 1965, as well as brief accounts of small mines and quite detailed descriptions of the geology of larger mines. These data have been supplemented by field mapping and structural investigations carried out in this study. The existence of such homogeneous and detailed databases was one reason for selecting the study areas.

Geological Background

Both study areas are centred on Archaean greenstone belts within the Zimbabwe Archaean craton.

The Masvingo-Mashava Area

The Masvingo-Mashava greenstone belt (Figs. 1, 2) contains mafic and ultramafic metavolcanics, banded iron formations and pelites, psammites, and carbonates of at least three ages, considered to be approximately 3,500 Ma ("Sebakwian"), 2,700 Ma ("Upper Bulawayan"), and 2,650 Ma ("Shamvaian"; Wilson 1979). The lithostratigraphic terminology of Wilson (1979) is followed here. A layered ultramafic intrusion, the Mashaba Igneous Complex, constitutes most of the greenstone belt to the east of Mashava. The Sebakwian greenstones are intruded by the Mushandike granite which has been dated at approximately 2,900 Ma (Moorbath et al. 1987), and lies in the central part of the study area.

The greenstones are surrounded to the east and north by granitoid gneisses of Sebakwian age, and to the southwest they are intruded by massive granites of the Chilimanzi suite dated at approximately 2,600 Ma (Taylor et al. 1991).

The major structure in the area is the Jenya-Mushandike Dislocation Zone, which probably had an early history of sinistral strike-slip, followed by late Archaean dextral strike-slip (Blenkinsop 1991, Tsomondo et al. 1994). A network of small strike-slip shear zones developed in the Mushandike granite during this movement (Blenkinsop et al. 1990). The base of the Mashaba Igneous Complex is a thrust that constitutes another regional structure. Other smaller structures include faults sub-parallel to the basal thrust of the Mashaba Igneous Complex, and NNE trending faults that cut the Mushandike granite and the Mashaba Igneous Complex.

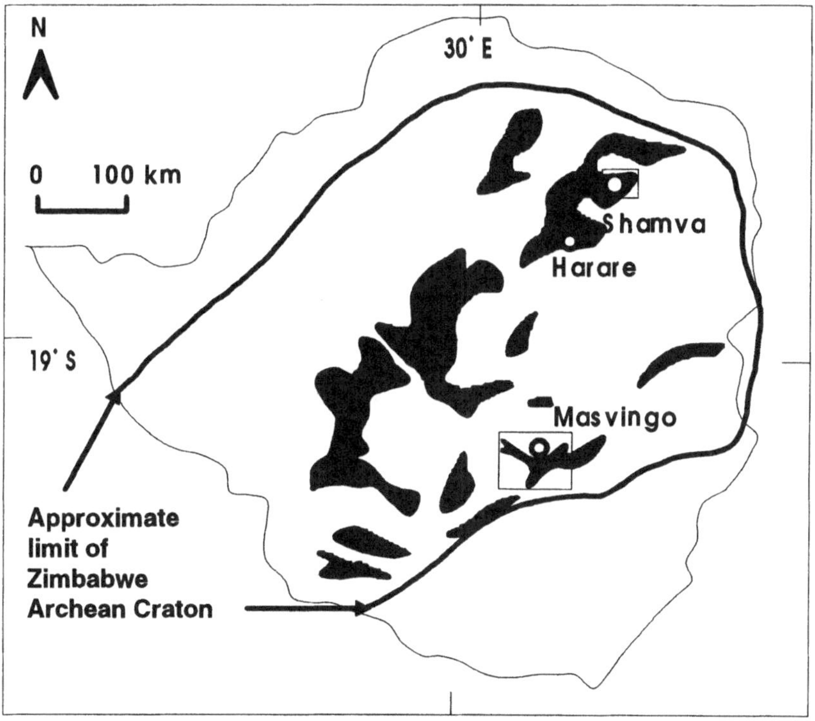

Fig. 1. Location of the study areas in the Zimbabwe Craton. Black areas are greenstone belts.

The Masvingo-Mashava area produced a total of just less than 7000 kg of gold during the study period; 147 mines could be located accurately in the study area and were used in the analysis (Fig. 2). Significant amounts of gold have been

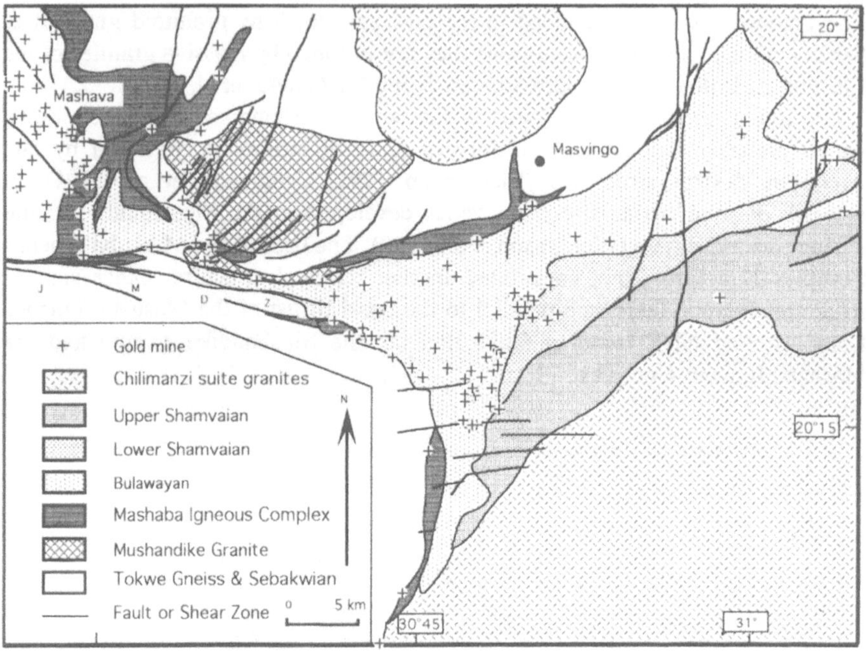

Fig. 2. Distribution of gold mines (+) and geology of the Masvingo-Mashava area. JMDZ - Jenya Mushandike Dislocation Zone.

produced from every single lithology in the area except the Chilimanzi granites. Large-scale associations between structure and mineralization are seen in the western part of the study area, where many mines lie in a broad linear trend parallel to the basal thrust of the Mashaba Igneous Complex and associated faults, along broad E-W trends parallel to the Jenya-Mushandike Dislocation Zone and on some of the strike-slip shear zones in the Mushandike granite. Localized deformation zones, fractures, and fold hinges clearly control mineralization on a mine scale. These observations have been interpreted to indicate that gold was deposited by fluid flow through structurally-controlled pathways (Foster 1985, Blenkinsop 1991).

The Shamva Area

The Shamva greenstone belt (Figs. 1, 3) consists of mafic and felsic metavolcanics with minor banded iron formations and carbonates of the Bulawayan, and metavolcanics and metasediments of the Shamvaian. The greenstone belt is surrounded by gneisses and granites of various ages, and intruded by a number of smaller granite stocks. The older Bulawayan rocks lie along the edges of the belt, surrounding the younger Shamvaian in the core, thus defining a major syncline on the scale of the whole belt. However, smaller scale folds or thrusts bring the Bulawayan to the surface as inliers within the Shamvaian (e.g. Tafuma Hill). Other major structures include a major normal shear zone along the southern margin of the belt, and a sinistral strike-slip shear zone near the central part of the Shamvaian. There are shear zones of uncertain kinematics along the northern margin of the belt and within the Bulawayan (Fig. 3, Jelsma et al. 1993). NNE-striking faults cross the axis of the greenstone belt.

The total production from the area was 56874 kg during the study period; 122 mines in the study area could be located accurately and used in the analysis (Fig. 3). As in the Masvingo-Mashava area, every major lithology has produced gold except the youngest granites. The largest mine in the belt (Shamva mine) lies on the Shamva shear zone near the centre of the Shamvaian outcrop (Jelsma et al. 1990). Gold production is also associated with the shear zones at the northern margin of the belt and in the Bulawayan. The notable concentration of mines in the southwest corner of the area lie on the Tafuma Hill inlier. On a small scale, structural controls similar to those described for the Masvingo-Mashava belt are evident in most mines.

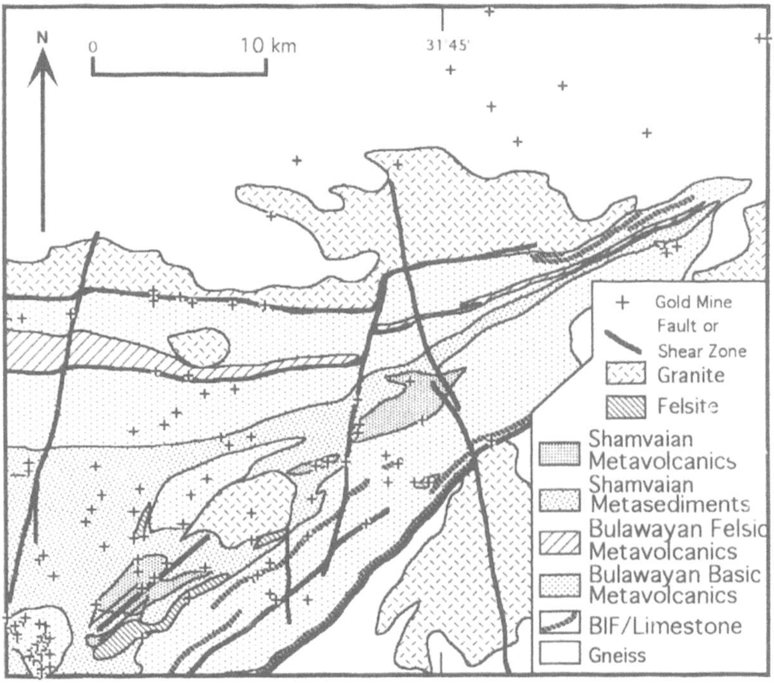

Fig. 3. Distribution of gold mines (+) and geology of the Shamva area.

Methods

Two methods of calculating the fractal dimension of the distribution of gold deposits have been applied in this study. In the method usually known as "box counting" (referred to as "square counting" here) the number of squares of side d necessary to cover all deposits is counted as a function of d, and the fractal dimension is D:

$$n(d) \propto d^{-D} \tag{1}$$

An alternative method of evaluating the fractal dimension is by counting the number of deposits N(r) in a circle of radius r, which are related as follows:

$$N(r) = C\,r^{D} \tag{2}$$

where C is a constant. Carlson (pers. comm. 1991) showed that the density of deposits, ρ(r), within a circle of radius r, and the density of deposits, d(r), at a distance r, could be derived from this equation to give the following relationships:

$$\rho(r) = C/\pi r^{D-2} \tag{3}$$

and

$$d(r) = D\,C/2\pi r^{D-2} = D/2\,\rho(r) \tag{4}$$

Carlson (1991) evaluated the fractal dimension by obtaining ρ(r) and then d (r) using equation 4 (the "density method"), but it is entirely equivalent and simpler to evaluate the fractal dimension from N(r) directly (equation 2). N(r) is evaluated for each deposit in turn and the results are averaged over all deposits to give the final relationship. This procedure has been applied to the data from both areas, and is referred to here as the "number-in-circle" method. D has been evaluated by standard linear regression and the error is given as the standard error of regression. In both the square counting and number-in-circle methods, the linear dimension (d or r) was varied from a minimum value larger than the measurement error (d = 300 m, r = 150 m) to a maximum value on the order of the dimensions of the study area (d = 29 km, r = 23 km).

Results

Graphs of n(d) vs d and N(r) vs r are shown in Figs. 4a and 4b respectively for both study areas over the total range d and r. The curves for both study areas show remarkably similar shapes in each figure. The square counting method especially shows a distinct change in fractal relationship at a value of d = 2.5 km.

There is clearly a better fractal relationship above d = 2.5 km. This is shown in more detail in Figs 5a and 5b where the data has been fitted by linear regression lines over the range d or r > 2.5 km for both methods.

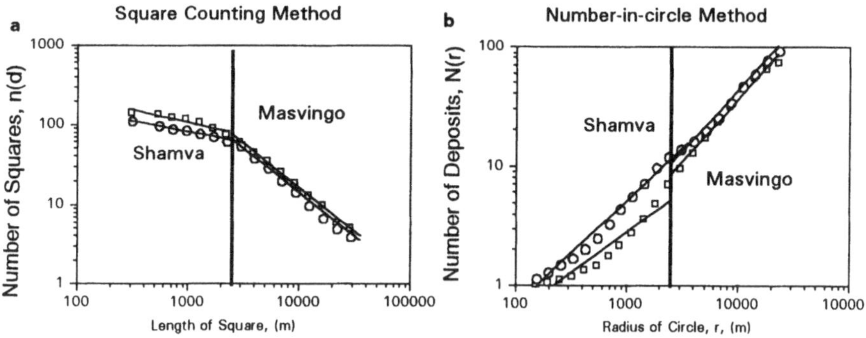

Fig. 4. a. Number of squares n(d) of length d (metres) necessary to cover all gold deposits as a function of d for the two study areas. Straight lines are derived from linear regression of the data below and above d = 2.5 km respectively with the parameters given in Table 1. **b.** Number of deposits N(r) within a circle of radius r (metres) as a function of r for the two study areas. Straight lines are derived from linear regression of the data below and above 2.5 km respectively with the parameters given in Table 2

Fractal dimensions, regression coefficients, and standard errors are given for the two methods in Tables 1 and 2 for the low range of d or r (< 2.5 km) and the high range of d or r (> 2.5 km). The tables also show Carlson's results from the Basin and Range study area. Economic implications of the results from the Masvingo-Mashava area are discussed in Blenkinsop (1994).

Table 1. Fractal dimension (D) of gold deposits by the square counting method for different ranges of length of square, d (km). R - correlation coefficient, E - standard error of regression, N - number of deposits. Data from this study (Masvingo and Shamva) are compared to results from Carlson (1991) in the Basin and Range.

Author	Range of d (km)	D	R	E	N
Masvingo (this study)	0.3 - 2.5	*0.32*	-0.947	0.0415	147
Shamva (this study)	0.3 - 2.5	*0.28*	-0.988	0.0169	122
Masvingo (this study)	2.5 - 29	*1.12*	-0.995	0.0379	147
Shamva (this study)	2.5 - 29	*1.18*	-0.999	0.0192	122
Basin &	1 - 15	*0.50*			4775
Range (Carlson 1991)	15 - 1000	*1.51*			4775

Table 2. Fractal dimension (D) of gold deposits by the number-in-circle and density methods for different ranges of circle radius, r (km). R - correlation coefficient, E - standard error of regression, N - number of deposits. Data from this study (Masvingo and Shamva) by the number-in-circle method are compared to results from Carlson (1991) in the Basin and Range by the density method.

Author	Range of r (km)	D	R	E	N
Masvingo (this study)	0.15 - 2.5	*0.69*	0.958	0.0595	147
Shamva (this study)	0.15 - 2.5	*0.87*	0.991	0.0347	122
Masvingo (this study)	2.5 - 22.5	*1.06*	0.994	0.0400	147
Shamva (this study)	2.5 - 22.5	*1.01*	0.997	0.0246	122
Basin &	1 - 15	*0.83*			4775
Range (Carlson 1991)	15 - 1000	*1.17*			4775

Discussion

Methods

The results show a number of differences between the square counting and number-in-circle methods. The square counting method gives lower values of D in the low range of d or r, and higher values of D in the high range of d or r, than the number-in-circle method, which has more consistent values of D over the whole range of length scale. This is because the former method is more sensitive to variations in fractal dimension with length scale, showing a distinct change in fractal relationship at d = 2.5 km. The number-in-circle method fits the entire range of data better with a single relationship. Both differences can be ascribed to the procedure of averaging N(r) for all deposits in the number-in-circle method.

The fractal dimensions for the high range of d or r given by the square counting method are greater than those given by the number-in-circle method. The error ranges of the fractal dimensions given by each method overlap for the Masvingo-Mashava area, but not for the Shamva area. The square counting method gives higher correlation coefficients and lower regression errors than the number-in-circle method.

These observations suggest that the square counting method is more useful than the number-in-circle method for detailed description of the fractal distribution over a range of length scales.

Geological Interpretation

Figures 4a and 4b show a distinct change in the fractal relationship at d or r = 2.5 km, which suggests that there is a higher density of deposits at small length scales, i.e. fewer boxes necessary to cover all deposits and more deposits in small circles than anticipated from the good fractal relationship for length scales above 2.5 km. This may be because exploration efforts are greater in the immediate vicinity of pre-existing mines. The true fractal relationship can only be established at larger length scales, and for a range of d or r of one order of magnitude. These are important limitations of the data that severely qualify the description of deposit distribution as "fractal".

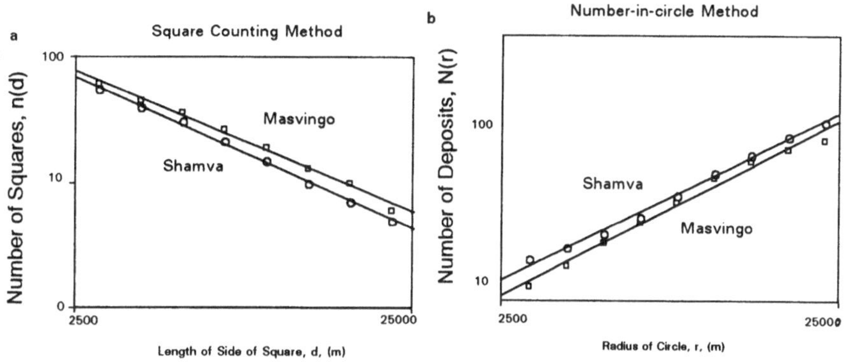

Fig. 5. a. Detail of part of Fig. 4a: Number of squares n(d) of length d (metres) necessary to cover all gold deposits as a function of d for the two study areas for the high range of d above 2.5 km. **b.** Detail of part of Fig. 4b: Number of deposits N(r) within a circle of radius r (metres) as a function of r for the two study areas for the high range of r above 2.5 km.

Nevertheless, within this length scale, the data give a very good power law relationship, which suggests that a single process could explain the spatial distribution of all the deposits. Furthermore, both study areas have the same fractal dimension within error for any one method, suggesting that the same process operated in both areas.

The geological evidence shows that there is no unique lithological control on mineralization (deposits are found in every lithology), but that the banded iron formations contain much higher concentrations of gold. There are clear structural controls on a large scale, and small-scale structural controls can be seen in most mines that have been described in detail. All these observations are consistent with the transport and deposition of gold in hydrothermal systems. Permeabilities of most undeformed rocks are too low to allow sufficient fluid flow to form significant deposits of gold (Thompson 1987). Fluid flow must have been localized in high permeability pathways in deformation zones. Fluid flow may have been concentrated in banded iron formations because they were more fractured during deformation than other less competent units, although Blenkinsop (1991) suggests that there may have been a primary syngenetic component of mineralization in the Sebakwian banded iron formations of the Masvingo-Mashava area.

Fractal spacing of deformation zones has been demonstrated for shear zones (Ord 1994) and for brittle fractures (Blenkinsop 1993, Kruhl 1993, Zhang & Sanderson 1994). The most simple explanation for the power law distribution of deposits is that they were formed in a fractal hydrothermal system with channelized flow in deformation zones.

Epigenetic mineralization at the end of greenstone belt evolution was suggested for this area by Wilson (1968). It has been proposed that late- to post-tectonic mineralization is a general pattern in greenschist-facies greenstone belt gold deposits on the basis of evidence from direct dating of mineralization (Robert et al. 1990). This work confirms that gold mineralization in these two greenstone belts is due to a common geological process, operating late in the evolution of the greenstone belts.

The fractal dimensions of the distribution of Archaean mineralization reported here are similar to results from the study of Mesozoic mineralization reported by Carlson (1991). Carlson identified a change in fractal relationship at a length scale of 15 km; the fractal dimension below this value is less than the fractal dimension for d or r between 2.5 km and 25 km in this study, but Carlson's dimensions for the length scale above 15 km are greater than the results reported here. The dimension of both studies over the length scale used here are therefore comparable. Carlson also suggested "fractal hydrothermal and fracture systems" as the control on mineralization for the length scale between 1 and 15 km. A fractal geometry with fractal dimensions of approximately 1 may be a fundamental property of precious-metal hydrothermal mineralizing systems.

This is only a preliminary study: a more complete analysis will incorporate quantities of gold as well as spatial distribution of deposits. Multifractals offer a promising approach to this problem (Evertsz, pers. comm. 1993). Future work should also endeavour to extend the range of length scales to several orders of magnitude before a true "fractal" distribution of gold deposits can be established.

Acknowledgements. Helpful discussions with Perter Monkhouse and with Carl Evertsz and Don Turcotte at the International Symposium on Fractals and Dynamic Systems in Geoscience are gratefully acknowledged.

References

Blenkinsop TG (1991) Controls on Archaean Gold Mineralization in the Mashava Area. In: African Mining '91, Institute of Mining and Metallurgy, Elsevier Applied Science, London and New York, pp. 320-336.

Blenkinsop TG (1993) Fracture spacing distributions in rock. Abstract, International Symposium on Fractals and Dynamic Systems in Geoscience, Johann Wolfgang Goethe-University, Frankfurt, Germany: 6-7.

Blenkinsop TG (1994) Fractal measures for size and spatial distributions of gold mines: Economic applications. Geological Society of Zimbabwe Special Publication No. 3, Proceedings of the Conference on Sub-Saharan Economic Geology, Balkema, Rotterdam, in press.

Blenkinsop TG, Dhilwayo J, Muranda SC (1990) Intracratonic shearing on shear zones of the Mushandike granite, Zimbabwe. Proceedings of the Second Symposium of Science and Technology, Research Council of Zimbabwe IIA: 396-421.

Carlson C A (1991) Spatial Distribution of Ore Deposits. Geology 19: 111-114.

Foster RP (1985) Major controls of Archaean gold mineralization in Zimbabwe. Transactions of the Geological Society of South Africa 88: 109-133.

Jelsma HA, Tomschi HP, Touret JLR, Kramers JD (1990) Gold mineralization at Shamva Mine, NE-Zimbabwe; An integrated structural magmatic control. Abstract volume, NUNA research conference on Greenstone Gold and Crustal Evolution, Geological Association of Canada/Society of Economic Geologists, Val d'Or, Quebec, Canada: 58-59.

Jelsma HA, Van der Beek PA, Vinyu ML (1993) Tectonic evolution of the Bindura-Shamva greenstone belt (northern Zimbabwe): progressive deformation around diapiric batholiths. J Struct Geol 15: 163-176.

Kruhl J H (1994) The Formation of Extensional Veins: An Application of the Cantor-Dust Model. In: Kruhl JH (ed) Fractals and Dynamic Systems in Geoscience. Springer, Berlin Heidelberg New York, pp 95-104 (this volume).

Mandelbrot BB (1983) The fractal geometry of nature. Freeman, New York.

Moorbath S, Taylor PN, Orpen JL, Treloar P, Wilson JF (1987) First direct radiometric dating of Archaean stromatolitic limestone. Nature 326: 865-867.

Ord A (1994) The Fractal Geometry of Patterned Structures in Numerical Models of Rock deformation. In: Kruhl JH (ed) Fractals and Dynamic Systems in Geoscience. Springer, Berlin Heidelberg New York, pp 131-155 (this volume).

Robert F, Phillips GN, Kesler SE (1990) Introduction. Abstract volume, NUNA research conference on Greenstone Gold and Crustal Evolution, Geological Association of Canada/Society of Economic Geologists, Val d'Or, Quebec, Canada: 6-10.

Stidolph PA (1977) The geology of the country around Shamva. Rhodesia Geological Survey Bulletin 78.

Taylor PN, Kramers JD, Moorbath S. Wilson JF, Orpen JL, Martin A (1991). Pb/Pb, Sm-Nd and Rb-Sr geochronology in the Archaean craton of Zimbabwe. Chemical Geology (Isotope Geoscience) 87: 175-96.

Thompson AB (1987) Some aspects of fluid motion during metamorphism. J Geol Soc 144: 309-312.

Tsomondo JM, Blenkinsop TG, Mandoreba P (1994) Strike-slip tectonics in south-central Zimbabwe; an intracratonic response to late Archaean Limpopo orogeny? J Geol Soc, in press.

Wilson JF (1964) The geology of the country around Fort Victoria. Bulletin of the Rhodesian Geological Survey 58.

Wilson JF (1968) The geology of the country around Mashaba. Bulletin of the Rhodesian Geological Survey 68.

Wilson JF (1979) A Preliminary Reappraisal of the Rhodesian Basement Complex. Special Publication of the Geological Society of South Africa 5: 1-23.

Zhang X, Sanderson DJ (1994) Fractal Structure and Deformation of Fractured Rock Masses. In: Kruhl JH (ed) Fractals and Dynamic Systems in Geoscience. Springer, Berlin Heidelberg New York, pp 37-52 (this volume).

Self-Organization of Mineral Fabrics

Karl-Heinz Jacob[1], Sabine Dietrich[1] & Hans-Jürgen Krug[2]
[1] Technische Universität Berlin, Institut für Angewandte Geophysik, Petrologie und Lagerstättenforschung
Ernst-Reuter-Platz 1, BH 4, D-10587 Berlin, Germany
[2] Technische Universität Berlin, Institut für Theoretische Physik, Arbeitsgruppe „Dissipative Strukturen"
Rudower Chaussee 5, Geb. 2.14, D-12489 Berlin, Germany

Abstract. Self-organization phenomena leading to the formation of patterned mineral fabrics are discussed in connection with electric field effects frequently occurring in the lithosphere. The different internal sources of electric potentials which may intensify ionic fluxes over long time scales are compiled. Model experiments with electrolysis in quartz sand basins between iron electrodes lead to the formation of Liesegang-like precipitate bands of iron hydroxides. Amplification of inhomogenities in the electric field or the capillary transport results in undulatory shapes of precipitation bands. Local breakthrough phenomena of ionic transport are observed leading to "boudinage" patterns along a horizontal precipitation band or "breccia"-like fabrics on top of vertical breakthroughs. Ripening effects within the primarily formed bands due to competitive particle growth lead to speckled or nodular patterns.

Several morphological similarities between these experimental findings and mineral fabrics are pointed out. Furthermore, the role of other non-electric effects, like the presence of diffusion barriers and capillar front instabilities ("Runge" pictures) in the evolution of geological patterns, is discussed.

Introduction

The formation of patterned mineral fabrics is caused by the interaction of driving forces, like the formation of any ordered structure in nature, via self-organization. In spite of the fact that gravitation is traditionally considered to be the fundamental order-creating force in geology, we must take into account further physical forces involved in reaction-transport feedbacks giving rise to in-situ structure formation (see: Ortoleva et al. 1987, Chen et al. 1990, Ortoleva et al. 1990):

1. *Reactive-infiltration instability*. This kind of instability may occur when a porous rock bearing a soluble mineral is infiltrated by a constant stream of reactive water. Small local differences in porosity give rise to the evolution of fingering or scalloping of the reaction-dissolution front.

2. *The supersaturation-nucleation-depletion cycle*. This is the basic mechanism for most of the observed isothermal and isobaric periodic precipitation processes

first reported by Liesegang (1896). In its simplest version, a precipitate is formed after the concentration product of a hardly soluble component attains supersaturation level. By the fast generation of nuclei and their subsequent growth the surrounding liquid concentration is rapidly depleted, thus constraining the precipitation process to separated rings or parallel bands.

3. *Autocatalytic particle growth (Ostwald ripening)*. The effect of competitive particle growth (CPG) may give rise to a further differentiation of Liesegang precipitation bands and is also capable of generating patterns from a continuous layer of precipitate.

4. *Stress-texture-solubility instability*. Due to an increasing solubility of minerals under stress, precipitation processes are fed back to their own stress-generating growth of grain particles. This feedback may also give rise to periodic pattern formation.

Furthermore, gravitation causes not only sedimentation processes, but may also lead to hydrodynamical instabilities resulting in "frozen-in" convection patterns observed in magma intrusions (Mc Birney & Noyes 1979).

In this chapter we want to treat Liesegang-like pattern formation due to a supersaturation-nucleation-depletion cycle and competitive particle growth under the influence of electric potentials. Besides effects on nucleation itself, the main role of electric potentials is here considered as intensifying ionic transport processes and enabling them to build up patterns over longer spatial scales than only by diffusion. To do this, electrolysis experiments were carried out to simulate qualitatively the ionar cross-flux processes within the lithosphere. The rhythmic precipitation patterns of metal hydroxides thus produced in quartz sand basins will also exhibit front instabilities like undulatory shapes and electrical "breakthroughs". Morphological similarities between such model patterns and natural mineral fabrics will be compiled to underline the role of internal electric fields in the lithosphere, which arise from many different sources.

Sources of Electric Potentials in the Lithosphere

In general, self-organization phenomena occur in open systems which are far from being in thermodynamic equilibrium. Continuous input or flux of energy or matter in a geological system prevents reaching a thermodynamic equilibrium and causes a continuous change of its order if a certain threshold of input flux is exceeded. This means that a chemically or physically produced geological body and its primary fabric will be altered as long as the flux of energy or matter persists.

In the laboratory experiments discussed below, an open system was maintained by a constant electric current delivering a continuous input of matter due to the continuous dissolution of the metallic anode.

The experiments are based on the actual knowledge of natural processes in the lithosphere, that are capable of generating an electrical potential. Klein (1990) investigated these processes and grouped them into the following six categories:

- Electrochemical potentials (Eh/pH)
- Tensor-induced potentials (pressure/temperature gradients)
- Potentials as a function of phase transformation (phase boundary)
- Potentials produced by induction (polarization)
- Electrokinetic potentials (electrophorensis/electro-osmosis)
- Electropotentials produced by radioactivity.

All factors are linked by linear and non-linear relations. According to Jacob (1988) and Klein (1990), the effects and influence of electric potential differences in nature are much underestimated in the geosciences, although numerous authors (Khayretdinov et al. 1987) have drawn attention to their immense significance over the last few decades.

Previous work on electric field effects on nonlinear waves was carried out by Feeney et al. (1981), Larter (1982), Feeney et al. (1983), Ortoleva (1979, 1981, 1993) comprehending bioelectric processes, chemical waves in the Belousov-Zhabotinsky-reaction and periodic precipitation processes. The main role of electric fields there was seen in altering the detailed physicochemistry of the wave process itself (e.g. leading to wave bifurcations), but not primarily in the intensifying of ionar transport processes. The latter phenomenon was considered to take place more frequently in porous media pervaded by aqueous flow (see: Sultan et al. 1990, Ortoleva et al. 1986) additionally giving rise to percolation instabilities.

One-dimensional electrolytic cross-flux experiments were described theoretically by Feeney et al. (1983).

Our experiments follow the same idea of ionar cross-flux, but their furthergoing aim is the investigation of two- and three-dimensional precipitation pattern formation occurring in greater basins involving undulatory instabilities of bands and also breakthrough phenomena.

Dissipative Structures Produced by Electrolysis Experiments

Fine quartz sand, as a homogeneous and non-structured medium, was placed between horizontal iron electrodes in cylindrical or square cross-section test vessels 15 to 20 cm across (Fig. 1). The pore space was occupied by sea water. A direct electric potential of 1.0 to 1.5 volts was applied; a potential of this order might be present also in many natural processes. In the test vessel, an electrolytic dissociation of sea water took place with the formation of two zones:

- the anode zone characterized by acidic ions (pH 2.0-2.5)
- the cathode zone characterized by alkaline ions (pH 10.5-11.5).

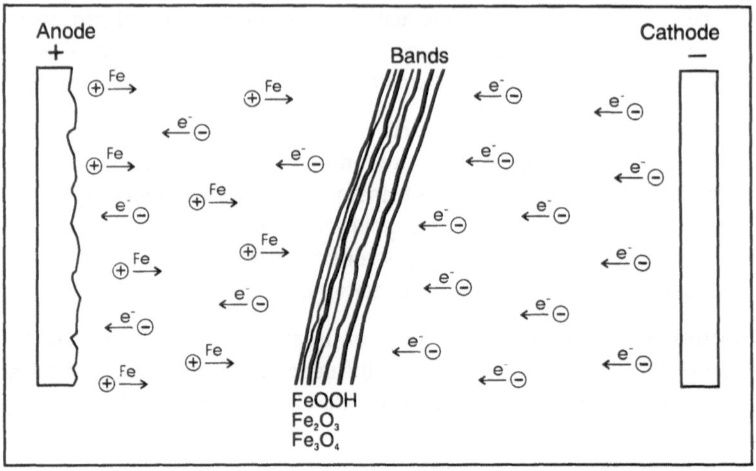

Fig. 1. Schematic arrangement of electrolysis experiments with vertical iron electrodes and wet quartz sand matrix.

As a consequence of the potential difference, a sharp boundary zone is developed in which anions and cations encounter and form a boundary zone at which a sudden change in pH from 2.5 up to 8 occurs. After reaching supersaturation conditions in this boundary zone, spontaneous precipitation of insoluble metal hydroxides and metal chlorides occurs at the isoelectrical point.

The following structures were observed, which result from the electrochemical reactions:

- Mainly rhythmic banding on planes at which pH changes abruptly
- Thin bands similar to secondary structures of Liesegang rings
- Undulatory and fold-like diagenetic banding
- Micro-diapirs
- Broken and/or vertically displaced precipitation horizons - ("boudinage" or "breccia")
- Discordant banding and patterns resembling "sedifluction"
- Formation of rosettes (cockades) of varying dimensions
- Fractional structures, such as dendrites or cauliflower structures
- Patchy and spotty patterns according to Ostwald ripening.

Experimentally produced structures as shown in the following figures are comparable phenomenologically to many natural structures and give an impression of the great variety of structure-forming forces (Jacob et al. 1992, Jacob & Zimmerle 1993).

The primary precipitation band created by electrolysis may be later accompanied with further parallel bands following Ostwald's classical supersaturation theory. Even within the primary band, a secondary structure can arise due to Ostwald ripening. The final form of the bands created will depend on the conductivity conditions within the diffusion matrix. The principle influence of diffusion barriers is demonstrated in the next pictures.

The first series of figures (Figs. 2a-c) shows vertically displaced structures which greatly resemble each other. A rhythmic bedding of sphalerite-galena ore combined with fine-sand tuffs is given in Fig. 2a. It is traditionally explained to be of sedimentary origin; the visible sharp displacement was hitherto explained mechanically through "subaqueous gliding". Fig. 2b shows the result of a classical Liesegang experiment 48 hours after silver nitrate application. Due to the separation of diffusion paths within the gel-containing test-tube two independent sequences of precipitation patterns could be established. Besides the separation within the gel, no further mechanical deviations were applied to the diffusion matrix. Different wavelengths in both series are due to unequal volumes of silver nitrate solution applied onto the tip of the gel.

The multiply displaced texture of "Pietra Paesina" or landscape marble (Fig. 2c) was in the past interpreted as the result of an external mechanical fracture. Pending proof to the contrary, Civitelli et al. (1970) were able to show that the visible bands are spreading over a homogeneous matrix of limestone only subject to vertical fissures later filled up with calcite impermeable to diffusion. The process of calcite intrusion into the fissures took place prior to the later

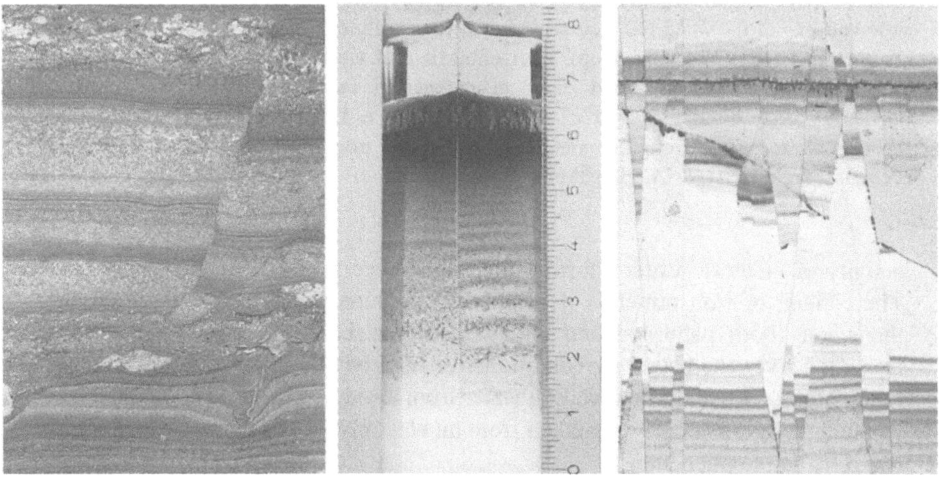

a b c

Fig. 2a-c. a. Rhythmic bedding of sphalerite-galena-ore of traditionally supposed sedimentary origin combined with fine-sand tuffs from Maccan mine, Peru. Height of section c 10 cm. **b.** Classical Liesegang experiment performed in a gel-containing test-tube with different diffusion paths separated by an impermeable sheet of polyethylene. Picture was taken 48 hours after start of the experiment. Initial recipe: 3.4 10-3 m K_2CrO_7 in gelatine, 0.5 m $AgNO_3$ in aqueous solution applied onto the tip of the test tube chambers. **c.** A piece of "Pietra Paesina" fine-grained limestone of Cretaceous-Eocene age. Height of section c 10 cm.

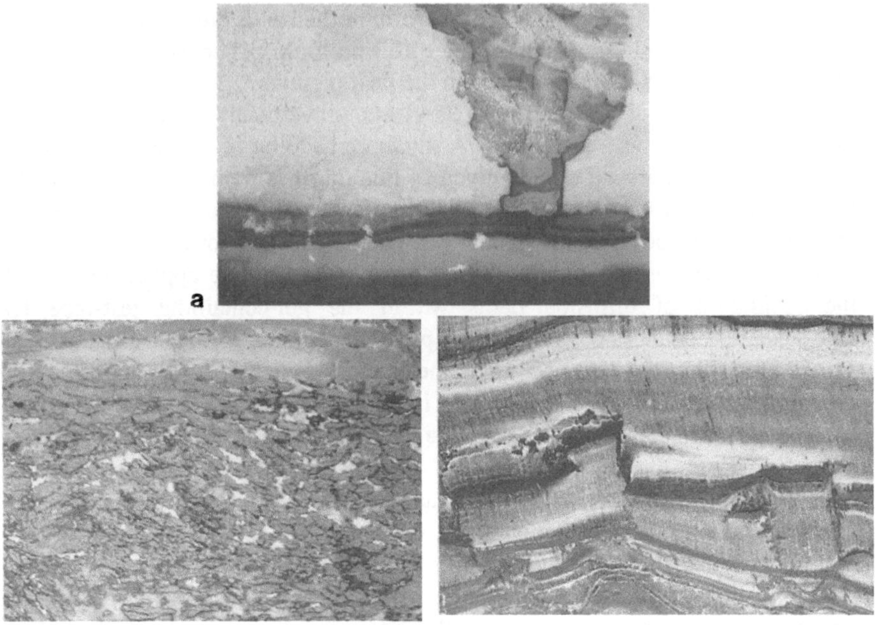

Fig. 3a-c. a. Experimentally produced sedimentary structure under the influence of a low-voltage (1.0 V) direct current with two different breakthrough phenomena: "boudinage" (left); "breccia" (top). Horizontal field of view c 20 cm. **b.** "Glide breccia" and "boudinage" causing load casts of compacted calcareous mud. Carboniferous formation. Saar district. Height of section c 10 cm. **c.** Displaced magnetite precipitate horizon in a sample from the metamorphic banded iron formation. Precambrian age. Postmasburg district, RSA. Height of section c 10 cm.

activity of meteoric water causing alteration processes between Fe^{++} and Fe^{+++}. The content of iron minerals itself thereby remains nearly constant throughout the whole (both patterned and unpatterned) matrix. Thus, the vertical calcite fissures deliver the boundary conditions for horizontally separated diffusion and redox processes caused by invading water from above.

The next figure (Fig. 3a) is taken from an electrolysis experiment which shows two remarkable phenomena.

1. Formation of a horizontal horizon of precipitate which is multiply perforated resembling the so-called "boudinage". Boudinage is an important term coined in geoscience denoting "a structure common in strongly deformed sedimentary and metamorphic rocks, in which an original continuous competent layer or bed between less competent layers has been stretched, thinned, and broken at regular intervals into bodies resembling boudins or sausages, elongated parallel to the fold axes." (Glossary of Geology p. 77)

2. A vertical breakthrough of the horizontal precipitation front with the subsequent formation of further horizons like "breccia".

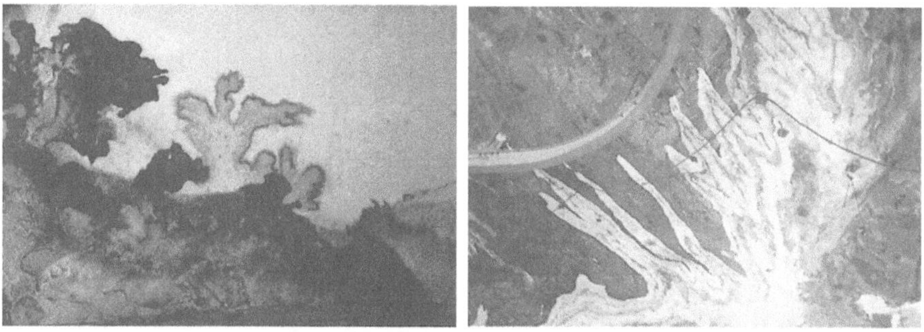

b

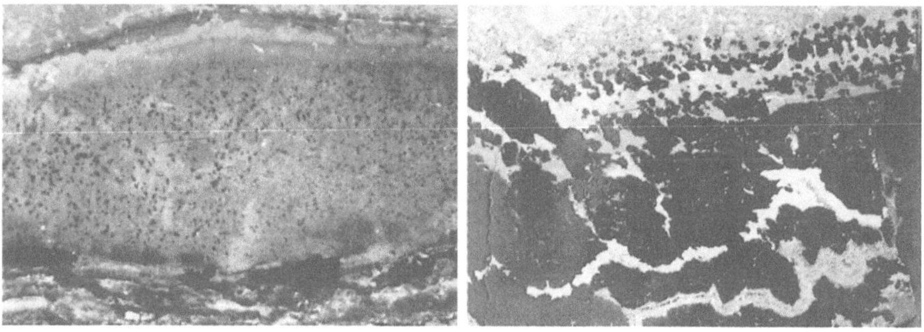

a b

Fig. 4a, b. a. Experimentally produced precipitation fabric of iron minerals after four weeks of constant low-voltage (1.0 V) electrolysis according to Fig. 1. Height of section c 6 cm. **b.** Secondary epsomite mineralization into kieseritic potassium salt layer. Zechstein formation, Werra district (Germany). Horizontal field of view c 10 m.

Fig. 5a, b. a. Experimentally produced coarse-grained sulphidic ore mineralization from formerly very fine-grained ore concentrate in the vicinity of the anode after two months of electrolysis. Height of section c 6 cm. **b.** Depositional lamination of idiomorphic galena (dark) and calcarenite (light). Wetterstein limestone. Bleiberg mine (Austria): Height of section c 6 cm.

Figs. 3b-c show mineral samples with similar faces with respect to boudinages and breakthrough phenomena which were hitherto still explained by mechanical brecciation or gliding of formerly continuous bands.

Scalloping structures both from electrolysis experiments and - on a larger scale - in a salt mine are to be seen in Figs. 4a-b. The undulated shape of the experimental pattern is due to a percolation instability occurring in the ionar transport driven by the low-voltage potential. A classical example of percolation instability is the so-called Runge pictures occurring in sole capillary transport of colloid matter (Runge 1850). The same kind of instability is expected to be responsible for the pattern in Fig. 4b, whereas the traditional explanations are based on mechanical, plastic deformations of the whole matrix (Halokinesis) or on a viscous flow process therein.

The grained galena structure in Fig. 5b is called idiomorphic galena. The reason for this grained structure was still assumed to be an accumulative

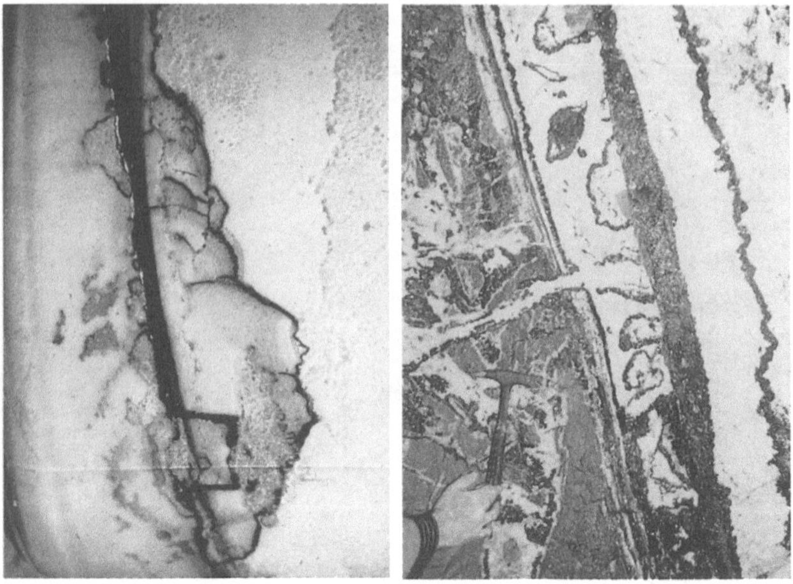

a b

Fig. 6a,b. a. Result of electrolysis experiment in a small test vessel similar to Fig. 1 (iron mineralization) exhibiting secondary horizons spontaneously arisen from the primary horizon on the right. Duration of the experiment: four weeks. Height of section c 10 cm. **b.** Hydrothermal vein mineralization exhibiting sphalerite (dark) and calcite (light). Bad Grund mine, Harz mountains (Germany). Horizontal field of view c 2 m.

crystallization during diagenesis without further detailed explanation given in geology. More probably nucleation and subsequent Ostwald ripening of precipitate took place as performed in our electrolysis experiment (Fig. 5a). This speckled pattern there evolved continuously in the space between precipitation bands.

A hydrothermal vein structure of sulphidic zinc ore in calcite is shown in Fig. 6b. Both components were created by precipitation from hot-water solution streamed up through a cleft formerly broken up mechanically. The process of precipitation there is assumed to be controlled by gradients of temperature, pressure and salinity of the solution. The ideal vein structure is twofold parallely banded, which is commonly taken as a proof for the assumption of upstreaming solutions. However, this ideal symmetric form of veins occurs only in exceptional cases.

More frequently, non-symmetric shapes with breakthrough phenomena, as shown in Fig. 6b, are encountered. Such phenomena unconstrainedly appear also in electrolysis experiments (Fig. 6a) suggesting that such secondary structure-forming processes should be taken into account in vein mineralization theories.

Conclusions

The experimentally created mineral fabrics demonstrate the capability of electric fields both to create periodic patterns by supporting the transport of ionar species and further to modify them with respect to undulatory or breakthrough phenomena.

Because electric potentials are also frequently present in the lithosphere and have often higher voltage values compared to our experiments, it is expected that these natural potentials bear a structure-creating power hitherto underestimated.

The detailed processes of nucleation and precipitation involved in some types of geological structure formation had already been discussed by Liesegang (1913, 1924) and Ostwald (1897) several decades ago. Modern theories - also involving electric field effects - were developed in the last decade (Ortoleva 1993), but the concept of self-organization has until now been accepted only to a small extent in the geological community. Further interdisciplinary efforts going into the detailed geological phenomena are still needed to compensate this lack of acceptance. Following Nicolis & Prigogine (1987, p. 48), several chapters of the geosciences should be rewritten, if the concept of self-organization were more broadly applied.

Acknowledgements. The authors greatly appreciate the many thoughtful suggestions of the referees. Research was supported by the Deutsche Forschungsgemeinschaft (DFG Kennzeichen Ja 262).

References

Chen W, Ghaith A, Park A, Ortoleva P (1990) Diagenesis through Coupled Processes: Modeling Approach, Self-Organization and Implications for Exploration. In: Meshri I and Ortoleva P (eds.) Prediction of Reservoir Quality Through Chemical Modeling, AAPG Memoir 49, pp 103-130.

Civitelli G, Funiciello R, Lombardi S (1970) Alcune considerazioni sulla genesi della « Pietra Paesina ». Geol Rom IX: 195-204.

Feeney R, Schmidt SL, Ortoleva P (1981) Experiments on Electrical Field-BZ Chemical Wave Interactions: Annihilation and the Crescent Wave. Physica 2D: 536-544.

Feeney R, Schmidt SL, Strickholm P, Chadam J, Ortoleva P (1983) Periodic Precipitation and Coarsening Waves: Applications of the Competitive Particle Growth Model. Jour Chem Phys 78: 1293-1311.

Glossary of Geology (1980), Second Edition, American Geological Institute, Falls Church, Virginia.

Jacob K-H (1988) Künstliche Bänderungen. Wissenschaftsmagazin d. TU Berlin: 75-78.

Jacob K-H, Zimmerle W (1993) Some diagenetic phenomena in carboniferous sedimentary rocks seen as rhythmic structures produced by energy dissipation in open systems. Zbl Geol Paleont 5: 437-459.

Jacob K-H, Krug H-J, Dietrich S (1992) Lagerstättenbildung durch Energiepotentiale in der Lithosphäre. Erzmetall 10: 505-513.

Khayretdinov IA, Kononenko AG, Belikova GI, Kononenko NP (1987) Structures, textures and shapes of geologic bodies of electrogeochemical origin. Int Geol Rev 29: 1073-1083.

Klein W (1990) Dissipative Gefügebildungen in Lockersedimenten durch Einwirkung elektrischer Felder; ein Beitrag zur Bedeutung des thermodynamischen Nichtgleichgewichts für lagerstättengenetische Vorgänge. Diss TU Berlin, Berlin.

Kuhnert L, Niedersen U (1987) Selbstorganisation chemischer Strukturen. Ostwalds Klassiker d exakten Wiss, Leipzig.

Larter R (1982) A Study of Instability to Electrical Symmetry Breaking in Unicellular Systems. Jour Theor Biol 96: 175-200.

Liesegang RE (1896) Liesegangs photographisches Archiv 801, Heft 21, Düsseldorf.

Liesegang RE (1913) Geologische Diffusionen. Steinkopf, Dresden Leipzig.

Liesegang RE (1924): Chemische Reaktionen in Gallerten. 2. umgearb. Aufl., Steinkopf, Dresden Leipzig.

Mc Birney AR, Noyes RM (1979) Crystallization and Layering of the Skaergaard Intrusion. J Petrology 20: 487-554.

Nicolis G, Prigogine I (1987) Die Erforschung des Komplexen. Piper, München.

Ortoleva P (1979) The Multifractal Family of the Nonlinear: Waves and Fields, Center Dynamics, Catastrophes, Rock Bands and Precipitation Patterns. In: Pacault A and Vidal C (eds) Synergetics Far from Equilibrium. Springer, Berlin Heidelberg New York, pp 114-127.

Ortoleva P (1981) Developmental Bioelectricity, Biological Effects of Nonionizing Radiation. In: ACS Symposium Series 157, Houston, TX, Am Chem Soc, Washington, D.C., pp 163-212.

Ortoleva P (1993): Geochemical Self-Organization. Oxford Univ Press, Oxford.

Ortoleva P, Auchmuty G, Chadam J., Hettmer J, Merino E, Moore C, Ripley E (1986) Redox Front Propagation and Banding Modalities. Physica 19D: 334-354.

Ortoleva P, Merino E, Chadam J, Moore CH (1987) Geochemical Self-Organization I: Reaction-Transport Feedbacks and Modeling Approach. Am J Sci 287: 979-1007.

Ortoleva P, Hallet B, McBirney A, Meshri I, Reeder R, Williams P, eds. (1990) Self-Organization in Geological Systems: Proceedings of a Workshop held 26-30 June 1988, University of California Santa Barbara. In: Earth Science Reviews 29. Elsevier, Amsterdam.

Ostwald W (1897) Lehrbuch der allgemeinen Chemie, 2. umgearb. Aufl., 2. Bd, 2. Tl, Verwandtschaftslehre. Engelmann, Leipzig, pp 777-780.

Runge FF (1850) Zur Farben-Chemie; Musterbilder für Freunde des Schönen und zum Gebrauch für Zeichner, Maler, Verzierer und Zeugdrucker. Mittler & Sohn, Berlin.

Runge FF (1855) Der Bildungstrieb der Stoffe, veranschaulicht in selbständig gewachsenen Bildern. - Oranienburg (Selbstverlag).

Sultan R, Ortoleva P , De Pasquale F, Tartaglia P (1990) Bifurcation of the Ostwald - Liesegang Supersaturation - Nucleation - Depletion Cycle. Earth-Sci Rev 29: 163-173.

The Formation and Fragmentation of Periodic Bands Through Precipitation and Ostwald Ripening

Hans-Jürgen Krug[1], Karl-Heinz Jacob[2] & Sabine Dietrich[2]
[1] Technische Universität Berlin, Intitut für Theoretische Physik, Arbeitsgruppe „Dissipative Strukturen"
Rudower Chaussee 5, Geb. 2.14, D-12489 Berlin, Germany
[2] Technische Universität Berlin, Institut für Angewandte Geophysik, Petrologie und Lagerstättenforschung
Ernst-Reuter-Platz 1, BH 4, D-10587 Berlin, Germany

Abstract. The formation of periodic band structures involving certain structural defects (lateral interruptions and branching of bands) through selforganization in colloidal media is demonstrated both by classical Liesegang experiments and by numerical simulation of a competitive particle growth model. Fragmentation or branching of bands can be explained as consequence of an instability during the precipitation process itself (Ostwald ripening) by excluding of any external mechanical influences. The detailed morphological similarity of these patterns with periodic bands in some mineral samples (zebra rock) suggests that the mineral pattern formation was in the same manner governed by self-organized precipitation processes.

Introduction

Periodic bands are omnipresent in the lithosphere. Their genesis was traditionally explained by sedimentation of sequentially infused matter. However, the sedimentary explanation does not suffice, if we encounter spherical concentric patterns or periodic patterns superposed on a chemically and structurally homogeneous background.

In the last decade, the alternative way of explanation by in-situ self-organization of structures in far-from-equilibrium systems has thus been discussed also in the geosciences (Ortoleva et al. 1987, Ortoleva 1994). Besides patterns formed by hydrodynamical convection (Mc Birney & Noyes 1979) or by mechanochemical interaction (Ortoleva et al. 1982b, Dewers & Ortoleva 1989, 1990a, b) periodic precipitation patterns created under isothermal and isobaric conditions are relevant for geological structure formation.

The discovery of periodic precipitation patterns by Liesegang (1896) initiated a huge series of subsequent experiments in colloid chemistry most of which were stimulated by the morphological similarity of these precipitation patterns to periodically occurring bands in geology and also in living matter. In two monographs entitled "Geologische Diffusionen" (Liesegang 1913) and "Die

Achate" (Liesegang 1915), Liesegang pointed out the possibility of periodic pattern formation due to precipitation processes in geological matter.

Older theories of Liesegang ring formation which still are valid and have been further developed since that time are Ostwald's supersaturation theory (Ostwald 1897a, b) and the principle of Ostwald ripening (Ostwald 1900). Further theories dedicated to other special aspects of the complex process of ring formation were hitherto only verbally formulated, e.g the theory of electrolyte flocculation (Freundlich & Schucht 1913), the adsorption theory (Bradford 1922), the diffusion-wave theory (Ostwald 1925), and the coagulation theory (Dhar & Chatterji 1925). Detailed reviews of these older theories and experimental findings are given by Hedges (1932), Stern (1954, 1967), and Kuhnert & Niedersen (1987).

Contemporary theories on Liesegang ring formation can be classified into two main groups:

1. Prenucleation theories (based on Ostwald's supersaturation theory, see: Prager 1956, Smith 1984, Dee 1986, Ortoleva et al. 1986, Sultan et al. 1990, Ortoleva 1994), and
2. Postnucleation theories (based on Ostwald ripening, see: Flicker & Ross 1974, Ortoleva et al. 1982a, Feeney et al. 1983, Ortoleva 1984, Sultan & Ortoleva 1993, Ortoleva 1994).

Detailed experimental investigations show that the phenomenon of ring formation cannot be explained exclusively by nucleation and depletion; their description must be accompanied by postnucleation effects (Feinn et al. 1978; Kai et al. 1982; Feeney et al. 1983, Kai & Müller 1985). The latter are also considered to be responsible for secondary band structures (Müller et al. 1982), formation of spirals and speckled patterns (Ortoleva 1984, 1987).

Meandric deformations of precipitation fronts (viscous fingering) can be explained by means of percolation instability (Ortoleva et al. 1986, Chadam 1987, Chadam et al. 1990, Chen & Ortoleva 1990a, 1990b) which is responsible for many geological curiosities (e.g. meandering gypsum or ore deposits in sandstones) or for the well-known phenomenon of Runge pictures (Kuhnert & Niedersen 1987), but not for discrete fragmentation of bands or pattern bifurcations (branchings) in two spatial dimensions. Besides flow-driven porous systems, meandered precipitation patterns and break-through instabilities may also be caused by electric fields present in the lithosphere, as recently demonstrated in ad-hoc model experiments (Jacob & Zimmerle 1990, Klein 1990, Jacob et al. 1992, 1994).

With respect to bifurcation of bands and fragmentation phenomena, we shall treat in this chapter two-dimensional instabilities revealed in a competitive particle growth model. Comparison of these results will be made with both mineral patterns (zebra spar) and classical Liesegang rings.

Results of Liesegang Experiments

The classical Liesegang experiment consists in diffusion of silver nitrate placed as a drop of aqueous solution onto a gel matrix containing potassium bichromate. The chemical reaction taking place is:

$$2(Ag^+ + NO_3^-) + 2K^+ + Cr_2O_7^{2-} \longrightarrow Ag_2Cr_2O_7 + 2(K^+ + NO_3^-)$$

Fig. 1 shows the patterns of a two-centered classical Liesegang ring experiment evolved 213 hours after application of the silver nitrate solution drops. Especially in the region between the drop centers remarkable bifurcations of precipitation rings occur, "alleys" of gaps within the rings and transitions to speckled patterns.

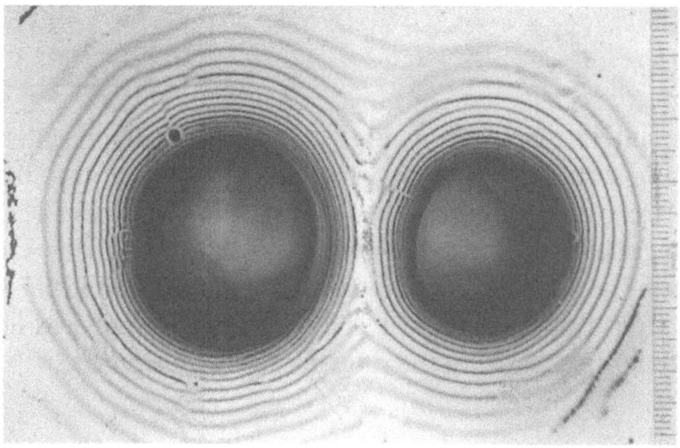

Fig. 1. Classical Liesegang experiments with diffusing silver ions in gelatine. Initial recipe: 1mm thick gelatine matrix in petri dish involving initially 3.4 10^{-3}M $K_2Cr_2O_7$. Initial concentration of $AgNO_3$ solution: 0.5 M. **a**: Evolution of patterns 213 hours after application of two separated drops of $AgNO_3$ solution onto gelatine.

Inspection of pictures taken before reveals that the gaps appearing within the precipitation bands or singular precipitation points are not the results of subsequent instabilities taking place in initially created continuous rings. Instead, discontinuities of rings already emerge in the first moment of precipitation. In the outer right part of the picture, the source of silver nitrate diffusion is almost concentric with the consequence of continuous ring formation whereas on the outer left-hand side little irregularities in the contour of the drop affect branchings, lateral interruptions and displacements of rings in their vicinity. This leads to the suggestion that band discontinuities are generally caused by non-circular or otherwise unevenly shaped diffusion sources (as in most realistic cases).

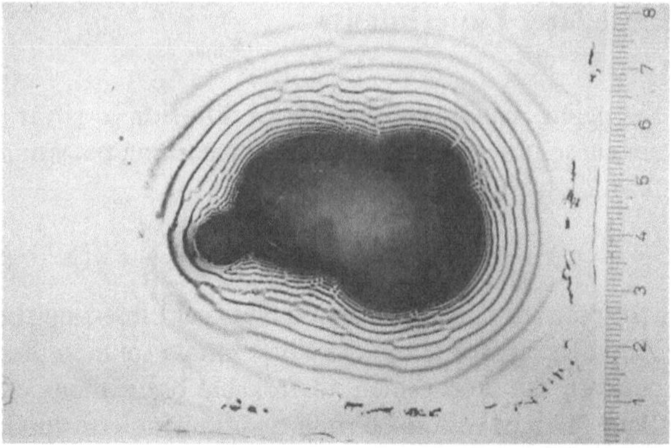

Fig. 2. Liesegang experiment with a pattern taken 313 hours after application of several single $AgNO_3$ solution drops having confluenced to one single center with undulatory contours. Recipe as in Fig. 1.

A similar result is displayed in Fig. 2, taken 313 hours after start of the experiment. Ring disturbances occur when the curvature of the silver ion source becomes undulatory or just unevenly shaped on a smaller spatial scale; bands will be formed as regular closed rings only when the source has both a nearly circular form and a smooth borderline. The fragmented patterns appearing in Fig. 2 were again present from the first moment of ring formation and are not due to any subsequent instability destroying the rings formerly precipitated.

The Competitive Particle Growth Model

For simulation of Liesegang experiment we use the competitive particle growth (CPG) model proposed by Feeney et al. in 1983. This model is based on a mechanism of a spatial instability in an Ostwald ripening process (see: Lifshitz & Slyozov 1961). It consists originally of two dimensionless variables only, the local particle radius ψ and the supersaturation σ. According to the Liesegang experiment, we assume formation of silver bichromate C as the result of

$$2\,A + B \xrightarrow{k_1} C \xrightarrow{\longleftarrow} \overline{C} \qquad \text{(solid precipitate)} \quad R1$$

where A stands for silver ions and B for bichromate ions. a and b are their dimensionless concentrations. Parameters β and G involve several material constants, q is derived from the velocity constant k_1. The dimensionless time is τ, ξ and η are the dimensionless spatial variables. For simplicity, we assume that the diffusion coefficients of all species are equal to each other. The equations read:

$$\begin{aligned}
&\frac{\partial \psi}{\partial \tau} = \sigma - g(\psi) \\
&\frac{1}{\beta}\frac{\partial \sigma}{\partial \tau} = \frac{\partial^2 \sigma}{\partial \xi^2} + \frac{\partial^2 \sigma}{\partial \eta^2} - \psi^2\left[\sigma - g(\psi)\right] + q\,a\,b \\
&\frac{\partial a}{\partial \tau} = \frac{D_b}{D_c}\beta\left[\frac{\partial^2 b}{\partial \psi^2} + \frac{\partial^2 a}{\partial \eta^2}\right] - 2\,q\,G\,a\,b \\
&\frac{\partial b}{\partial \tau} = \frac{D_b}{D_c}\beta\left[\frac{\partial^2 b}{\partial \xi^2} + \frac{\partial^2 b}{\partial \eta^2}\right] - \;q\,G\,a\,b
\end{aligned} \tag{1}$$

with $g(\psi) = \dfrac{2\psi^2}{2\psi^3 + \psi_c^3}$ being the equilibrium function for a particle with radius ψ surrounded with monomer of supersaturation σ. Using the scaling procedure of the original CPG model after Feeney et al. (1983) we have for the variables τ, ξ, η, a, b, σ and ψ, and for the matter parameters β, G, and q the detailed expressions as follows:

Particle radius	$\psi = R / \overline{R}$	
Supersaturation	$\sigma = \overline{R} / \Gamma$;	$S = [C] / [C]^{eq} (\infty) - 1$
		$\Gamma = 2\gamma_\infty / \rho \nu R_G T$
Concentrations	$a = [A] / [C]^{eq} (\infty)$	
	$b = [B] / [C]^{eq} (\infty)$	
Time	$\tau = t / \bar{t}$;	$\bar{t} = \dfrac{\overline{R}^2 \rho}{K[C]^{eq}(\infty)\Gamma}$
Space variables	$\xi = x / L$	
	$\eta = y / L$;	$L^2 = D / 4\pi \overline{R}^2 \bar{n} K)$
Matter parameters	$\beta = \overline{R} G / \Gamma$	
	$G = 4\pi \overline{R}^3 \rho \bar{n} / [C]^{eq}(\infty)$	
Velocity constant	$q = k_1 \bar{t}\, [C]^{eq}(\infty) / G$	

R, t, x, and y were the real particle radius, the real time, and the real space variables. The matter constants involved above are in detail:

$[C]^{eq} (\infty)$	saturation concentration of C above the plain crystal
γ_∞	surface tension of the plain precipitate crystal
ρ	molar density of the solid phase
ν	stoichiometric factor
R_G	gas constant

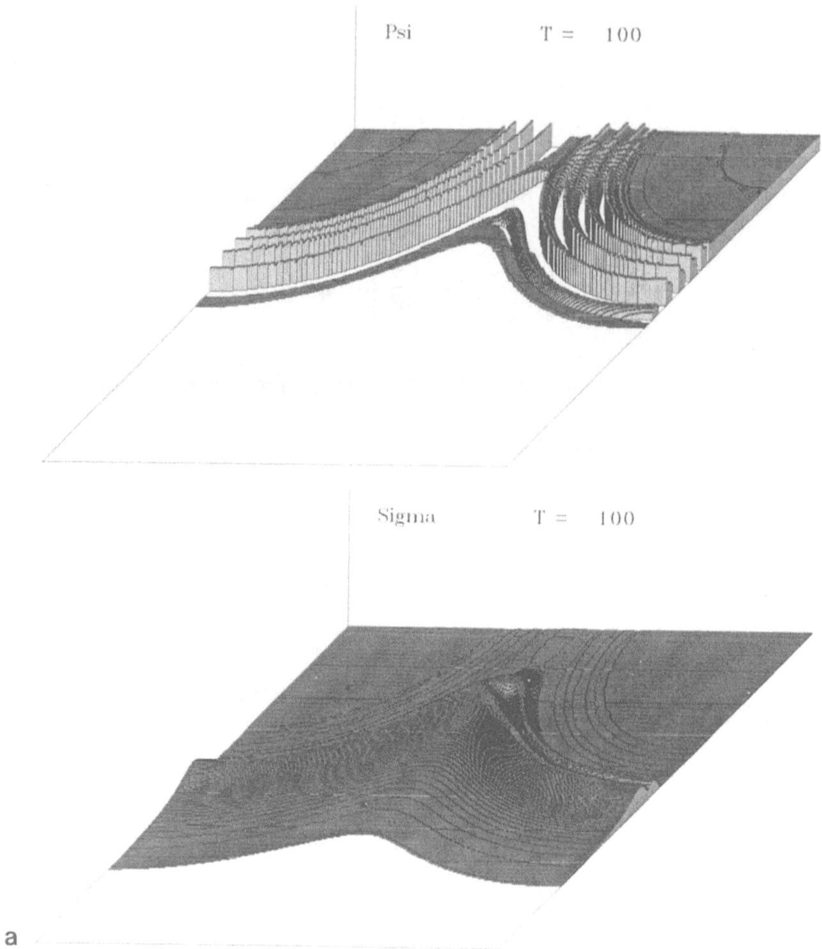

Fig. 3 a-c. Simulation of the Liesegang experiment shown in Fig 1 using the competitive particle growth (CPG) model. Pictures were taken at τ = 100 (**a**), τ = 150 (**b**), and τ = 1000 (**c**). Parameters: β = 0.1, G = 0.1, q = 1. Lengths of the spatial axes: L = 25, length of the ψ - axis: 72, length of the σ - axis: 15. Initial conditions: b_0 = 17.6 anywhere and a_0 = 2590 within the drop application areas with r_1 = 3.8 (upper left corner) and r_2 = 4.5 (upper right corner). Scaled concentrations used in the CPG model calculations are multiples related to the saturation concentration of $Ag_2Cr_2O_7$ (1.93 10^{-4}M), compare recipe given in capture of Fig. 1.

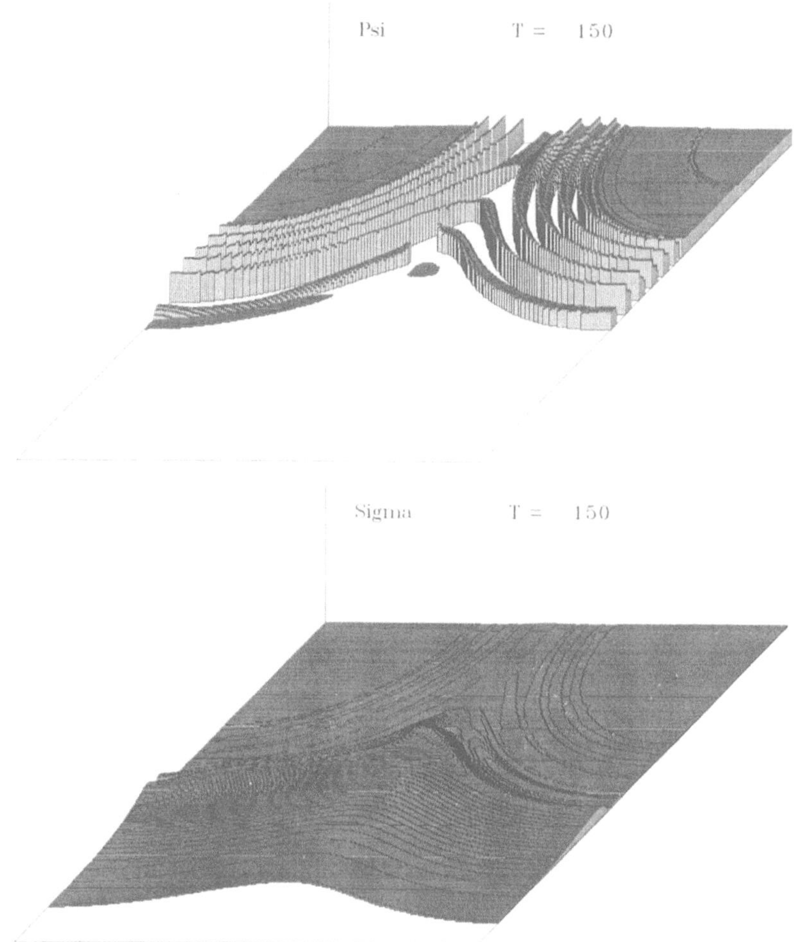

b

Fig. 3b1 (upper) and 3b/2 (lower). Legend see p.274.

T	temperature
K	growth rate constant of precipitate particles
D	diffusion coefficient (equal diffusion of A, B, and C assumed)
$\overline{n}$	"typical" particle number density
$\overline{R}$	"typical" particle radius

To avoid numerical problems in integration of the system (1), velocity constant q was set equal to 1. Increasing values of q (more corresponding to the fast velocity of R1) did indeed not alter the qualitative shapes of simulated Liesegang patterns.

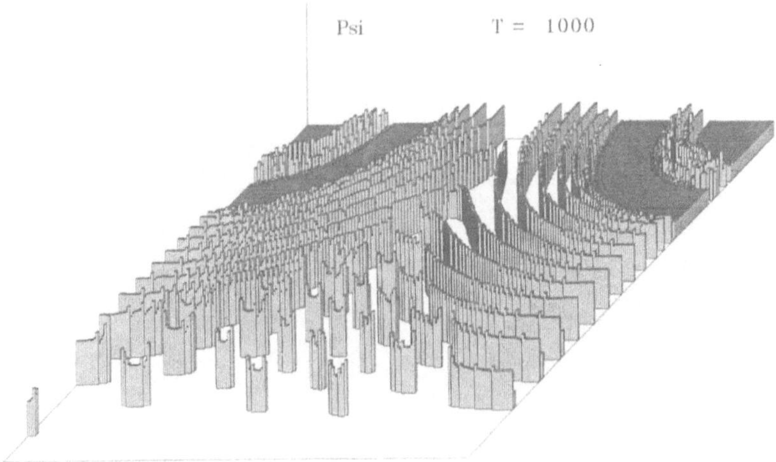

c

Fig. 3c. Legend see p.274.

Results of Numerical Simulation

Fig. 3a-c shows a sequence of pictures from simulation of a Liesegang experiment according to Fig. 1. In Fig. 3a-b are presented both the local particle radius ψ and the supersaturation σ of the monomer. After a distinct time, an evenly distributed precipitate is formed within and around the area of initial application of the silver nitrate drop. The radius of the circular area of even precipitate is finally almost twofold of the initial spreading radius of the simulated silver ion drop. The first rings created outside the centers of even precipitate are closed and show no disturbances (Fig. 3a). The second ring will also be created continuously by both diffusion centers. The first lateral discontinuity will occur in the third ring (Fig. 3b). The occurrence of this gap within the precipitation front is due to the fact that the supersaturation "wall" formed prior to precipitation is always slightly undulated if non-centric matter sources are present instead of being "isoconcentric".

This lateral undulation of the crest line of supersaturation σ may give rise to an instability outlined as follows: precipitation of a part of the whole ring will take place where σ exceeds the critical limit $\sigma_c = max\ [g\ (\psi)]$ within a certain interval of time. Further growth of the precipitated particles leads to a depletion of monomer in both radial and lateral direction. Also at the outer endings of a precipitated band the fast depletion of monomer continues, thus permitting further continuation of the ring in lateral directions. Formation of a next-neighbouring band to the left or to the right takes place only behind a persisting gap as illustrated by Figs. 3b/1 and 3b/2. Fig. 3c shows a picture of a later stage of pattern evolution. Beginning with the third ring created by superposition of both diffusion centers, each succeding ring exhibits an increasing number of gaps. With increasing distance to the diffusion sources, the formerly only partly perforated rings turn over to speckled patterns.

The simulation of the second Liesegang experiment (Fig. 2) is shown in Figs. 4a-b. The instability of band formation is again initiated by the undulatory curvature of the silver ion source. The occurrence of the first band branching at $\tau = 120$ is shown in Fig. 4a. Further branchings appear within later evolving bands (Fig. 4b, $\tau = 600$). Notice that - according to the observation in Fig. 2 - there occur very regular "alleys" of gaps within the sequence of bands and also bifurcations or branching of bands. Further continuation of the numerical simulation reveals a very slow decay of the periodic pattern once formed into a speckled pattern due to Ostwald-ripening (Fig. 4b, $\tau = 7500$).

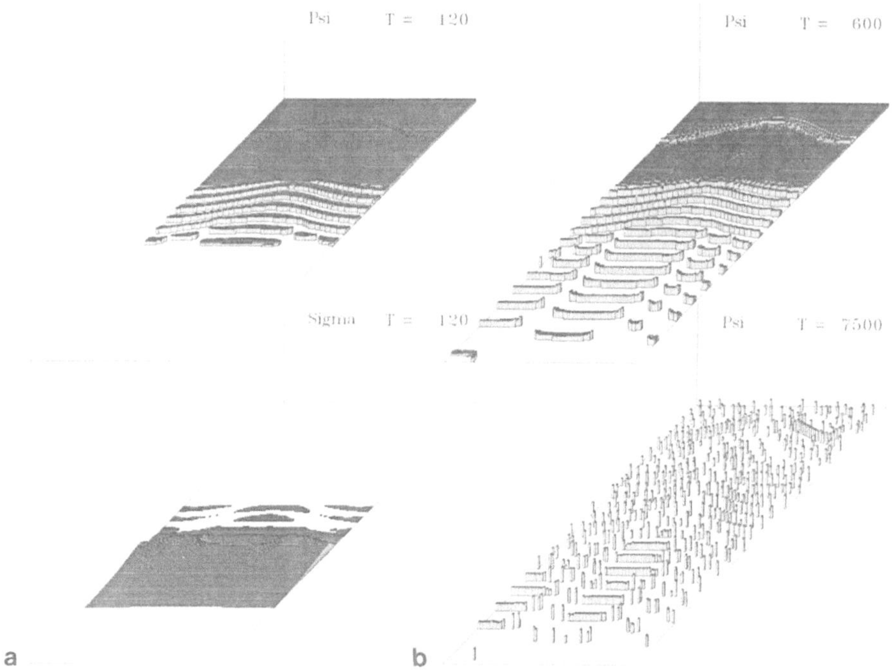

Fig. 4a-b. Simulation of the Liesegang experiment from Fig. 2. Horizontal length scale here: $L = 12.5$; length of the ψ - axis: 265, length of the σ-axis: 20 (Fig. 4a). Initial conditions: $b_0 = 17.6$ elsewhere, $a_0 = 2590$ within the curvature $\xi \leq 3.0 + 1.0 \cos(2\pi\, \eta/25)$, where ξ is the vertical, and η is the horizontal spatial coordinate. Other parameters and scales are according to Fig. 3. Pictures were taken at $\tau = 120$ (Fig. 4a), and at $\tau = 600$, and $\tau = 7500$ (Fig. 4b).

Discussion

Irregular forms of rhythmic bands are often encountered in the geosciences. Whereas the formation of regular patterns by means of in-situ self-organization became more accepted in the last decade, disturbances of patterns are still considered to be caused by subsequent tectonic effects leading to brecciated structures or by landslides of slack sediments. This mechanical argument begins to fail when irregular patterns occur on the ground of a structurally unaffected mineral matrix or are localized in the vicinity of undisturbed bands.

One non-tectonic explanation of brecciated patterns consists in considering of local barriers impermeable to diffusion. They may be built up by microscopic fissures within the rock later occupied with impermeable material (calcite) as reported on the "Pietra Paesina" limestone (Jacob et al. 1994).

Further disturbances like lateral band gaps, transition to speckled pattern or bifurcations (splitting or branching of one band into two others in lateral direction) can be explained even without the presence of any local irregularities in the diffusion matrix as we tried to show in the chapters above by demonstration of Liesegang ring experiments and model simulations. In the following, three examples of zebra spar or zebra rock exhibiting the named phenomena will be presented (Figs. 5-7).

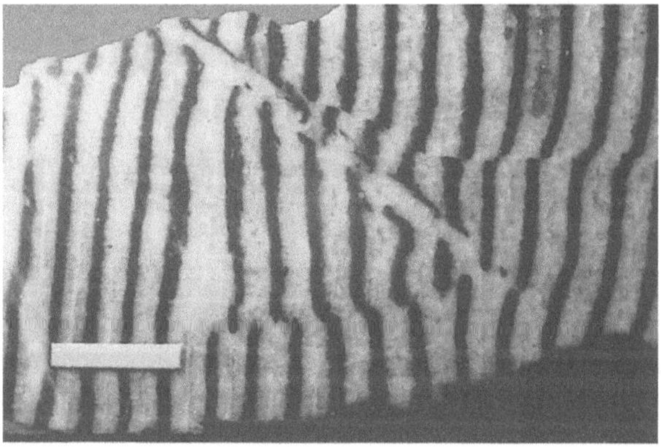

Fig. 5. Fluorite-rhythmite or zebra spar from the Sierra de Baza (Spain). Scale: 1 cm.

Figure 5 shows a fluorite or zebra spar from Sierra de Baza (Spain). This mineral consists of calcium fluorite in both the white and dark bands; the dark bands contain impurities of organic clay minerals. The regular rhythm of parallel bands is disturbed with a precipitate-free "alley" diagonally crossing the piece and deviations of bands finally turning over to band branchings. These are quali-

Fig. 6. Zebra spar (dark dolomite bands alternating with pure calcium carbonate) from the Sierra de Baza (Spain). Horizontal field of view: c 15 cm.

tatively the same patterns we encountered both in Liesegang experiments (Figs. 1-2) and their numerical simulation (Figs 3-4).

Similar pattern irregularities were also observed on a much smaller spatial scale (periodicity lengths of 1 μm order of magnitude) in rhythmic bands of W-As-Au skarn of Salanfe (Switzerland) recently reported by Chiaradia (1993).

A piece of zebra spar with dark dolomite bands (including organic clay minerals) alternating with pure calcium carbonate (Fig. 6) exhibits similar irregularities as the former example: precipitate band branchings and alleys perpendicularily crossing the rhythmic train of bands.

Fig. 7. Fluorite-rhythmite or zebra spar from the Sierra de Baza (Spain). Horizontal field of view: c 15 cm.

The last picture (Fig. 7) shows a fluorite-rhythmite with a rather dissolved band structure with still visible residuals of alleys and band branchings. This pattern exhibits similarities to the Liesegang pattern (Fig. 1) and also to the corresponding simulated structure (Fig. 3c) created at greater distance to the source of infused matter. On the other hand, this dissolved band structure of Fig. 7 may also be caused by a continued decay of a formerly only slightly fragmentated pattern as given in Figs. 5-6 by subsequent long-time Ostwald ripening as performed in numerical simulation (Fig. 4b, $\tau = 7500$).

The formation of zebra spars was formerly assumed to be due to "discrete-rhythmic" sedimentation in connection with a carboniferous mud facies. In a later phase, a purification-crystallization with formation of pure white and impured black fluorite or dolomite bands is assumed to have proceeded (Jacob 1974).

Much more convincing and generally accepted is the well-understood mechanism of Liesegang-ring formation (Ostwald-Liesegang supersaturation-nucleation-depletion cycle with subsequent Ostwald ripening). Considering especially the various types of structural defects appearing in rhythmic band trains of mineral pieces, diffusion experiments and in model calculations, they may now be explained rather by the process of competitive particle growth than by ad-hoc introduction of external mechanical effects.

References

Bradford SC (1922) Die Adsorptionstheorie der geschichteten Niederschläge. Kolloid-Z 30:364-367.

Chadam J (1987) Reaction-Percolation Instability. In: Nicolis C, Nicolis G (eds) Irreversible Phenomena and Dynamical Systems Analysis in Geosciences. Reidel, Dordrecht, pp 523-532.

Chadam J , Peirce A, Ortoleva P (1990) Stability of Reactive Flows in Porous Media: Coupled Porosity and Viscosity Changes. SIAM J Appl Math 51: 684-692.

Chen W, Ortoleva P (1990a) Self-Organization in Far-From-Equilibium Reactive Porous Media Subject to Reaction Front Fingering. In: Walgraef D, Ghonlem NM (eds) Patterns, Defects and Materials Instabilities. NATO ASI Series 183. Reidel, Dordrecht, pp 203-220.

Chen W, Ortoleva P (1990b): Reaction Front Fingering in Carbonate-Cemented Sandstones. Earth-Sci Rev 29: 183-198.

Chiaradia M (1993) Arsenopyrite Geothermometry and As_2 Conditions in the W-As-Au Skarn of Salanfe (Valais, Switzerland): Tectonic Implications in the Skarn Formation Processes. Mineralium Deposita, (submitted).

Dee GT (1986) Patterns Produced by Precipitation at a Moving Reaction Front. Phys Rev Lett 57: 275-278.

Dewers T, Ortoleva P (1989) The Self-Organization of Mineralization Patterns in Metamorphic Rocks through Mechano-Chemical Coupling. J Phys Chem 93: 2842-2848.

Dewers T, Ortoleva P (1990a) Geochemical Self-Organization III: A Mechano-Chemical Model of Metamorphic Differentiation. Am J Sci 290: 473-521.
Dewers T, Ortoleva P (1990b) A Coupled Reaction / Transport / Mechanical Model for Intergranular Pressure Solution, Stylolites, and Differential Compaction and Cementation in Clean Sandstones. Geochim Cosmochim Acta 54: 1609-1625.
Dhar NR, Chatterji AC (1925) Theorien der Liesegangringbildung. Kolloid-Z 37: 89-97.
Feeney R, Schmidt SL, Strickholm P, Chadam J, Ortoleva P (1983) Periodic Precipitation and Coarsening Waves: Applications of the Competitive Particle Growth Model. J Chem Phys 78: 1293-1311.
Feinn D, Ortoleva P, Scalf W, Schmidt S, Wolff M (1978) Spontaneous Pattern Formation in Precipitating Systems. J Chem Phys 69: 27-39.
Flicker MR, Ross J (1974) Mechanism of Chemical Instability for Periodic Precipitation Phenomena. J Chem Phys 60: 3458- 3465.
Freundlich H, Schucht E (1913) Über die Geschwindigkeit des Adsorptionsrückgangs bei der Umwandlung des Quecksilbersulfids aus der amorphen Form in eine mehr kristallinische. Z phys Chem 85: 660-680.
Hedges ES (1932) Liesegang Rings and other Periodic Structures. Chapman & Hall, London.
Jacob KH (1974) Deutung der Genese von Fluoritlagerstätten anhand ihrer Spurenelemente - insbesondere an fraktionierten seltenen Erden. Thesis, Techn Univ Berlin, Dep of Mining and Geosciences.
Jacob KH, Zimmerle W (1990) Diagenetische Phänomene in Sedimentgesteinen des Karbons, gesehen als rhythmische Gefügebildungen durch Energiedissipation in offenen Systemen. Z Förd Bergb Hüttenw der Techn Univ Berlin 2: 12-18.
Jacob KH, Krug HJ, Dietrich S (1992) Lagerstättenbildung durch Energiepotentiale in der Lithosphäre. Erzmetall 45: 505-513.
Jacob KH, Dietrich S, Krug HJ (1994) Self-Organization of Mineral Fabrics. In: Kruhl JH (ed) Fractals and Dynamic Systems in Geoscience. Springer, Berlin Heidelberg New York, pp 259-268 (this volume).
Kai S, Müller SC (1985) Spatial and Temporal Macroscopic Structures in Chemical Reaction Systems: Precipitation Patterns and Interfacial Motion. Science on Form (Tokyo) 1: 9-39.
Kai S, Müller SC , Ross J (1982) Measurements of Temporal and Spatial Sequences of Events in Periodic Precipitation Processes. J Chem Phys 76: 1392-1406.
Klein W (1990) Dissipative Gefügebildungen in Lockersedimenten durch Einwirkung elektrischer Felder. Thesis, Techn Univ Berlin, Dep of Mining and Geosciences.
Krug HJ, Jacob KH (1993) Genese und Fragmentierung rhythmischer Bänderungen durch Selbstorganisation. Z Dt Geol Ges 144: 451-460.
Kuhnert L, Niedersen U (1987) Selbstorganisation chemischer Strukturen. (Ostwalds Klassiker der exakten Wissenschaften 273). Geest & Portig, Leipzig.
Liesegang RE (1896) A-Linien. Lieseg photogr Arch 21: 321-326.
Liesegang RE (1913) Geologische Diffusionen. Steinkopf, Dresden Leipzig.
Liesegang RE (1915) Die Achate. Steinkopf, Dresden Leipzig.
Lifshitz IM, Slyozov VV (1961) The Kinetics of Precipitation from Supersaturated Solid Solutions. J Phys Chem Solids 19: 35-50.
Mc Birney AR, Noyes RM (1979) Crystallization and Layering of the Skaergaard Intrusion. J Petrol 20: 487-554.

Müller SC, Kai S, Ross J (1982) Curiosities in Periodic Precipitation Patterns. Science 216: 635-637.
Ortoleva P (1984) From Nonlinear Waves to Spiral and Speckle Patterns. Non-Equilibrium Phenomena in Geological and Biological Systems. Physica 12 D: 305-320.
Ortoleva P (1987) Modeling Geochemical Self Organization. In: Nicolis C, Nicolis G (eds) Irreversible Phenomena and Dynamical Systems Analysis in Geosciences. Reidel Publishing Comp., Dordrecht, pp 493-510.
Ortoleva P (1994): Geochemical Self-Organization. Oxford University Press, Oxford.
Ortoleva P, Chadam J, El-Badewi M, Feeney R, Feinn D, Haase S, Larter R, Merino E, Strickholm P, Schmidt S (1982a) Mechanisms of Bio- and Geo- Pattern Formation and Chemical Signal Propagation. In: Reichl LE, Schieve WC (eds) Instabilities, Bifurcations and Fluctuations in Chemical Systems. University of Texas Press, Austin.
Ortoleva P, Merino E, Strickholm P (1982b) Kinetics of Metamorphic Layering in Anisotropically Stressed Rocks. Amer J Sci 282: 617-643.
Ortoleva P, Auchmuty G, Chadam J, Hettmer J, Merino E, Moore CH, Ripley E (1986) Redox front Propagation and Banding Modalities. Physica 19 D: 334-354.
Ortoleva P, Merino E, Chadam J, Moore CH (1987) Geochemical Self-Organization I: Reaction-Transport Feedbacks and Modeling Approach. Amer J Sci 287: 979-1007.
Ostwald Wi (1897a) A-Linien von R. E. Liesegang (Review). Z phys Chem 23: 365.
Ostwald Wi (1897b) Lehrbuch der allg. Chemie, 2. umgearb Aufl, II, 2.: Verwandtschaftslehre. Engelmann, Leipzig, pp 777-780.
Ostwald Wi (1900) Über die vermeintliche Isomerie des roten und gelben Quecksilberoxyds und die Oberflächenspannung fester Körper. Z phys Chem 34: 495-503.
Ostwald Wo. (1925) Zur Theorie der Liesegang'schen Ringe. Kolloid Z 36: 380-390.
Prager S (1956) Periodic Precipitation. J Chem Phys 25: 279-283.
Smith DA (1984) On Ostwald's Supersaturation Theory of Rhythmic Precipitation (Liesegang's Rings). J Chem Phys 81: 3102- 3115.
Stern KH (1954) The Liesegang Phenomenon. Chem Revs 54: 79-99.
Stern KH (1967) Bibliography of Liesegang Rings (2nd ed.). U.S. Governement Printing Office, Washington D.C.
Sultan R, Ortoleva P (1993) Periodic and Aperiodic Macroscopic Patterning in Two Precipitate Postn-Nucleation Systems. Physica D 63: 202-212.
Sultan R, Ortoleva P , De Pasquale F, Tartaglia P (1990) Bifurcation of the Ostwald - Liesegang Supersaturation-Nucleation-Depletion Cycle. Earth Sci Rev 29: 163-173.

Agates, Geodes, Concretions and Orbicules: Self-Organized Zoning and Morphology

Peter Ortoleva[1,2], Yueting Chen[2] & Wei Chen[3]
[1]Departments of Chemistry and of [2]Geological Sciences, Indiana University
Bloomington, IN 47405, USA
[3]Silicon Graphics
1 Cabot Road, Hudson, MA 01749, USA

Abstract. A variety of objects growing in lithified or unlithified environments exhibit self-organization. These include single crystals and polycrystalline or amorphous solids such as agates, geodes, concretions and orbicular granites. Here we discuss mechanisms for their genesis and quantitative reaction-transport-(and possibly)mechanical (RTM) equations that can account for many of their most spectacular characteristics.

The main two classes of phenomena are internal compositional or textural patterns and morphological patterns. These phenomena can occur under a wide range of conditions of geochemical change: igneous, metasomatism, metamorphosis, contact metamorphosis, diagenesis and weathering. Conditions of change, and more precisely of strong displacement from thermodynamic equilibrium, are required for self-organization.

Mechanisms of self-organization fall into two classes. Growth instabilities occur during the growth of the inclusion from components in the host medium. In contrast to these "open system" phenomena, there are mechanisms which rely only on material in the inclusion (already formed) in its pre-self-organized state in the matrix. Both types of models are considered, contrasted and formulated as quantitative RTM models.

Self-organization dynamics of four types are examined. In the transport-limited, open case the inclusion grows from an initial nucleus or seed as a low or zero porosity body with a feedback arising from the growth rate mediated by the state of the surface of the body and that of the host medium near the body surface. Another open case involves growth of isolated crystals in a nucleation zone from mobile components derived outside the original zone; growth of these crystals may be mediated by their surface composition and that of the host near their surface and may proceed to the removal of host grains within their interior by force of crystallization effects. A third mechanism involves the replacement of an initial inclusion phase by a late phase such that minimal mass exchange between the inclusion and host medium is required. Finally, growth of the body may be under open or closed conditions but involves Ostwald ripening.

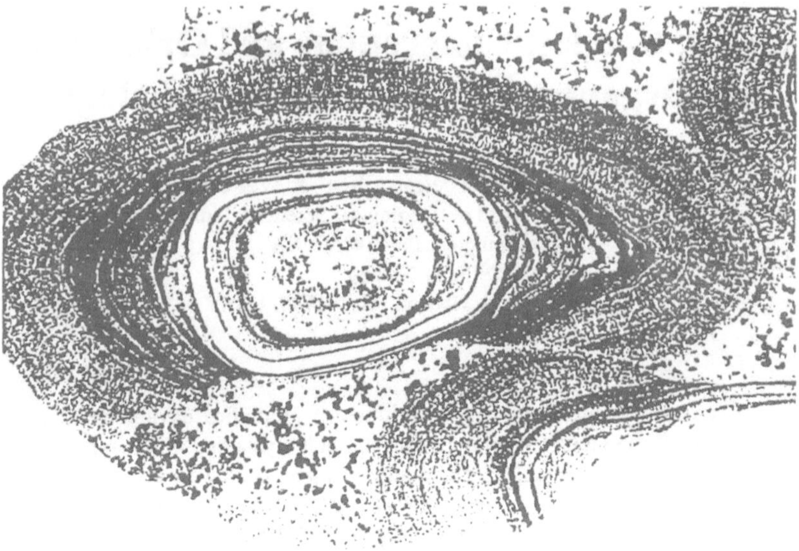

Fig. 1. Patterns of concentric rings of alternating light and dark minerals constitute "orbicules" in otherwise homogeneous igneous rocks: Orbicular quartz diorite from southern Alaska. Courtesy of A. McBirney (from McBirney and Noyes, 1979).

Geodes, Concretions, Agates and Orbicules

The Phenomena

Roughly spheroidal objects of supragrain dimension are observed to develop under a variety of conditions including igneous, metamorphic, and diagenetic. Examples include orbicular granite, concretions, geodes, and agates. They are noteworthy from the point of view of self-organization in that they may have morphologies or internal compositional distributions which suggest that autonomous patterning processes have played a role.

Orbicular Granites

Orbicular granites are patterns of spheroidal geometry involving concentric shells alternating in mineral content. They may involve one or many shells (Fig. 1). The darker shells are generally phyllosilicate rich, whereas the lighter ones are predominantly quartz and feldspar rich. It has been claimed that their core shows evidence for being a remnant of the country rock into which the magma has been intruded. Evidence also exists that some may have formed while the magma was in a liquid or gel-like state.

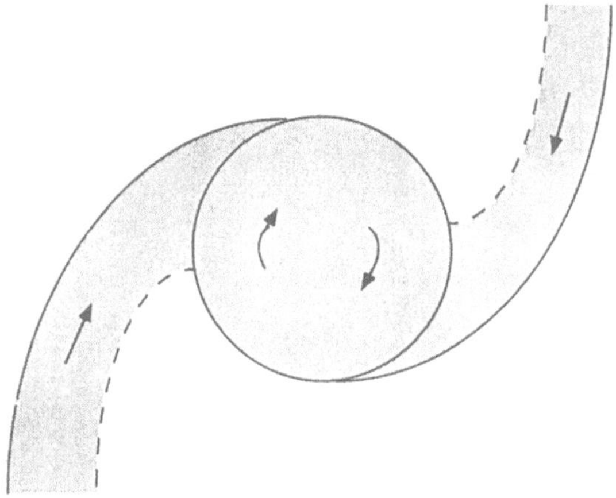

Fig. 2. Suggested genesis of a spiral garnet.

Metamorphic Spheroids and Spiral Garnets

Spheroidal (Loberg 1963, Fisher 1970) or spiraled (Rosenfeld 1970) patterns of (apparently) metamorphically differentiated mineralization are observed (Fig. 2). The types and ranges of mineral segregations are similar to those observed for metamorphic layering as reviewed in Dewers and Ortoleva (1990a) and Ortoleva (1994a Chapter 11). It is tempting to suggest that they have common origins, perhaps differing only in the geometry of the shear or other imposed conditions on the system during differentiation.

Agates

Agates are microcrystalline SiO_2 inclusions that have grown in a host medium. They contain low levels of impurities that are distributed in a variety of beautiful patterns, as suggested in Fig. 3 (Quick 1963, Frondel 1985). The compositional zonation seen in some agates is related to oscillatory variations in OH^- or other components contained in structural sites within quartz fibers. In the case of concentric-sphere patterns, the color contrast may be coordinated with alternations of crystallographic axes of the SiO_2 crystallites or may be due to variations in pigmenting impurities. The crystals in neighboring bands may also differ in that those in one set of bands are twisted in shape. The tightness of the agates and the orientational and twist ordering suggest a possible role of mechanochemical coupling.

Concretions and Geodes

Concretions and geodes are domains containing one or more minerals in a host sedimentary rock of contrasting mineralogy (Pettijohn 1975, Maliva & Siever 1988, Bjorkum & Walderhaug 1990). They have been observed to grow in young, unlithified sediment (Coleman & Raiswell 1993). They may also form by precipitation with concomitant dissolution of the host rock to make room for the

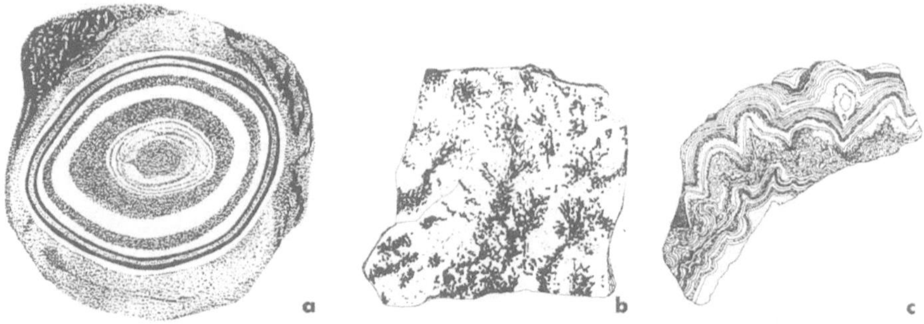

Fig. 3a-c. Sketches of three types of commonly observed agates: **a** orbicular, **b** mossy, and **c** scalloped.

growing inclusion. Frequently these features appear to start around an initial inclusion or inhomogeneity but may grow to many times the size of the original feature. A typical case is the growth of quartz (or other SiO_2 minerals) from sponge spicules or other high SiO_2-free energy source in a host carbonate rock (Dewers & Ortoleva 1990b). Alternatively, concretions may be formed from carbonate minerals. Because there does not seem in many cases to be evidence that these inclusions grow by fracturing a lithified host, it seems that the host, if lithified during growth, dissolves out by pressure solution or other mechanisms.

The question arises as to the nature of the driving force for growth against the opposing stress of a lithified host and about the morphological stability of the growing inclusion. Concretions and geodes may be roughly spherical but may be

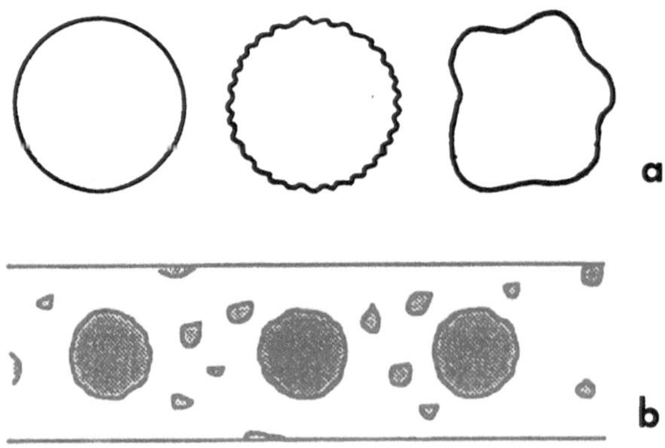

Fig. 4a, b. a Schematic view of concretion or geode morphologies showing spherical, bumpy, and large-amplitude outgrowth forms. **b** Schematic drawing of a roughly periodic distribution of large geodes in a bed containing many smaller geodes seen in the Harrodsburg limestone of South Central Indiana.

elongate or display multiple bumps or other morphologies, as suggested in Fig. 4a. The nonspherical shapes may be the result of a morphological instability of the spherical growth mode, as was the case for reaction fronts studied in Chadam et al. (1986, 1987, 1988, 1990), Ortoleva et al. (1987b), Chen & Ortoleva (1990c) and Ortoleva (1994a Chapter 7). Concretions are sometimes observed to occur in roughly periodic distribution in space within a certain formation, as suggested in Fig. 4b. There may be a second spatial scale of differentiation operating during their genesis.

Geodes may be fine or coarse grained, may range in diameter from a millimeter to a meter, and may be solid or hollow. If hollow, they may even exist as a very thin shell. Although predominantly constituted of quartz, they may have other minerals in their core. When found situated in the host rock, they often are surrounded by a layer of clay or other residue that appears to have been left behind as the inclusion grew and dissolved out the soluble components in the host.

The carbonate comprising the cements in concretions commonly fills pores in a host sandstone in the spheroidal domain defining the concretion. In many cases the quartz host has been removed, possibly by some pressure-of-crystallization mechanism (Dewers & Ortoleva 1990b). They may be quite large (occurrences 5 m in diameter are known).

Both geodes and concretions have apparently grown from some high-free-energy source material. Sources include sponge spicules in the case of SiO_2 and shells for $CaCO_3$. Growth scenarios that involve the import of mass are also possible. For example, Ca^{2+} could come from local feldspars and CO_2 might be imported from underlying petroleum reservoirs.

In both cases grain size may vary considerably within the inclusion. The most common pattern for geodes involves an increase in grain size as the center of the inclusion is approached.

Origins

A number of scenarios for the development of these inclusion phenomena and their internal or external structure may be conjectured. Let us now cite some of them and make suggestions for their potential applicability to the phenomena listed earlier. It must be emphasized that this is not intended to be a careful examination of the known data. The remarks to follow are only intended to demonstrate that the preceding phenomena are expected to follow directly from future studies of reaction-transport (and perhaps mechanochemical) models.

Replacement Versus Mechanically Uninhibited Growth

There are several limiting cases of inclusion growth. If the host medium is unconsolidated or of sufficiently low "viscosity," the inclusion can grow by pushing the host medium out of its way. If, on the other hand, the host medium is of high viscosity, the inclusion can only grow as the host material dissolves. The latter case is expected to involve pressure-of-crystallization effects (Dewers & Ortoleva 1990c).

Next, consider the growth of an inclusion in a porous rock. The mineral grains constituting the inclusion may in principle grow and reproduce themselves by nucleation of crystallites at their surface centrifugally from the inclusion core so as to fill the pore space further and further from the core. This could even lead to the replacement of the host matrix grains within the inclusion by pressure of crystallization effects. For example, the growth of a calcite inclusion in a sandstone from high-$CaCO_3$ free-energy shell material may cause the dissolution of quartz grains trapped within the inclusion by pressure-of-crystallization.

In the preceding mechanisms one must address the question as to why the mineral growth was confined to the inclusion and did not occur pervasively. It seems likely that there must be some role for a nucleation threshold to prevent pervasive growth and, furthermore, some type of catalyzed nucleation within or at the surface of the inclusion. In the latter case it appears that existing grains sometimes can serve as a heterogeneous nucleation site for new grains.

Another growth mode may arise through the same types of mechanochemical feedback that lead to metamorphic layering (Dewers & Ortoleva 1989b, 1990a, 1993, Ortoleva et al. 1982a). In this case, the stress on a grain (and therefore its free energy) can depend on the mechanical properties and therefore the volume percent of various minerals in its vicinity. This effective dependence of equilibrium constants on mineral modes leads to a self-organizational feedback (Dewers & Ortoleva 1988).

Internal Compositional Self-Organization

The internal compositional zonation in some inclusions could have developed either in association with growth or after growth had ceased. Concentric orbicular composition patterns could arise via a mechanochemical instability, as noted earlier. In the case of growth in a low-viscosity medium, a mechanism analogous to the autocatalytic surface attachment kinetics of crystal grain zoning (Haase et al. 1980, Ortoleva 1990, 1994a Chapter 3) could operate.

Internal compositional patterning may also occur in the postgrowth period. One mechanism is a mechanochemical differentiation, as occurs for metamorphic layering (Ortoleva et al. 1982a, Dewers & Ortoleva 1989a, 1990a, Ortoleva 1994a Chapter 11). An unstable ripening dynamic (Feeney et al. 1983, Ortoleva et al. 1982b, Ortoleva 1994a Chapter 9) could also lead to banding or mosaic patterns. Alternatively, if exchange of mass between the host and inclusion takes place, then Liesegang (cross coprecipitate gradient) phenomena (Ortoleva 1994a Chapter 8) could operate. In the latter case, the mineralization of the transformed inclusion could be quite distinct from what it was originally.

Inclusion Morphology

The growth of the inclusion need not be stable morphologically. For example, the state of spherical growth could become unstable through the development of bumps or other distortions, suggested in Fig. 4a. In this case the ultimate inclusion morphology would be determined by the opposition of tendencies toward the development of bumps and some stabilizing effect that tends to repress bump formation. An example of the latter for growth in a very viscous host is discussed in Section "Morphological Instability of a Growing Inclusion".

Agates and Orbicules

Liesegang banding (Liesegang 1913, Hedges & Meyers 1926, Ortoleva 1994a Chapter 8 and references therein) are banded, concentric ring or other repetitive patterns of precipitate content. They were originally believed to arise from the Ostwald/Prager supersaturation-nucleation-depletion cycle (Sultan et al. 1990, Ortoleva 1994a Chapter 8). More recently it has been shown that these phenomena can also arise through the instability of a uniform sol to an inhomogeneous ripening effect (Ortoleva 1978, 1994 Chapter 8, Feinn et al. 1978, Sultan & Ortoleva 1993). These two phenomena need not act independently i.e., in some systems they can act in concert lending to multiple scale banding; "secondary banding" is apparently due to a supersaturation-nucleation-depletion band becoming unstable to the formation of shorter scale banding within it.

The Classic O-P Banding at Reactive Infiltration

These have been discussed via modeling studies in the context of diagenesis (Ortoleva et al. 1986) and more generally (Sultan et al. 1990, Ortoleva 1994a Chapter 8). They involve bands of one or more species that form normal to the imposed reactive flow. This may be at the heart of banded skarns, Mississippi Valley ores, and iron oxide beds associated with redox fronts and banded patterns of contact metamorphism. An alternative type of banding not involving nucleation or ripening phenomena but rather the kinetic interplay of minerals in multimineralic reaction fronts has also been pointed out (Ortoleva 1994a Chapter 6).

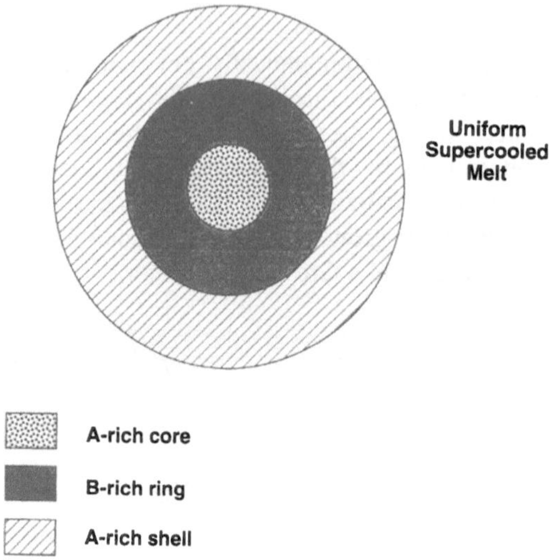

Fig. 5. Generation of orbicules showing mineral A-rich core surrounded by induced B-rich core surrounded by induced B-rich shell followed by an induced A-rich shell.

Orbicular Patterns in Igneous Rocks

These could form in a supercooled melt. Take a simple case of a two solid component A, B melt. A nucleus (such as a fragment of country rock) favors nucleation of A at its periphery. As A crystals grow they deplete the melt in the vicinity of the A nucleation core of components needed to grow A. But this factor increases the concentration of the B-forming components. This can proceed until such a time that B nucleation is induced in a ring surrounding the A-rich core. As B grows, A-forming components will be enriched in a third new shell and the cycle repeats. Thus many shells alternatingly rich in A and B may be produced (Fig. 5).

Contact Metamorphic Patterning Through Ripening Phenomena

Rapid heating in country rock adjoining an intrusion, fragments of country rock fallen into a magma or any other induced rapid change in pressure and temperature or chemical environment through rapidly infiltrated silicate or other fluid can cause the production of extensive and pervasive nucleation of new phases. The resulting sol, because of the particle size dependence of the equilibrium constant, can be unstable to the evolution of spatial nonuniformities. As a result the uniform sol may become unstable to the development of mottled patterns, spots, bands, rings or orbicules. Some examples of phenomena that one may illustrate using such reaction-transport-particle growth/dissolution models are shown in Fig. 6. Observed phenomena include

a) Alternating bands of larger and smaller (or even dissolved-out) crystals (Ortoleva 1978, Feinn et al. 1978, Feeney et al. 1983) (Fig. 6).
b) Bands where the majority of particles have survived and grown alternating with bands where most crystals have been dissolved out but a few grow to be the largest in the system (Ortoleva et al. 1982, Strickholm & Ortoleva 1985)–see Fig. 7.
c) For cases of a mixed sol of A-type and B-type crystals (A and B being either different minerals or the same mineral with contrasting impurity content or habit), it is found that the A bands can alternate with B bands or in other cases be coordinated with them--see Fig. 8 (Sultan & Ortoleva 1993). Under other circumstances, bands containing both large A and B crystals alternate with bands containing both small (or dissolved out) A and B crystal regions (Sultan & Ortoleva 1993).
d) Same as (c) except that the A and B growth bands may occur in a chaotic pattern (Sultan & Ortoleva 1993).
e) Imposed gradients of temperature, composition or flow or other vector can regularize the patterns as in (a)-(d) to be essentially periodic, as can imposed initial nonuniformity.
f) Initial nonuniformity in the sol growth tends to induce a regular banded, spotted or other pattern but as the pattern advances into the sol it becomes mottled and may completely lose the coherence of the pattern of initial data or boundary perturbation.

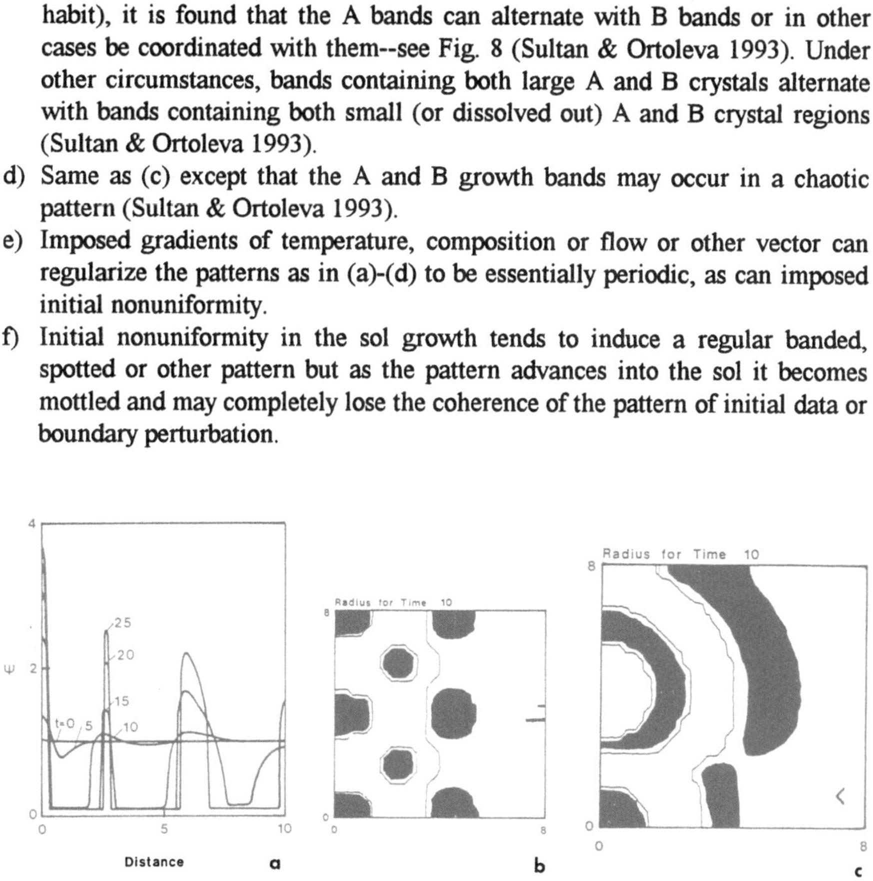

Fig. 6a-c. a Simulation of the CPG model in a one-spatial-dimensional system with no flux ends. Note the induction of a number of satellite bands by the initial, small-amplitude disturbance at the left. **b** Same as in frame (a) except for a two dimensional domain with the initial nonuniformity being a set of discrete (small amplitude) heterogeneities along the left wall. **c** Same as in frame (b) except for a small maximum in the lower left corner and a small minimum at the middle of the left wall (in particle size).

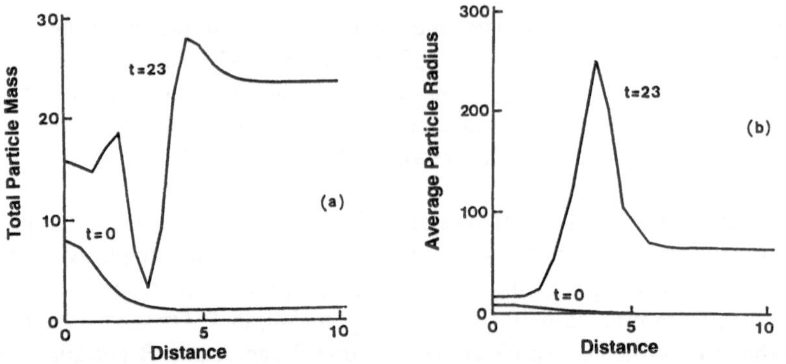

Fig. 7. Numerical simulation of a particle size distribution model showing the evolution of a local disturbance in average local particle radius and total mass contained in all particle size classes.

The above phenomena can be described by a model as follows (see the above cited references for details)

$$\frac{\partial c_\alpha}{\partial t} = -\vec{\nabla} \bullet \vec{J}_\alpha + \sum_k \nu_{\alpha k} W_k(\underline{c}) + \sum_i 4\pi n_i R_i^2 G_i(\underline{c}, R_i) \tag{1}$$

$$\frac{\partial R_i}{\partial t} = G_i \tag{2}$$

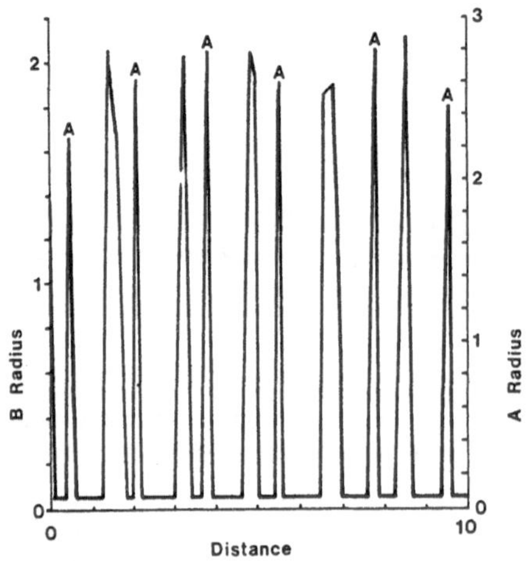

Fig. 8. Simulation of equations of two-salt CPG dynamics showing the evolution of patterns from random initial noise. The pattern is anticorrelated (maxima in the radius of A and B do not overlap).

Here c_α is the molar concentration of solute α and $\vec{J}_\alpha$ is the α flux; R_i is the radius of an i-type particle and G_i and n_i are its radial growth rate and number density, respectively. W_k is the rate of the k-th reaction in the fluid filling the interparticle space; $\nu_{\alpha k}$ and $\omega_{\alpha i}$ are stoichiometric coefficients. The particle type index may distinguish between particles of the same mineral but different size, thereby allowing an analysis of the dynamics of the particle size distribution (Ortoleva et al. 1982a, Strickholm & Ortoleva 1985). Alternatively, it can distinguish between mineral types in a multi-mineralic system (Sultan & Ortoleva 1993). The size dependence of the equilibrium constant (key to the present phenomenon) is manifest in the R_i-dependence of G_i. Other formulations involving nucleation effects and a continuum of particle sizes are outlined in Ortoleva (1994a Chapters 8 and 9).

A key element in the self-organizing dynamics embedded in the above competitive particle growth (ripening) dynamics is the dependence of the growth rates $G_i(\underline{c}, R_i)$ on the R_i due to surface energy (radius of surface curvature) effects on the equilibrium constant. A second dynamic can also operate due to nucleation. In this case G_i has been modeled via the nucleation law

$$G_i = \theta G_i' \tag{3}$$

where G_i' is a $\underline{c}$ (and possibly R_i) dependent classical rate law and the factor θ is given by

$$\theta = \begin{cases} 0, R_i = 0 \text{ and critical nucleation} \\ \quad \text{saturation for mineral i has} \\ \quad \text{not been obtained} \\ 1, \quad \text{otherwise} \end{cases} \tag{4}$$

For further details see Ortoleva et al. (1986), Sultan et al. (1990) and Ortoleva (1994a Chapter 7).

Morphological Instability of a Growing Inclusion

Let us now analyze a simple model of geode or concretion growth in a lithified host rock and pose the question of its morphological stability and the effects of pressure of crystallization. Consider an SiO_2 inclusion growing as a solid body of negligible porosity in a limestone containing sponge spicules. The SiO_2 in the spicules is high in free energy relative to that in the quartz (or other SiO_2 mineralization) constituting the growing inclusion. Assume that the inclusion is initially spherical. As it grows at the expense of the high-free-energy SiO_2 source, the latter will dissolve out and a region depleted in high-free-energy SiO_2 source material will be created around the inclusion (Dewers & Ortoleva 1990c). Within

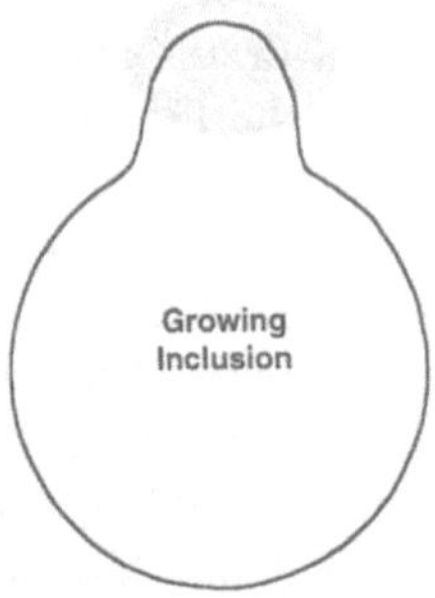

Fig. 9

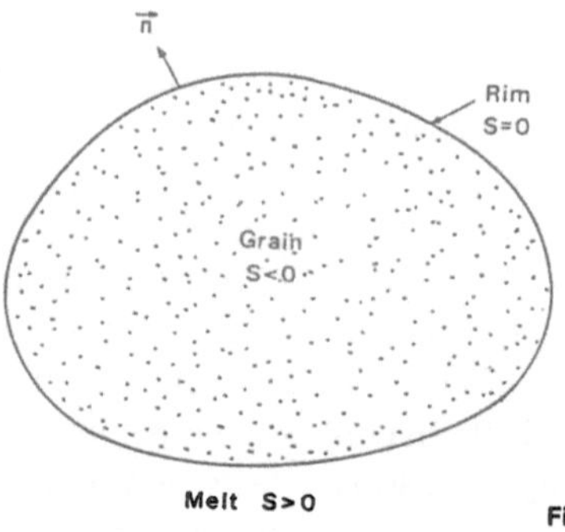

Fig. 10

Fig. 9. Inclusion with bump growing from relatively high concentration in a matrix producing solutes from which the inclusion is growing. The region of increased stress near the growing tip (shaded domain) tends to slow down bump growth while increased stress in the host near the outgrowth promotes dissolution of the host matrix. **Fig. 10.** Growing grain illustrating normal ($\vec{n}$) and bounding surface ($S = 0$).

this region a gradient of SiO_2 (aq) concentration will develop in the pore fluid, increasing [SiO_2 (aq)] with distance from the inclusion. This situation allows for the spherically symmetric growth state to become unstable in analogy with the Mullins and Sekerke morphological instability of a growing amorphous particle (Dandekar 1969, Chadam et al. 1987).

In Figure 9 we see a schematic of a growing spherical SiO_2 inclusion with a bump. At the tip of the bump [SiO_2 (aq)] is larger than at its flanks and hence the bump tends to elongate as the normal velocity of advancement increases with [SiO_2 (aq)]. However, as the bump grows, it butts against the host, causing increased stress within the bump and in the host near the bump (as suggested by the shading in Fig. 9). The extra stress in the bump over that at the larger radius of curvature domains of the inclusion increases the free energy of the quartz in the bump and hence tends to slow down its rate of growth. Thus, the pressure dependence of the solid free energy tends to stabilize the spherical growth mode. Interestingly, however, the extra stress induced in the host near the bump promotes dissolution of the host to allow for further bump growth. It is seen that the growth of concretions or geodes poses interesting questions of morphological stability. To address these problems one must invoke equations of grain growth/dissolution, solute reaction and migration, and rock mechanics. The criterion for morphological instability will then involve grain growth rate parameters, elastic and/or plastic properties of the host and inclusion, and initial lithology of the host. Interesting effects might also enter as a result of stresses applied to the host far from the inclusion.

The morphological instability of a growing inclusion may be modeled via the phenomenology of Ortoleva (1994a Chapter 10). The simplest case is when the body grows at negligible porosity and contains only one mineral. An example of this is quartz or chalcedony growing in a limestone host. We consider the single mineral, low-porosity inclusion briefly here. An alternative case in which a quartz concretion starts to grow so that its interior initially contains both quartz and calcite with calcite subsequently dissolved out by pressure of crystallization is presented in Dewers & Ortoleva (1990c).

Let the growing body be contained within a mathematical surface defined by the set of points $S(\vec{r},t)=0$ as in Fig. 10. Assume the medium comprising the inclusion deforms with velocity $\vec{u}'(\vec{r},t)$ ($S<0$). Consider a point $\vec{r}_0$ on the surface of the growing body at time t_0 (i.e., $S(\vec{r}_0,t_0)=0$). Let μ be the normal growth velocity and let $\vec{n}$ be the unit normal vector (Fig. 10). Then in a short time δt the surface of the body at $\vec{r}_0$ at time t_0 has moved to $\vec{r}_0+(\vec{u}'+\mu\,\vec{n})\delta t$. Therefore $s(\vec{r}_0+(\vec{u}'+\mu\,\vec{n})\delta t, t_0+\delta t)=0$. Because δt is small but arbitrary, this implies

$$\frac{\partial S}{\partial t}+(\vec{u}'+\mu\,n)\bullet\vec{\nabla}S=0 \tag{5}$$

Assume that there is no slip of the growing inclusion relative to the host medium. If $\vec{u}$ is the deformation velocity of the host,

$$\left.(\vec{u}'+\mu\vec{n})\right|_{0^-}=\left.\vec{u}\right|_{0^+} \tag{6}$$

where 0^+ and 0^- are positive and negative infinitesimals, respectively, and in (Eq.6) they indicate evaluation at $S=0^+$ and 0^-, respectively.

The modeling proceeds by assuming a rock rheology in each domain as outlined in Ortoleva (1994a Chapter 10), for example. Because the interface $S=0$ moves at finite speed, the stresses must be continuous across $S=0$. In particular, the mean stress P^m is continuous:

$$\left.P^m\right|_{0^-}=\left.P^m\right|_{0^+} \tag{7}$$

Assuming the inclusion to be growing from a single solute of concentration c, we adapt the simple growth law

$$\mu=k\,[c-c^{eq}(P^m)], \tag{8}$$

for example; here k is a rate constant and c^{eq} is an equilibrium constant that depends on the stress just inside the body ($S=0^-$) at the given point on the surface of the growing body. In the interior of the body, incompressibility requires

$$\vec{\nabla}\bullet\vec{u}'=0,\ S<0 \tag{9}$$

if no grain growth takes place inside the body. Outside the growing body grain growth/dissolution may be contributing to c. If reactions are taking place in $S>0$, we have

$$\mathrm{tr}(\dot{\underline{\underline{\varepsilon}}}) = \sum_{i=1}^{M} 4\pi\, n_i R_i^2 G_i \tag{10}$$

where $(\dot{\underline{\underline{\varepsilon}}})$ is the rate of strain tensor, n_i is the number of grains per unit volume and R_i and G_i are their volume-effective radii and growth rates, respectively (Ortoleva 1994 Chapter 13).

The preceding model contains many of the features we expect to play a role in determining the morphology of inclusions whose porosity was negligible during growth. For example, the P^m dependence of the growth law (Eq.8) will insure that a rapidly growing bump will be somewhat slowed down by a rise in solubility at the growing tip caused by higher stress (P^m) there. Because the rate laws G_i for the reactions in the host depend on stress we see that the effect is accounted for because a growing bump in the body can cause stress-induced dissolution of minerals in the host. Criteria for morphological instability can be obtained once the spherically symmetric state of growth is determined. The methods used to determine morphological stability of growing amorphous solid particles would carry over most directly to the present problem (Mullins & Sekerke 1963, 1964, Dandekar 1969, Chadam et al. 1987). Finally, if the host medium is unlithified or otherwise much more ductile, then the growing inclusion, the aforementioned stress effects would be minimized, and the problem would reduce to the classical morphological instability of a growing solid except for the possibility of the grain dissolution reactions in the host medium.

Simulation of Metamorphically Differentiated Orbicular Patterns

For illustrative purposes, consider patterning in two dimensions for a two-mineral system. Take quartz and muscovite or garnet to be the two participating minerals. Neglect quartz dissolution for simplicity. Thus the radius of quartz grains R_{qz} is constant if constant initially. With this, we adopt the chemical model (for muscovite)

$$\text{Muscovite} + 3H_2O + H^+ = K^+ + 3X + 3Y \tag{11}$$

where X is $SiO_2(aq)$ and Y is $Al(OH)_3(aq)$. If $[H^+]$ and $[K^+]$ in the pores are relatively large, then they can be taken to be constant. Furthermore, if X = Y everywhere initially, then X = Y everywhere always if their diffusion coefficients are equal ($D_X = D_Y$). Thus the system is reduced to the effective reaction

$$A \Leftrightarrow 6X \tag{12}$$

for the mineral A and aqueous species X. Using the formalism of Dewers and Ortoleva (1989a, 1990a) and Ortoleva (1994a Chapters 10 and 11), the model reduces to the determination of X, $\vec{u}$, P^m, R_A, n_A, and n_{qz}. Finally, we adopt the simple pressure solution model of Ortoleva (1994a Chapter 10) where the molar free energy μ_A of an A grain is given by

$$\mu_A = \mu_A^0 + \frac{\overline{V}_A P^m \kappa_A}{\phi_A \kappa_A + \phi_{qz} \kappa_{qz}}, \tag{13}$$

where μ_A^0 is the molar free energy of muscovite in the unstressed state and $\overline{V}_A$ is the molar volume of muscovite (taken to be constant).

The simulations were carried out in a rectangular domain with periodic boundary conditions to reduce boundary effects as discussed further in Ortoleva (1994a, Chapter 11). In Fig. 11 we see the result of the evolution of a small initial maximum of muscovite volume fraction in the center of the system under no imposed shear (i.e., $\vec{u} = \vec{0}$ at the top and bottom boundaries). Note that muscovite differentiation in the center proceeds to 100%. A muscovite depletion zone is also created. Further out, a halo ring of muscovite concentration is also induced. Apparently, the halo induction phenomenon will continue, ultimately yielding a number of rings, resulting in a mechanochemically differentiated orbicular structure.

Imposition of a shear yields two interesting variants on the development of the depletion zone and halos. In Fig. 12 we see the evolution of a muscovite spot under an imposed shear. Note the induction of the asymmetry in the depletion zone. This asymmetry is similar to that shown in the sample of Fig. 2a. After long times under imposed shear a spot pattern evolves into essentially parallel bands. The effect of shear is not simply to stretch the muscovite segregation ellipse. In time, one ellipse is brought sufficiently close to another that halo

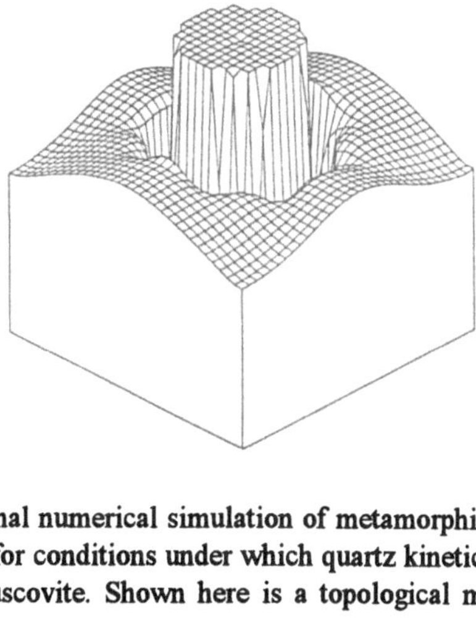

Fig. 11. Two-dimensional numerical simulation of metamorphic spot differentiation in a quartz-muscovite rock for conditions under which quartz kinetics are assumed to be slow relative to those of muscovite. Shown here is a topological map of muscovite volume percent.

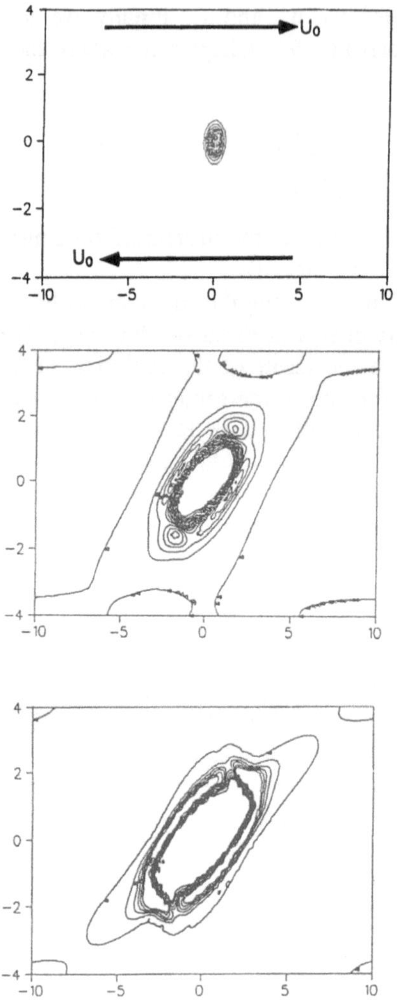

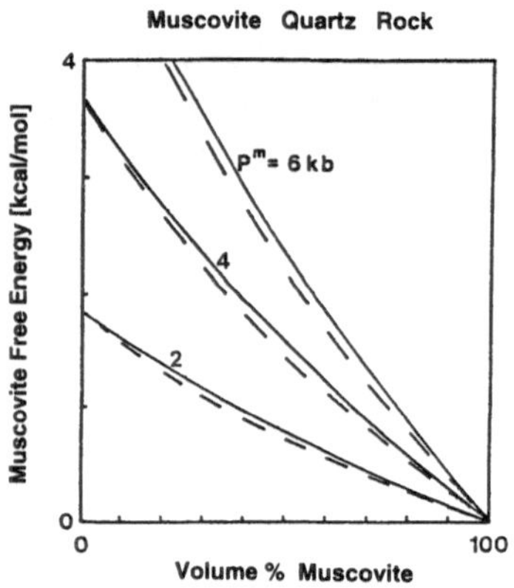

Fig. 12

Fig. 13

Fig. 12. Same as Fig. 11 except that the process takes place under an imposed shear. Note the resemblance to the differentiated spots of Fig. 2.

Fig. 13. Molar free energy of muscovite in a low-porosity quartz-muscovite system as a function of increasing muscovite volume fraction. The solid curve is obtained by the MFEK approach whereas the dashed curve is for the volume fraction-averaged Voight formula (from Dewers & Ortoleva 1989).

induction is interfered with and more complex patterns are produced. This is due to the competition between the natural induction length for a halo and the proximity to other ellipses induced by the imposed shear.

The classic vortex of Taylor occurs when a fluid is subjected to a shear. In that case the inertia of the fluid tends to keep it rotating once rotation commences. Viscosity tends to dissipate such rotation whereas an applied shear tends to sustain it. The critical condition for the formation of a rotation is thus that the applied rate of shear strain must exceed a critical value that increases with viscosity and decreases with fluid mass density (the inertial factor). This condition is never obtained in metamorphic rocks, and hence spiral garnets are not an imprint of a Taylor vortex.

One aspect of spiral garnets can be understood in terms of the rotation of a rigid body in a sheared viscous fluid. As the fluid flows past the body, it tends to rotate it. Indeed, a garnet/quartz rock appears to have increasing viscosity with increasing garnet content. However, this does not explain the arm structure of the spiral garnet suggested in Fig. 2b. Neither does rigid body rotation explain the tendency of the garnet to differentiate out of the undifferentiated medium.

The preceding discussion indicates that an interesting dynamic of rotation and differentiation could underlie the genesis of spiral garnets. The incompressibility of quartz relative to that of garnet suggests that the free energy of garnet in a garnet/quartz rock decreases with garnet volume fraction as for muscovite in Fig. 13. Furthermore, the increase of viscosity with garnet volume fraction can mediate another mechanism of differentiation under compressive stress. Enhanced compressive stress probably exists in the medium around the spiral garnet core as the bulk medium is forced around it due to the applied shear.

With the preceding factors, a scenario exists for differentiating a spiral garnet from an initially uniform medium. Suppose there is a mechanism whereby garnet can segregate from quartz so that spots of garnet can form. This was the case for the pressure solution-mediated segregation of muscovite spots as in the simulation discussed above (Figs. 11 and 12). Imposition of a shear flow will tend to elongate the spot as seen in Fig. 12. However, as the spot intensifies, the viscosity rises, becoming largest in the core. The latter then will tend to rotate when the viscosity contrast becomes sufficiently large. The two elongated ends of the spot are then wound in with the noninertial vorticity as suggested in Fig. 2b. These extensions are more viscous, being of higher garnet content than the ambient medium and hence tend to be drawn in by the rotating core faster than the "lubricating" quartz-rich environment on their flanks. Thus the elongation of a garnet spot creates putative spiral arms spontaneously.

Patterned Agates

The concentric ring and other complex agate patterns of Figs. 3a-c may have developed during the growth of the inclusion. Leaving aside stress considerations as noted in the previous section, one might set forth a surface attachment model analogous to that for zoned crystal growth of Ortoleva (1990, 1994a Chapter 3).

As the quartz lattice is rather resistive to the incorporation of impurities, we expect that a minority species model such as that of Ortoleva (1994a Chapter 3, Ortoleva & Chen 1994a) will capture this phenomenon. In this way the analysis of Ortoleva (1990, 1994a Chapter 3) can be used to study the mosaic and other more complex patterns of Figs. 3a-c. This is clearly a rich area for future research.

Inclusion Replacement Through Inward Growth

A single crystal or polycrystalline mass or amorphous object can be replaced through either an inward advancing front or pervasive reaction. Replacement can occur simultaneously throughout the body due to a uniform change of pressure and temperature or the fact that the initial state was thermodynamically unstable with respect to another state of aggregation, for example, amorphous silica → quartz. Alternatively, the replacement may start at the surface of the initial inclusion and advance inward. This may be because of a favorable site for the nucleation of the new phase being at the contact between the original inclusion and the host. Alternatively, the medium surrounding the original inclusion may contain reactant species which, when combined with the chemical constituents of the original inclusion, make the new phase(s).

Closed System Inward Replacement

Here the initial inclusion is assumed to contain all the chemical components needed to make the new body. Thus the normal flux of all components from the interior of the inclusion vanishes at the contact between the inclusion and the host. Let us first assume that the replacement occurs through an inward propagating front separating the new solid from the original inclusion material.

If $\vec{n}$ is the unit normal pointing in the advancement direction then the flux of component a to the new solid, $-\vec{n} \bullet J_\alpha$ and the influx due to incorporation of material already present, $c_\alpha u$, is equal to the net incorporation rate:

$$-\vec{n} \bullet J_\alpha + c_\alpha u = \sum_{k=1}^{N_a} \omega_{\alpha k} G_k . \tag{14}$$

For dilute solutions all c_α are small relative to the solid molar density(ies). In this case, the $c_\alpha u$ term can be neglected. If this is the case, however, then there is far from sufficient material in the original inclusion for the replacement to be significant, i.e., the "replacement" would be as a thin coating on the bounding surface of the original inclusion – a factor apparently not recognized in Wang & Merino (1990).

An interesting self-organization problem in agate genesis is their common concentric ring composition/color patterns, see Fig. 3a. These are impurities in the solid SiO_2 matrix. Let the impurity be Y. Then consider the simple model

$$SiO_2(aq) \Leftrightarrow SiO_2(s) \equiv A(s) \qquad (15)$$
$$Y(aq) \Leftrightarrow B(s).$$

Then the model takes the same form as that of equation (8) of Ortoleva (1994b, see also Ortoleva 1990). This is in the spirit of Wang & Merino (1990) except that it corrects a number of their misconceptions. For example, in the present notation, they assume $-\vec{n}\bullet\vec{J}_y = \eta\, G_A$ where η is a measure of the number of impurities released as SiO_2 polymers are incorporated into the replacing solid lattice. This would imply that if A is dissolving ($G_A < 0$) then Y is being incorporated. Equally unphysical is the fact that impurity would be created even if it had been depleted by earlier incorporation. Furthermore, their model does not account for their assumption that the original material was SiO_2(gel) or perhaps SiO_2(amorphous solid). The problem is that in this case the majority of SiO_2 is not mobile. To correct these and other misconceptions consider the following type of model

$$Y_m(SiO_2)_n(gel) \Leftrightarrow mY + nSiO_2(aq) \qquad (16)$$

$$SiO_2(aq) \Leftrightarrow SiO_2(s) \equiv A(s) \qquad (17)$$

$$Y(aq) \Leftrightarrow B(s) \qquad (18)$$

$$SiO_2Y(aq) \Leftrightarrow SiO_2 + Y(aq) \qquad (19)$$

With this, (Eq.16) accounts for the breakup/dissolution of the immobile gel matrix and (Eq.19) accounts for a Y-SiO_2 complexing so that if SiO_2(aq) is incorporated into the new solid then the equilibrium (Eq.19) (assuming it is fast) will release Y(aq).

The ramifications of this more realistic model can be profound on the self-organization character of the evolving agate. First, the compositional (i.e., A/B ratio) zoning requires Y incorporation, not just release via the artificial boundary condition (Wang & Merino 1990). Secondly, the stability analysis and the potential for oscillatory growth or morphological instability depends sensitively on the gradient of the species at the growing solid that are key to the feedback. For example, if Y is not incorporated and is a minor constituent ($m \ll n$) then for diffusive flux we have $\vec{n}\bullet\nabla Y = 0$ if one neglects the uY term as in Wang & Merino (1990) and hence the profile of Y is normal to the advancing surface. Thus Y would keep building up at the surface, never decreasing due to incorporation and hence there would be no oscillation.

If the agate is to completely or predominantly fill the volume of the original inclusion then the molar density of SiO_2 (as gel, free SiO_2(aq) or complex) must be near the molar density of the SiO_2(s). Therefore the gradients of SiO_2

concentration ahead of the replacement interface will be minimized. This would certainly not support the very small scale fibrils as being a manifestation of morphological instability, another claim of Wang & Merino (1990).

As agates are often solid or have only a very small void yet still have fibrils and concentric rings, the above considerations are a key to understanding their genesis. Furthermore, the zoning patterns of agates can be very complex; of particular note are the mottled patterns (termed mossy agates--see Fig. 3b) or patterns with various domains within which bands are parallel and the orientation of bands in one such domain has no obvious relation to the periphery of the agate or to bands in neighboring domains (Quick 1963). What seems needed, at least for many systems, is an agate model that can account for both internally generated patterns as well as ones correlated to the boundary.

Open Replacement Systems

In his consideration of agates, Liesegang (1913) attributed their patterning to his interdiffusion precipitation phenomenon. More recent work on Liesegang banding has shown that it can also occur as a post nucleation phenomenon through the instability of the uniform state of Ostwald ripening to the formation of banded, mottled or other patterns (Ortoleva 1978, Feinn et al. 1978, Lovett et al. 1978, Feeney et al. 1983, Sultan et al. 1990, Ortoleva 1994 Chapter 9). Can this competitive particle growth (CPG) model provide an explanation for agate phenomena?

Consider the original inclusion to be a gel or matrix of mineral grains different from the final SiO_2(s) replacement. The broad view of such a system can be as follows. The initial SiO_2(gel) (or other material) is thermodynamically unstable to the nucleation of lower free energy forms such as quartz or impurity-bearing fibrillular SiO_2(s). According to the CPG model the uniform state of this "sol" is unstable to the formation of patterns of domains of larger and smaller particle size. If there is more than one crystallite type, then there may also be a pattern of domains alternatingly dominant in the size of one or the other crystal type (Sultan 1986, Sultan & Ortoleva 1990).

If gradients of composition or other variable are absent, the pattern would appear random (mottled or mossy). If a concentration of SiO_2(aq) is imposed because of the influx of SiO_2(aq) from outside the inclusion (or similarly for other kinetics-affecting agent), the pattern could become well correlated with respect to the inclusion periphery. Thus both mossy and banded agates could have a unified explanation (Sultan 1986, Ortoleva 1994 Chapter 9).

Other details of observed agates are also implied by an unstable coarsening model. The constant influx of SiO_2(aq) from the outside can cause ongoing growth until the sol evolves into a very low porosity grain network. Strong forces (pressure of crystallization) can reorient and coordinate grains into tightly locked structures wherein fibers can be parallel and, if twisted, can have their twist coordinated. Occasional misoriented crystals could recrystallize into correctly oriented ones. Compositional zoning of the tightly packed structure would reflect that of the primary pattern in the sol. Particle size would tend to increase inward because near the inclusion periphery $SiO_2(aq)$ concentration would have been highest initially allowing for the nucleation and survival of many grains. Grains in the interior would be fewer because the nucleation event took place at a somewhat lower $SiO_2(aq)$ concentration. Furthermore, in the interior the Ostwald ripening (leading to fewer but larger crystals) would not be mediated against by the high $SiO_2(aq)$ coming from the inclusion periphery. All these suggestions are consistent with simulations of the CPG model for monodisperse, particle size-distributed and multi-mineralic systems (Ortoleva 1978, Feinn et al. 1978, Lovett et al. 1978, Feeney et al. 1983, Sultan et al. 1990, Ortoleva 1994 Chapter 9--see also Fig. 7).

Acknowledgements. This work was supported by a grant from the Office of Basic Energy Sciences, U.S. Department of Energy (Grant #DE-FG02-91ER14175), a contract with the Gas Research Institute (No. 5092-260-2443), and grants from Sun Microsystems and Intel Supercomputer Systems Division.

References

Bjorkum PA, Walderhaug O (1990) Geometrical arrangement of calcite cementation within shallow marine sandstones. Earth Science Reviews 29: 145.

Chadam J, Ortoleva P, Sen A (1986) Reactive infiltration instabilities. IMA J Appl Math 36: 207-221.

Chadam J, Howison S, Ortoleva P (1987) Existence and stability for spherical crystals growing in a supersaturated solution: IMA J Appl Math 39: 1.

Chadam J, Ortoleva P, Sen A (1988) A weakly nonlinear stability analysis of the reactive infiltration interface. SIAM J Appl Math 48: 1362-1378.

Chadam J, Peirce A, Ortoleva P (1990) Stability of reactive flows in porous media: coupled porosity and viscosity changes. SIAM J Appl Math 51: 684-692.

Chen W, Ortoleva P (1990c) Self-organization in far-from-equilibrium reactive porous media subject to reaction front fingering. In: Walgraef D and Ghoniem NM (eds) Patterns, defects and materials instabilities. NATO ASI Series, Vol. 183, pp 203-220.

Coleman ML, Raiswell R (1993) Microbial mineralization of organic matter: mechanisms of self organization and inferred rates of precipitation of diagenetic minerals. Phil Trans Roy Soc Lond A 344: 69-87.

Dandekar DP (1969) Variation in the elastic constants of calcite with temperature. J Appl Phys 39: 3694.

Dewers T, Ortoleva P (1988) The role of geochemical self-organization in the migration and trapping of hydrocarbons. Appl Geochemistry 3: 287-316.

Dewers T Ortoleva P, (1989a) Mechano-chemical coupling in stressed rocks. Geochim Cosmochim Acta 53: 1243-1258.

Dewers T, Ortoleva P (1989b) The self-organization of mineralization patterns in metamorphic rocks through mechano-chemical coupling. J Phys Chem 93: 2842-2848.

Dewers T, Ortoleva P (1990a) Geochemical self-organization III: a mechano-chemical model of metamorphic differentiation. Am J Sci 290: 473-521.

Dewers T, Ortoleva P (1990b) A coupled reaction/transport/mechanical model for intergranular pressure solution, stylolites, and differential compaction and cementation in clean sandstones. Geochim Cosmochim Acta 54: 1609-1625.

Dewers T, Ortoleva P (1990c) Force of crystallization during the growth of siliceous concretions. Geology 18: 204.

Dewers T, Ortoleva P (1993) Formation of stylolites, marl/limestone alternations, and clay seams through unstable chemical compaction of argillaceous carbonates. In Wolf KH, Chilingarian GV (eds) Diagenesis Vol. IV, in press.

Feeney R, Schmidt SL, Strickholm P, Chadam J, Ortoleva P (1983) Periodic precipitation and coarsening waves: applications of the competitive particle growth model. J Chem Phys 78: 1293-1311.

Feinn D, Scalf W, Schmidt S, Wolff M (1978) Spontaneous pattern formation in precipitating systems. J Chem Phys 69: 27.

Fisher GW (1970) The application of ionic equilibria to metamorphic differentiation: an example. Cont Min Petrol 29: 91.

Frondel C (1985) Systematic compositional zoning in the quartz fibers of agates. Am Min 70: 975.

Haase S, Ortoleva P, Feinn D, Chadam J (1980) Oscillatory zoning in plagioclase feldspar. Science 209: 272.

Hedges ES, Myers JE (1926) The problem of physico-chemical periodicity. Longmans-Green, New York.

Liesegang RE (1913) Geologische Diffusionen. Steinkopff, Dresden Leipzig.

Loberg B (1963) The formation of a flecky gneiss and similar phenomena in relation to the migmatite and vein gneiss problem. Geol Foren Stockholm Forh 85: 2.

Lovett R, Ortoleva P, Ross J (1978) Kinetic instabilities in first order phase transitions. J Chem Phys 69: 947.

Maliva R, Siever R (1988) Mechanisms and controls of silicification of fossils in limestones. J Geol 96: 387.

McBirney AR, Noyes RM (1979) Crystallization and layering of the Skaergaard Intrusion. J Petrol 20: 487-554.

Ortoleva P (1978) Selected topics from the theory of physico-chemical instabilities. In Eyring H (ed) Theoretical chemistry, periodicities in chemistry and biology, Vol. IV Academic Press, 235-286.

Ortoleva P (1990) Role of attachment kinetic feedback in the oscillatory zoning of crystals grown from melts. In Ortoleva P, Hallet B, McBirney A, Meshri I, Reeder R, Williams P (eds) Self-organization in geological systems: proceedings of a workshop held 26-30 June 1988, University of California Santa Barbara. Earth Science Reviews 29: 3-8.

Ortoleva P (1994a) Geochemical self-organization. Oxford University Press, New York.

Ortoleva P (1994b) Self-organized zoning in crystals: free boundaries, matched asymptotics and bifurcation. In Proceedings of the workshop on application of pattern formation, Fields Institute, March 23-28 1993, to appear.

Ortoleva P, Merino E, Strickholm P (1982a) Kinetics of metamorphic layering in anisotropically stressed rocks. Am J Sci 282: 617.

Ortoleva P, Chadam J, El-Badewi M, Feeney R, Feinn D, Haase S, Larter R, Merino E, Strickholm P, Schmidt, S (1982b). Mechanisms of bio- and geo-pattern formation and chemical signal propagation. In Reichl LE and Schieve WC (eds) Instabilities, bifurcations and fluctuations in chemical systems. Univ. of Texas Press, Austin.

Ortoleva P, Auchmuty G, Chadam J, Merino E, Moore C, Ripley E (1986) Redox front propagation and banding modalities. Physica 19D: 334-354.

Ortoleva P, Merino E, Chadam J, Moore CH (1987a) Geochemical self-organization I: reaction-transport feedbacks and modeling approach. Am J Sci 287: 979-1007.

Ortoleva P, Merino E, Chadam J, Sen A (1987b) Geochemical self-organization II: the reactive-infiltration instability. Am J Sci 287: 1008-1040.

Pettijohn FJ (1975) Sedimentary rocks, 3rd ed. Harper and Row, New York.

Quick, L. (1963) The Books of Agates (Radnor, Chilton).

Rosenfeld JL (1970) Rotated garnets in metamorphic rocks. GSA Special Paper No 129.

Sultan R (1986) macroscopic waves and patterning in nonequilibrium physico-chemical systems. Ph.D. thesis, Indiana University.

Sultan R, Ortoleva P (1993) Periodic and aperiodic macroscopic patterning in two precipitate post-nucleation systems. Physica D 63: 202-212.

Sultan R, DePasquale F, Tartaglia P, Ortoleva P (1990) Bifurcation of the Ostwald-Liesegang supersaturation-nucleation-depletion cycle. In Ortoleva P, Hallet B, McBirney A, Meshri I, Reeder R, Williams P (eds) Self-organization in geological systems: proceedings of a workshop held 26-30 June 1988, University of California Santa Barbara. Earth Science Reviews 29: 163-173.

Wang Y, Merino E (1990) Self-organizational origin of agates: banding, fiber, twisting, composition, and dynamic crystallization model. Geochim Cosmochim Acta 54: 1627-1638.

The Formation of Manganese Dendrites as the Mineral Record of Flow Structures

Juan M. García-Ruiz[1], Fermín Otálora[1], Antonio Sanchez-Navas[2] & [3]Francisco J. Higes-Rolando
[1]Instituto Andaluz de Geología Mediterránea. CSIC-Universidad de Granada. Fuentenueva s/n. Granada 18002. Spain
[2]Departamento de Mineralogía y Petrología, Universidad de Granada. Fuentenueva s/n. Granada 18002. Spain.
[3]Departamento de Química Inorgánica. Universidad de Extremadura. Campus Universitario. Av. de Elvas s/n. Badajoz 06071. Spain.

Abstract. Manganese oxide patterns known as "pyrolusite dendrites" are explained as the result of the mineral record of flow structures in porous media. This interpretation is supported by a) fractal characterization, b) Mn profiles across the mineral pattern and the matrix rocks, c) structures reminiscent of flow pattern observed at the scale of the grain size of the matrix rocks, d) the lack of long-range order of the manganese oxide particles and e) the existence of clays and quartz grains of colloidal size intimately linked to the manganese particles. This interpretation also explains other fractal and non-fractal patterns accompanying the beautiful treelike fractal forms associated with these manganese oxide patterns.

Introduction

As pointed out by Potter & Rossman (1979), pyrolusite dendrites is a somewhat misleading name for the well-known tree-like patterns formed by manganese oxide minerals displayed by all Natural History museums. Because of their very common occurrence, the knowledge of the genetic conditions of the manganese and iron dendrites would be of great practical interest in understanding geological environments. Nevertheless, they are considered today to be rather meaningless structures in the deciphering of geological environments, probably due to the uncertainty of their genesis.

These amazing tree-like patterns of manganese dendrites are fractal objects, i.e. scale-invariant structures (Mandelbrot 1982). In fact, the non-trivial dilation symmetry of these structures was suggested by Swartzlow (1934) sixty years ago when he realized that as for megascopic observation, "the microscope reveals a more detailed splitting and bifurcation of the branches as well as more details of the irregularities of the edges". This author suggested that dendrites are formed by distribution of suspended material due to surface tension of evaporating waters in a epigenetic environment and Schoedler (1851 cited by Swartzlow 1934) pointed out that the pattern obtained by pressing a clay layer confined between two glass microslides is reminiscent of dendrites. Recently, Chopard et al. (1991) proposed that dendrites can be explained by a diffusion controlled reaction

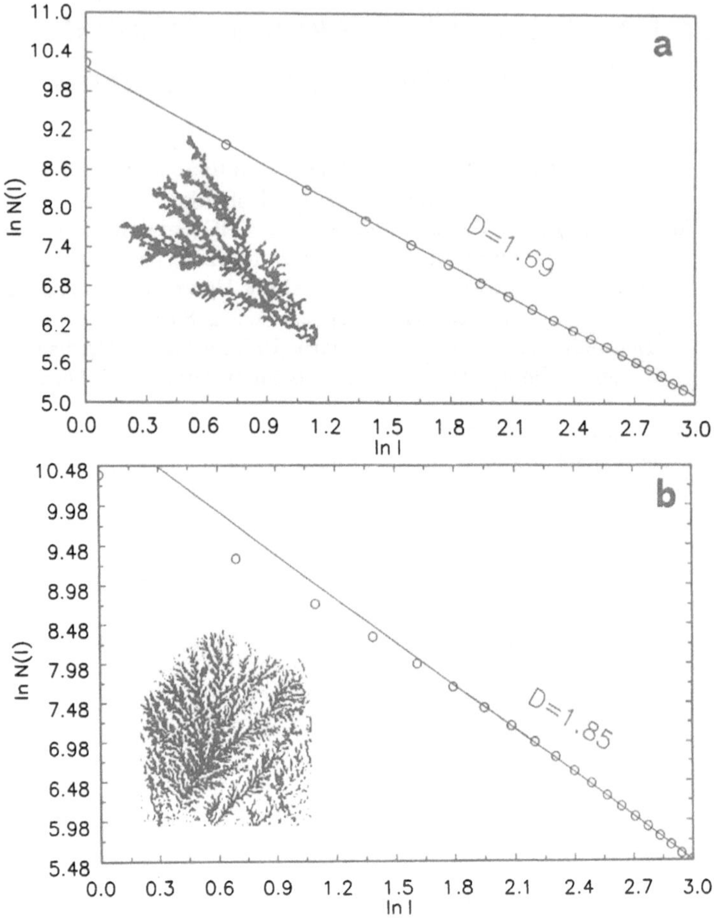

Fig. 1. Two digitized images of manganese dendritic patterns and their respective fractal dimension obtained by the box-counting method. The number of boxes N(l) of side length l (pixels) is plotted versus l. Sample a) is a dendritic pattern observed between two laminae of fine-grained calcarenite from a flysh facies. Sample b) was found on a cracked surface of a limestone. Deviation from the straight line for low values of l in Fig 1b are interpreted as the effect of a inner cut-off at $l = 4.5$ pixels.

mechanism. In this chapter we present data supporting the idea that manganese and iron dendrites are the mineral record of flow instabilities. The driving force for the flow may arise from any type of geological pressure gradient (for instance in a hydrothermal field) but in the absence of such an external driving force for injection, we propose a self-fed mechanism operating during early diagenesis and we discuss the geological plausibility of this process.

Pattern Characterization

We list below several observational and analytical facts relevant to "pyrolusite dendrites". The following account was made based upon our observations in Sierra Elvira (Granada, Spain) and the flysch of Tarifa (Cádiz, Spain) and from descriptions collected from the literature:

1. As stated above, pyrolusite dendrites are fractal objects. We have selected for this study two dendrite-like patterns from Tarifa (southern Spain) and one specimen kindly supplied by Prof. Michael Russell from the University of Glasgow. They were digitized (512 by 512 pixels) and their fractal dimension D was measured by the box-counting method (Barnsley, 1988; Vicsek, 1989) using an *ad hoc* computer program. The D value obtained from these selected specimens was 1.69±0.01 (Fig. 1a) which practically matches that for Laplacian growth patterns. This result agrees with Van Damme's conjecture (1989) that the general appearance of the manganese dendrites could be the result of either a diffusion limited aggregation (Witten & Sander 1983) or a viscous fingering (Vicsek 1989) process. It is important to note that the above specimens were selected by a previous visual inspection of the branching structures and chosen as typical manganese dendrites. However, these beautiful patterns are usually accompanied by other fractals in which the width of the branches and the branching behaviour are different and there are even non-dendritic and non-fractal patterns associated with the beautiful dendrites having a D value of 1.70. When the above procedure of selection is eliminated, other values for D are currently found (Figure 1b and Chopard et al. 1991). It follows that any explanatory theory of manganese pattern formation must explain such a diversity of patterns.
2. The infrared study carried out by Potter & Rossman (1979) showed that "pyrolusite dendrites" are formed by different mineral associations, their composition being neither singular nor universal. According to these authors, ring-structure manganese oxides such as Romanecheita $(Ba,H_2O)_2Mn_5O_{10}$ and those of the Hollandite ($A_{1-2}Mn_8O_{16}$; with A mainly Ba,K,Na,Pb, Mn^{4+},Mn^{2+}) group are the main components of manganese dendrites. Accessory minerals, mainly carbonates and silicates, are not characteristic and in most of the cases are the same as those appearing in the host rocks. These authors also find that iron oxides, when present, are minor components. However, it has to be noted that non-crystalline iron minerals also form dendritic tree-like patterns.
3. Manganese dendrites are low-crystallinity products. This conclusion was reached using several techniques. The black scab (crust) of manganese minerals forming the dendritic patterns was removed with a graver. The collected material was placed on a micro-sample holder and then analyzed by the X-ray diffraction powder method using filtered copper radiation. The bands corresponding to manganese oxides are broad and were investigated at

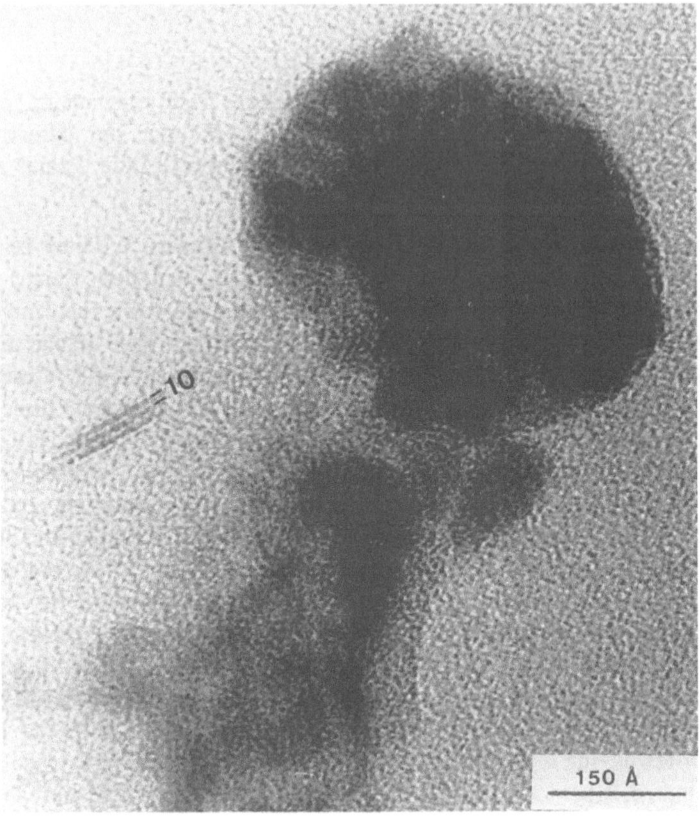

Fig. 2. High resolution electron microscopy view of Mn oxide particles showing the lack of long-range order. The small particle showing lattice structure is a clay mineral with a 10 Å interplanar d-spacing.

slow exploration (static record during ten seconds for every 0.02°). When attempting to obtain the degree of crystallinity using the line profile method (Sanchez-Navas 1989) we found a crystallinity degree below the range of resolution for that method. Therefore, in order to obtain the coherent domain size we used high resolution transmission electron microscopy. For this purpose a suspension of the manganese minerals was made and part of the solid was dispersed on a holey carbon grid. A transmission electron microscope operated at 100 kV, with an objective aperture of 90 μm was used in order to add 000 transmitted beam and diffusion haloes. Low beam current and minimum exposure were used to avoid beam damage. The study showed the absence of even short range three-dimensional periodical structure and the clear colloidal nature of the Mn dendrites (Figure 2). This means that the kinetics of the precipitation process has been fast and irreversible enough to prevent any internal reorganization of the growth units into a periodical lattice.

4. In agreement with 3), scanning electron microscopy views (Figure 3) clearly show that there is no crystallographic continuity in the precipitate. The scanning electron microscope was operated at 30 kV in back-scattered electron (BSE) mode and at 15 kV in secondary electron (SE) mode. To ensure that we were scanning on the Mn oxides, we previously obtained a BSE image of the specimen and then in the same place we turned to SE images. Under SEM, the material forming the dendrite shows a colloidal aspect and appears coating the mineral grains forming the matrix rock. Figure 3 also shows the cracked surface of the colloidal phase and its penetration through intergranular joints.
5. Chemical analyses show that the Mn-concentration profile has a negative gradient towards the host rock over a narrow band (a few micrometers) and then attains a zero-value outside the precipitate. This information was obtained by using a electron microprobe operated at 30 kV in BSE mode. Figure 4 shows an example of the results reported above.
6. Manganese (and iron) dendrites appear on or inside many different types of rocks, such as limestone and sandstone which suggests that the mineralogy of the host rock is not a limiting factor.
7. Unlike desert varnish (Perry & Adams 1978, Carlton 1978) manganese dendrites are commonly associated with cracks and sedimentary laminations, or in general with quasi two-dimensional spaces.

Statement of the Problem

Crystalline dendritic patterns are ordered fractals where the anisotropy of the crystal structure governs the growth process. On the contrary, the so-called random dendritic fractals are patterns where the fluid structure created in the growth environment dominates crystal anisotropy. Therefore, it is a requirement for the formation of the random dendrites that the kinetics of the phase transition be faster than the kinetics of the pattern formation and consequently, the precipitation process must be a highly irreversible one occurring at conditions far from equilibrium. It follows from this consideration that the precipitate formed must be a low crystallinity product and that the whole growth pattern must lack any geometrical relationships derived from the crystal structure. According to observational facts 1) 3) and 4), the so-called manganese dendrites belong to this group of random fractals. Beyond systematic terminology, this distinction is important in building a model of their morphogenetical mechanisms.

According to observational fact 1), some of the patterns displayed by manganese dendrites have a visual appearance and a fractal dimension which are typical of those patterns derived from the Laplace equation. Today we know

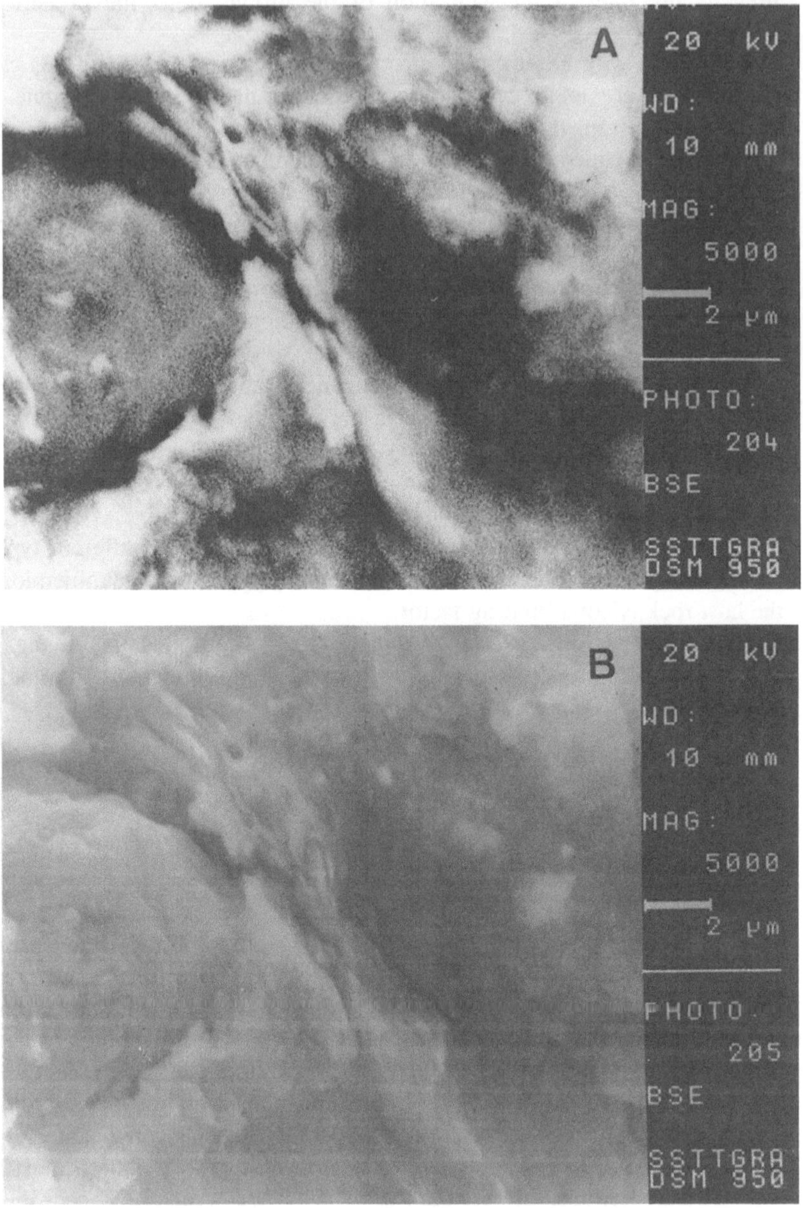

Fig. 3. Scanning electron microscopy views of Mn oxihydroxide minerals into the intergranular space between two calcitic grains. To ensure that we were scanning the Mn oxides, we obtained previously a back scattered electron image of the specimen (a) and then we turned to secondary electron mode (b). Note that the material forming the dendrite (the whitish regions in 3a) has a colloidal appearance, showing the penetration of the Mn coated clay minerals (top-left part of the image) through intergranular joints.

several chemical and physical Laplacian growth processes (Vicsek 1989) but only two of them are relevant to the formation of the Mn dendrites. The first of these processes leading to two-dimensional objects with fractal dimension around 1.7 are irreversible growth processes governed by the diffusional control of the nutrient phase, the so-called diffusion-limited aggregation (DLA) model (Witten & Sander 1983). The second one is the viscous fingers (VF) process (Van Damme 1989, Feder 1988, Vicsek 1989) which is known to produce DLA-type patterns when the less viscous fluid invades a more viscous and non-Newtonian one. Despite the fact that one of the most important findings of the new physics of pattern formation is that both DLA and VF display the same fractal dimension and that they can be described by the same formal equations (Daccord et al. 1986, Paterson 1984), it is important to note that from a phenomenological viewpoint, DLA and VF are clearly different. As suggested by Van Damme, to reveal the origin of the pyrolusite dendrites reduces to the choice of one of these growth phenomena.

DLA Versus VF Mechanism

A diffusion limited aggregation mechanism requires a starting point or linear sink of energy to create a positive concentration gradient in the surrounding fluid, which must be maintained throughout the growth process. As the driving force for precipitation here is the redox potential leading Mn^{2+} ions to be oxidized, such a condition is difficult to fulfil in geological environments. Furthermore, because of the two-dimensional geometry of the manganese dendrite, the chemical potential gradient must be created and maintained in planes parallel to the bedding surfaces where chemical gradients are difficult to imagine in a geological environment which, in accordance with observational fact 6), must be a general one. Another argument against the DLA-type mechanism is the Mn concentration profile around the manganese oxide precipitate (observational fact n° 5, Figure 4). It is clear that a pure DLA mechanism cannot explain a sharp negative concentration gradient outside the dendrite. It must also be taken into account that the manganese dendrites are in Tarifa flysch associated with cracks and always parallel to sedimentary laminations. Consequently, the supply of Mn for such a mechanism should be placed in the planar region between cracks, that is, into the region where no Mn has been detected. Recently, Chopard et al. (1991) proposed a genetic mechanism supported by fractal characterization and computer simulations. They suggest a model in which two chemical species counterdiffuse and interact, forming a precipitate after an induction time during which the reaction products can still diffuse. Chopard et al. were able to obtain the different values of fractal dimension they measured for natural dendrites but have not yet translated the computer simulation into a geochemical model. The colloidal structure of the manganese oxides, the manganese profile concentration and the coexistence of well-differentiated Mn and Fe dendrites seems difficult to explain by a reaction-controlled precipitation with diffusional transport. Nevertheless, the existence of banding structures and isolated DLA-like clusters suggests that the Chopard et al. model is also worth developing and that further studies on the interaction between physical instabilities and chemical reactions, that is, on the mineral record of flow patterns, are required. Another mechanism for the formation of "pyrolusite dendrites" was proposed by Swartzlow (1934) supported by laboratory experiments of evaporation of solutions under capillary forces. This mechanism led to the type of pattern later studied by Hazlehurst (1941) who named them"structural precipitates", which are interesting fractal aggregates but in our opinion this evaporation mechanism cannot explain the observational facts described above.

Viscous fingering is a pattern created when one fluid pushes another of higher viscosity. The experimental device for studying two-dimensional viscous flow is the Hele-Shaw cell, i.e., two glass plates separated by a gap typically of two millimetres. Pionering experiments using this geometry were carried out by Schoedler (1851 cited by Swartzlow 1934) to create air dendrites on clays. The

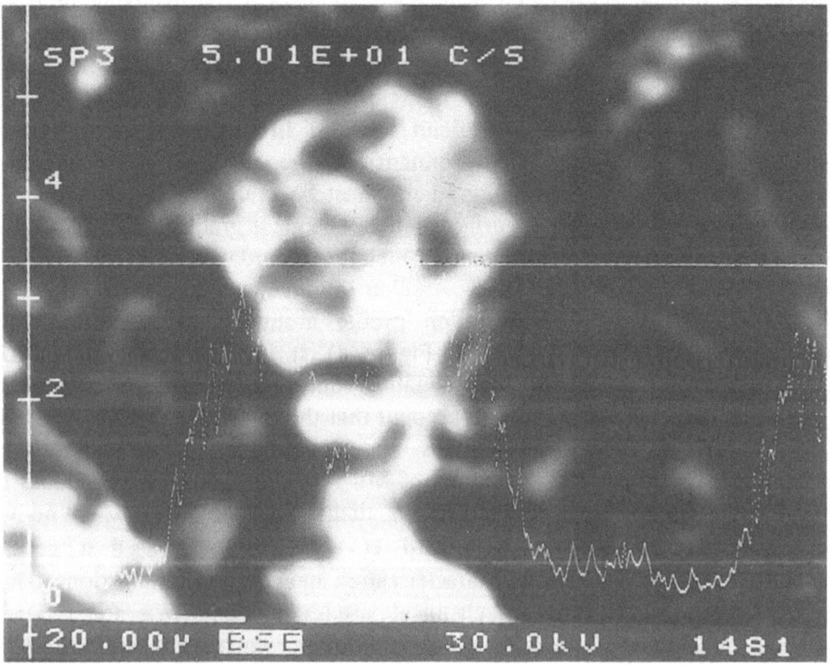

Fig. 4. Mn profile across a dendrite obtained by electron microprobe chemical analysis. Image width is around 70 μm.

unstable interface between the two fluids bifurcates iteratively and therefore a branched treelike structure reminiscent of manganese dendrites is obtained. The main parameter controlling the interfacial pattern is the critical wavelength

$$\lambda = \frac{b^2}{\varpi^2} * \frac{\gamma}{\nu V},$$

where b and ϖ are the thickness and the width of the cell respectively, μ is the difference in fluid viscosity, γ the surface tension and V the velocity of the injected fluid. Today it is well known that viscous fingering have a fractal behaviour when either the pushed fluid is a non-Newtonian one or it occurs into porous media. The type of fractal obtained depends on the applied flow rates and the surface tension between the fluids (Lenormand & Daccord 1988). In porous media, for very low values of V (or according to the previous equation, for larger cell thickness) the fluid provokes an invasive percolation pattern with a calculated fractal dimension $D = 1.82$ (Stauffer 1985). This pattern changes for larger V values and after a crossover it becomes a fractal structure with a $D = 1.70$, visually reminiscent of DLA patterns (Lenormand & Daccord 1988). This situation accounts very well for the geometrical properties of manganese dendrites and their related structures. First, as described above, the fractal dimension of typical Mn dendrites with a DLA-like morphology matches the one expected for viscous fingers at high injection rates. Because of their beauty, Mn dendrites are by far the best known patterns, but as stated in observational fact n°1, there are many other accompanying deposits which show other kinds of patterns. Among them, invasive percolation and other patterns with fractal dimensions and visual geometry difficult to explain by aggregation under diffusional control are the most developed. These accompanying non-dendritic patterns also can be explained by the viscous fingering mechanism. For a given value of the injection flow, the parameter b may adopt small local changes but large enough to produce the well-known change of regime provoked by variations in V. Thus, the VF mechanism explains not only the beautiful treelike fractal growth, but also the fractal and non-fractal patterns accompanying them, which constitute the majority of manganese oxides. Unlike a DLA-type mechanism, the viscous fingering approach is consistent with the manganese profiles across the rock matrix containing the dendrites (observational fact n°5). Moreover, the SEM views (Figure 3) of Mn dendrites suggest the existence of flow patterns surrounding the quartz and calcite grains of the rock and high resolution electron microscopy shows that the Mn oxides closely adhere to clay particles and quartz grains with a colloidal size which also suggests that a high viscosity colloidal fluid was pushed by the Mn^{2+}-rich injected fluid.

The above discussion indicates a genetic mechanism for the formation of manganese dendrites based on the mineralization of fluid structures. These fluid structures occur during the invasion of sedimentary discontinuities or in general, low attachment surfaces, by a fluid rising through cracks. The cracks (or sedimentary laminations) are the conduits for Mn^{2+} and Fe^{2+}, and the surfaces of low cohesion associated with them allow the development of narrow gaps which in many cases are the *loci* for pattern formation. It is therefore reasonable to think that the existence of rough impermeable pairs of quasi-parallel surfaces with a gap between them led to the formation of natural Hele Shaw cells.

The most intuitive source (and probably the most general one) of Mn^{2+} and Fe^{2+} is the rise of enriched solutions through the cracks. However, even in the absence of such an "external" source, we propose a genetic model using a self-fed system that could account for those geological scenarios where the existence of an external source of manganese and iron solutions is difficult to support. We start from a sedimentary material undergoing fracturing and compaction during early diagenesis. It follows that the host rock body to be injected must be at least partially cemented and must preserve some bedding discontinuities. Also, the injected fluid must displace another of higher viscosity. Thus, the pattern formation should be generated during diagenesis, when a colloidal suspension fills the interbedding laminations and cavities of the sedimentary materials. Under these conditions, the injection of a rising solution rich in Mn^{2+} into a natural Hele Shaw cell with rough surfaces containing a colloidal suspension will produce either the dendritic pattern with D=1.70, the invasive percolation pattern with D=1.82, or an intermediate pattern somewhere in the range between these two values (Lenormand & Daccord 1988). The width and thickness of the gap, the roughness of the surfaces, the pressure of the injection fluid and the viscosity of the pushed fluid are the parameters controlling the great variety of dendritic and non-dendritic patterns observed in field studies and extending the range of possible D values. From the Eh-pH diagram for Mn and Fe oxides, sediments containing Mn^{4+} and Fe^{3+} compounds are reduced after being buried below the anoxic-oxic interphase, and thus enriched in Mn^{2+} and Fe^{2+}. Compaction and cracking events provoke the upward injection of these solutions which travel through the cracks of the sedimentary body and invade the upper alkaline and oxygenated zone, pushing the colloidal suspension that fills the sedimentary laminations and consequently forming the dendritic patterns. Under such conditions Mn^{2+} will precipitate as γ-MnOOH (via $Mn(OH)_2$ Mn_3O_4 plus ß-MnOOH) and Fe^{2+} as γ-FeOOH (Giovanoli 1980), both in the form of low crystallinity colloidal precipitates. The Mn^{3+} oxides are intermediate phases that later convert to manganese of higher oxidation states. This mechanism of formation also explains the Mn-concentration profile obtained by electron microprobe chemical analysis (Fig. 4), which has a negative gradient towards the host rock over a narrow band of a few micrometers and attains a zero-value outside the precipitate (a similar behaviour was found for Fe^{3+}). Finally, preservation of the pattern can be produced by three mechanisms working simultaneously: a) the cementation of the host rocks which obstructs the intraporous diffusion of reducing agents, b) the plugging of the system by the formation of colloidal iron oxides, and c) the exhaustion of the source of reducing agents. We have simulated this mechanism in the laboratory. A reservoir was filled with a colloidal suspension of γFeOOH and γMnOOH oxide particles.

Three 5 mm thick glass disks were piled and the surfaces between them coated with tooth paste, a material made 99% of colloidal silica. The pile was immersed in the manganese solution and then a set of fractures was provoked by impacting a hammer on the glass pile. After a couple of minutes, dendritic patterns were observed in the silica thin layer (Fig. 5). A different chemical system was also tried: the colloidal silica was replaced by a suspension of clay with BaO solution and the pile was now immersed in a true solution of Fe^{2+} and Mn^{2+}. Details on the similarities and differences of these laboratory model and real manganese dendrites will be published elsewhere.

Fig. 5. Laboratory analogous of the self-fed mechanism proposed in this work.

Acknowledgements. This work was carried out under financial support from CICYT Project PB92-1137 and the Junta de Andalucía. We acknowledge Tamás Vicsek, Eugene Stanley and Michael Russell for their useful comments on this work.

References

Barnsley M (1988) Fractals Everywhere. Academic Press Inc, Boston.
Carlton CA (1978) Desert varnish of the Sonoran desert. Optical and electron probe microanalysis. J Geol 86: 743-752.
Chopard B, Herrmann HJ, Vicsek T (1991) Structure and growth mechanisms of mineral dendrites. Nature 353: 409-412.
Daccord G, Nittmann J, Stanley, H.E. (1986) Radial viscous fingers and diffusion limited aggregation: fractal dimension and growth sites. Phys Rev Lett 56: 336-339.
Feder J (1988) Fractals. Plenum Press, New York.
Hazlehurst TH (1941) Structural precipitates: the silicate garden type. J Chem Education June: 286-290.
Giovanoli R (1980) On natural and synthetic manganese nodules. In: Varentsov IM and Grasselly Gy (ed) Geology and Geochemistry of Manganese. Schweizerbart, Stuttgart, pp 191-199.
Lenormand R, Daccord G (1988) Flow patterns in porous media. In: Stanley HE and Ostrowsky N (ed) Random Fluctuations and Pattern Growth. Kluwer Academic Publishers, Dordrecht, pp 69-74.
Mandelbrot BB (1982) The fractal geometry of Nature. Freeman, San Francisco.
Paterson L (1984) Diffusion limited aggregation and two fluid displacements in porous media. Phys Rev Lett 52: 1621-1624.
Perry RS, Adams JB (1978) Desert varnish: evidence for cyclic deposition of manganese. Nature 276: 489-491.
Potter RM, Rossman GR (1979) The tetravalent manganese oxides: Identification, hydration and structural relationships by infrarred spectroscopy. Am Miner 64: 1199-1218.
Sanchez-Navas A (1989) Aplicación del análisis de perfil de línea al estudio de la cristalinidad mineral. PhD. Thesis. Universidad de Granada, Granada
Stauffer D (1985) Introduction to Percolation Theory. Taylor and Francis, London.
Schoedler C (1851) Elements of Geology and Mineralogy. Joseph Griffin & Co, London, p 56.(Cited by Swartzlow, 1934).
Swartzlow C (1934) Two dimensional dendrites and their origin. Am Miner 19: 403-411.
Van Damme H (1989) Flow and interfacial instabilities in newtonian and colloidal fluids. In: Avnir D (ed) The Fractal Approach to Heterogeneous Chemistry. John Wiley & Sons, Chichester, pp 199-226.
Vicsek T (1989) Fractal Growth Phenomena. World Scientific, Singapore.
Witten TA, Sander LM (1983) Diffusion Limited Aggregation. Phys Rev B 27: 5686-5697.

Structure and Fractal Properties in Geological Crystallization Processes Due to Nucleation and Growth

Ingo Orgzall & Bernd Lorenz
Hochdrucklabor bei der Universität Potsdam
Telegrafenberg A43, D-14473 Potsdam, Germany

Abstract. Based on a continuum model of nucleation and growth in two and three dimensions we present results for grain size and cluster size distributions for different nucleation and growth mechanisms. The percolative properties of these systems are determined with respect to percolation threshold and appropriate exponents for volume, surface, and hull of the clusters. The universal behaviour of these systems is confirmed. The present results are applicable for the description of crystallization processes in igneous rocks as well as for understanding the structure of pore spaces in sedimentary rocks.

Introduction

Crystallization and precipitation from the melt or a solution play an important role in many geologic processes as, for instance, in magmatism or in phase separations in solutions. The physical properties of the new phase are influenced by the microstructure resulting from this phase transition. A usual description of these processes is given by the nucleation and growth model for first order phase transformations. Critical nuclei of the thermodynamically stable phase are generated randomly and grow at a distinct rate. Where they impinge on one another larger and larger clusters form. The microstructures as described e.g. by grain and cluster size distributions are accessible by experiments as e.g. by investigations of thin sections of rock material, and are characteristic for the different physical processes that control the transition. Quantities that influence these structures are first of all the nucleation and growth rates. Thus, the investigation of characteristic features occurring during such transition could give some important insights into the underlying physical mechanisms and geologic formation conditions.

At a certain stage of the transformation - the percolation threshold - one cluster spans through the system. Here, the physical properties of the whole system change drastically, especially pronounced for electrical transport and liquid permeability or mechanical stability in the case of fracture propagation. The properties like volume, surface, and hull of the clusters become fractal at this point. These structures are usually described by continuum percolation models.

On the other hand the continuum models are also relevant for describing different phenomena in the geosciences. Random void models are frequently used to describe the pore space and the structure of sedimentary rocks. Geologic processes occurring within a larger rock matrix are often influenced by structure, size distribution, and extension of the pore space. Especially transport and storage properties of oil and gas reservoirs are determined by volume and surface of these voids and their connectivity (Thompson et al. 1987). The mechanical behaviour and the stability depend on the occurrence and propagation of cracks and have essential influence on seismic activities.

The Model

Crystallization as appearing in rock formation processes from the melt is described by the Avrami model (Christian 1965). The random nucleation of nuclei proceeds with rate I(p,T,t), where p, T and t denote pressure, temperature, and time, respectively. The nuclei grow with linear velocity v(p,T,t). If different nuclei impinge on one another during their growth larger clusters are formed. If the largest cluster spans through the system, percolation occurs and the physical properties of the system change qualitatively. The structures and the size distributions of grains and clusters strongly depend on both physical quantities I and v. While the onset of percolation depends on the model, the structure of the system exhibits fractal properties that are equal for different models belonging to the same universality class.

Under fixed thermodynamic conditions, i.e. p and T are kept constant, the transformation degree Φ follows from the Avrami equation

$$x(t) = 1 - \exp\left[-\alpha \int_0^t I(t_1) \left[\int_{t_1}^t v_{t_1}(t_2)\,dt_2\right]^D dt_1\right] \tag{1}$$

where $v_{t1}(t_2)$ denotes the growth velocity at time t_2 of nuclei that were created at time t_1. D gives the spatial dimension and α is a geometrical constant depending on the shape of the nuclei.

The following models for nucleation and growth kinetics represent typical situations:

(i) Continuous nucleation
Nuclei of the growing phase form continuously during the whole transformation time in the untransformed volume. The nucleation rate is given by $I = I_0 =$ const.

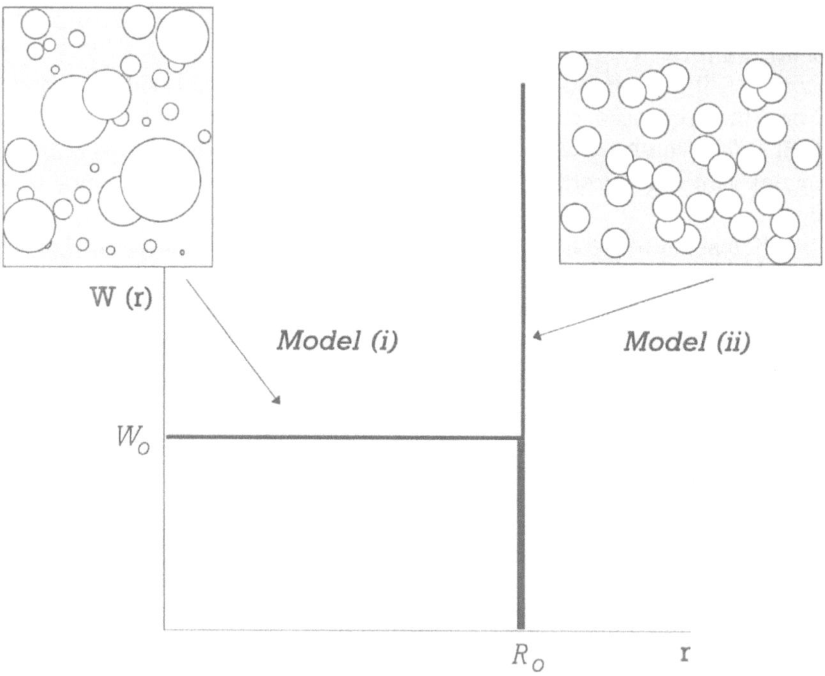

Fig. 1. Radius distributions for models (i) and (ii).

(ii) Simultaneous nucleation
All nuclei are formed instantaneously at the beginning of the transition (or in a small time interval, short compared with the transformation time), i.e. $I = n_0 \delta(t)$, where n_0 gives the density of the nuclei and δ the Dirac-δ-function.

(iii) Heterogeneous nucleation at grain boundaries
If nucleation is restricted to grain boundaries Eq. (1) is no longer valid due to nonhomogeneous nucleation. At the beginning of the transition ($t \ll 1$) the system behaves like a continuous one. At the end ($t \gg 1$) the preferred nucleation sites are exhausted and one-dimensional growth processes dominate.

The general features of models (i) and (ii) concerning the radius distribution and the related structures at early times are illustrated in Fig. 1.

In the models (i) to (iii) the growth velocity v(t) is assumed to be equal for all growing grains and constant with time. This represents the case of interface limited growth and fixed thermodynamic conditions. However, crystallization processes under isochoric (Lorenz et al. 1991) or adiabatic conditions often are accompanied by changing pressures and/or temperatures due to volume changes and latent heat production, which is more realistic for rock formation processes (Spohn et al. 1988). In this case for the theoretical and numerical analysis of the crystallization process even more sophisticated methods should be applied (Lorenz & Orgzall 1991).

Moreover, if the diffusion of growth units to grain surface is sufficiently slow the growth mechanism changes qualitatively. This results in a dispersion of growth velocities which depend on the grain size. For each individual grain v decreases with time according to $v \sim (t-t_o)^{-0.5}$. $(t-t_o)$ denotes the life time of the grain.

To differentiate between the various influences of nucleation rate and growth velocity and to extract the characteristic differences, the various models are evaluated for the nucleation and growth kinetics discussed above.

The continuum model also describes the pore structure in sediments where the size distribution of the individual pores is determined by the diagenesis.

Numerical Procedure

The simulation process of the random system is performed using a specially developed algorithm (Orgzall & Lorenz 1988, Lorenz et al. 1993) and the microstructures are evaluated with special emphasis to the grain and cluster size distributions, the percolation thresholds, and the fractal cluster properties.

A configuration is generated by randomly placing discs (two dimensional model - 2D) or spheres (three dimensional model - 3D) into a unit volume until the volume fraction Φ is covered. The special nucleation and growth rates were chosen according to the models outlined in the previous section.

The generated structure is evaluated numerically using the technique of Hoshen & Kopelman (1976) extended to continuum problems by Gawlinski & Stanley (1981). After labelling the individual clusters in the system the volume and surface of each cluster and the volume of each grain belonging to this cluster are calculated numerically within a relative error of less than 0.5%. Additionally, in the 2D models the outer surface (perimeter or hull) of the clusters is determined by an exact integration. Periodic boundary conditions are applied for a better simulation of the infinite system.

For the estimation of the percolation threshold Φ_c the extension of the largest cluster is taken into account derived for periodic as well as nonperiodic boundary conditions. This method yields a considerably higher accuracy in determining Φ_c.

This procedure was implemented on a RISC processor system. Systems up to 250 000 discs (2D) and 100 000 spheres (3D) were simulated.

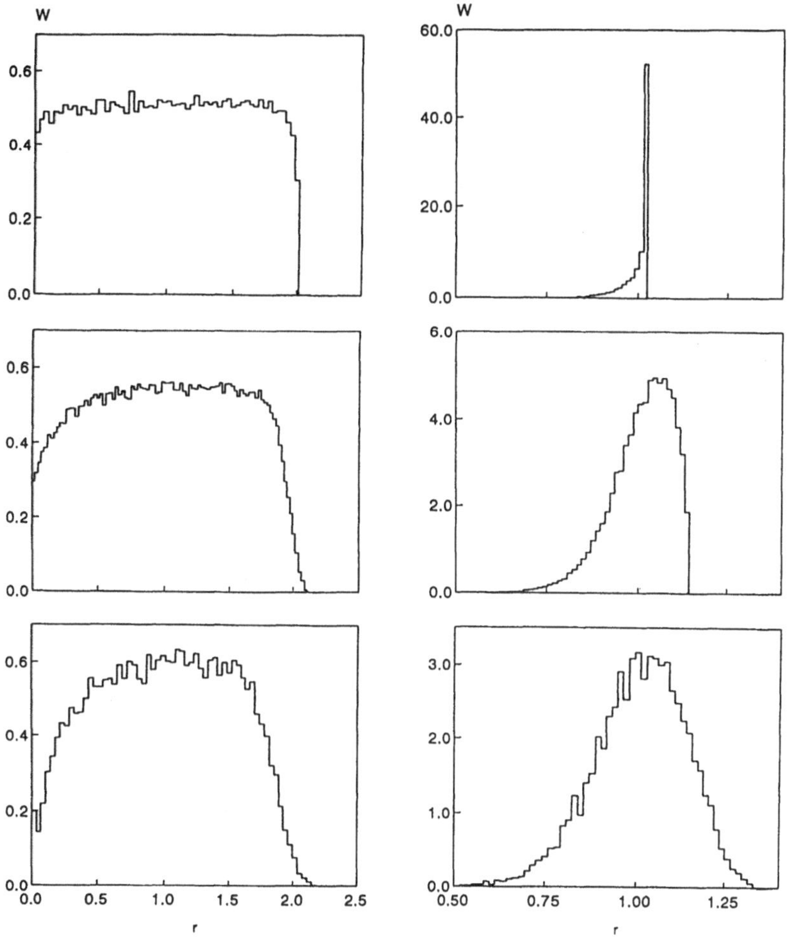

Fig. 2. Evolution of the grain size distribution for models (i) **left** and (ii) **right**. Three different stages of the transition ($\Phi = 0.1$ - above, $\Phi = 0.5$, and $\Phi = 0.9$ - below) are shown.

Grain Size Distribution (GSD)

We shortly summarize the results for 3D models taking into account different kinds of nucleation kinetics. Fig. 2 illustrates the evolution of the grain size distribution W(r) for different stages of the phase transition characterized by the value of Φ ($\Phi = 0.1$, 0.5, 0.9). The grain size is given in units of the average grain length and the distributions are normalized to 1.

Continuous Nucleation

In the limit of small Φ the GSD tends to a rectangular one, since grains of different lengths up to the maximum grain size are present with nearly equal amount. With increasing Φ the GSD rounds mainly at the edges and the width becomes smaller. In general, in the case of continuous nucleation the microstructures are characterized by a rather broad distribution even for transformation degrees $\Phi \approx 1$.

Simultaneous Nucleation

If all nuclei are generated instantaneously at the beginning of the transition, the initial GSD is characterized by a δ-like peak with zero width for $\Phi \to 0$. Increasing Φ leads to impingement of the grains and cluster formation. The narrow peak is broadened until it becomes approximately belt shaped for $\Phi \to 1$. However, compared with the results of continuous nucleation, the width of W(r) remains considerably smaller.

Besides the GSD's, further characteristic quantities (e.g. the total number of grains per unit volume, the average grain volume, the number of nonimpinged single grains, their volume fraction, the degree of impingement etc.) have been shown to exhibit a distinct behaviour for both models discussed above (Lorenz 1989a).

Scaling Behaviour of the Structure Function

The GSD's W(R,t) have a considerable influence on the structure function evaluated from scattering experiments (e.g. neutron scattering, Axe & Yamada 1986). The relevant grain autocorrelation function is given by:

$$G(R,t) = \frac{\pi}{6} \int_R^{2vt} (r-R)^2 (r + \frac{1}{2} R) W(r,t)\, dr \tag{2}$$

2vt is the largest possible grain diameter. It has been shown that this function fulfils a scaling relation (Lorenz 1989b) and for both nucleation mechanisms the determined scaling lengths agree well with simple scaling assumptions, $L = [\Phi(t)]^{1/D+1}$ (continuous case), and $L = c[\Phi(t)]^{1/D}$ proposed by Axe & Yamada (1986). The average grain size, however, deviates from the calculated scaling length which means that it is not the proper quantity to determine the characteristic length scale in the growing system.

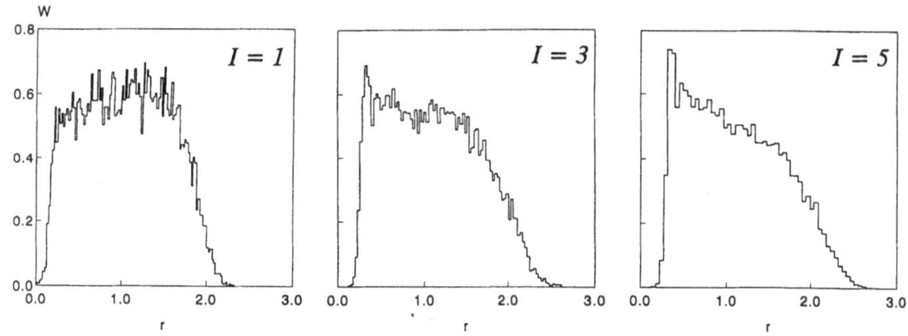

Fig. 3. Grain size distribution for grain-boundary nucleation at $\Phi = 0.9$ for different nucleation rates.

Heterogeneous Nucleation at Grain Boundaries

The simulation of heterogeneous nucleation at grain boundaries yields qualitatively different results in the late stage of the transition process. The shape of the GSD now becomes a function of nucleation rate I, as shown in Fig. 3 for $\Phi = 0.9$. Comparing the results with those of model (i) it is obvious that with increasing rate I the GSD becomes asymmetric with a significant shoulder in the range of small grain sizes. This asymmetry develops due to a change in the kinetic process. If the nucleation processes are fast enough the preferred nucleation sites are soon exhausted and only growth will dominate in the last stages of the transition leading to a broadening of the distribution as first discussed by Lorenz (1990).

Cluster Size Distribution (CSD)

General differences between the CSD's determined for the two nucleation mechanisms (i) and (ii) are shown in Fig. 4. The distributions of Fig. 4 are derived from the numerical simulation of 3D models and similar results are also obtained for $D = 2$ (Orgzall & Lorenz 1988). The CSD's are given in reduced length variables.

The cluster size distribution for continuous nucleation is a decreasing function of the reduced length and is characterized by a sharp decrease in the range of the largest single grains. This distribution becomes broader with increasing transformation degree and the shoulder is shifted towards larger radii. The numerical simulation results agree well with an analytical approach (Lorenz 1987, broken line in Fig. 4).

The CSD for simultaneous nucleation is characterized by a number of pronounced peaks in agreement with the analytical results (Lorenz 1987). The

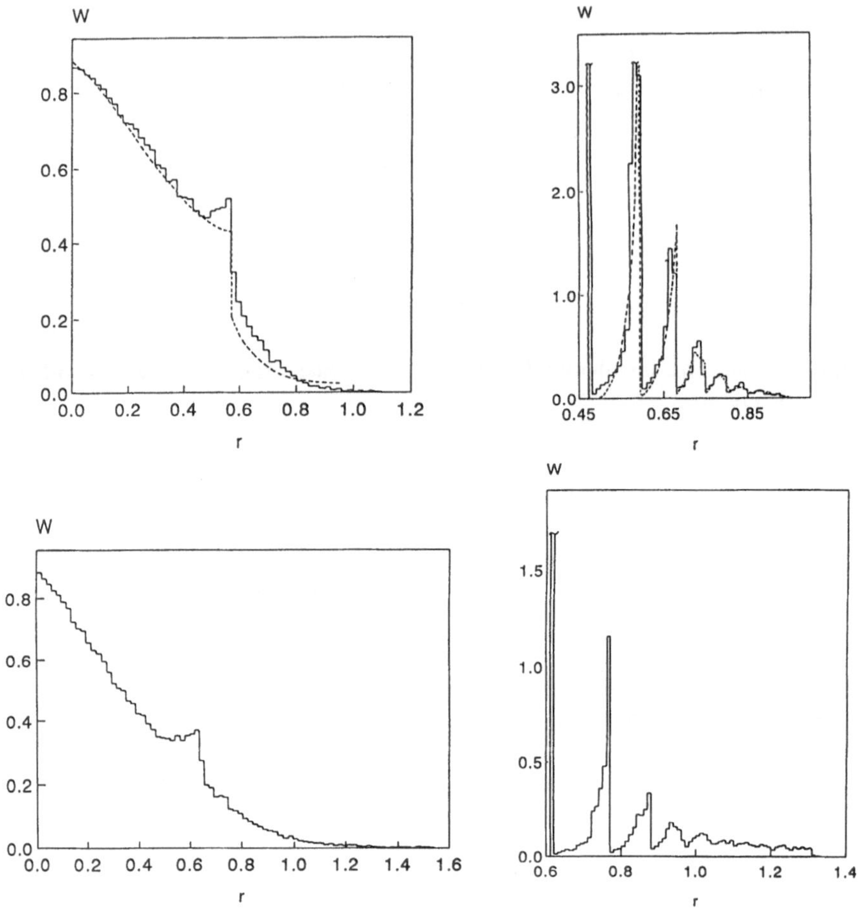

Fig. 4. Cluster size distribution for models (i) ($\Phi = 0.1$ - above, and $\Phi = 0.15$) and (ii) ($\Phi = 0.1$ - above, and $\Phi = 0.2$).

position of the n-th peak is roughly given by $r_n = n^{1/3} r_1$, where r_1 determines the size of the single grains. For growing Φ the peak heights decrease and the weight of the distribution is shifted to larger sizes.

To investigate the influence of different growth mechanisms on the features of the CSD we performed simulations for interface and diffusion controlled (i.e. size dependent) growth. The results are summarized in Fig. 5 for two different transformation rates Φ. Thereby, it has been assumed that nucleation is continuous and homogeneous in space. The CSD for size dependent growth rates shows a pronounced maximum and vanishes for small and large sizes r. This qualitative behaviour is in clear contrast to the interface limited growth process (constant growth rate).

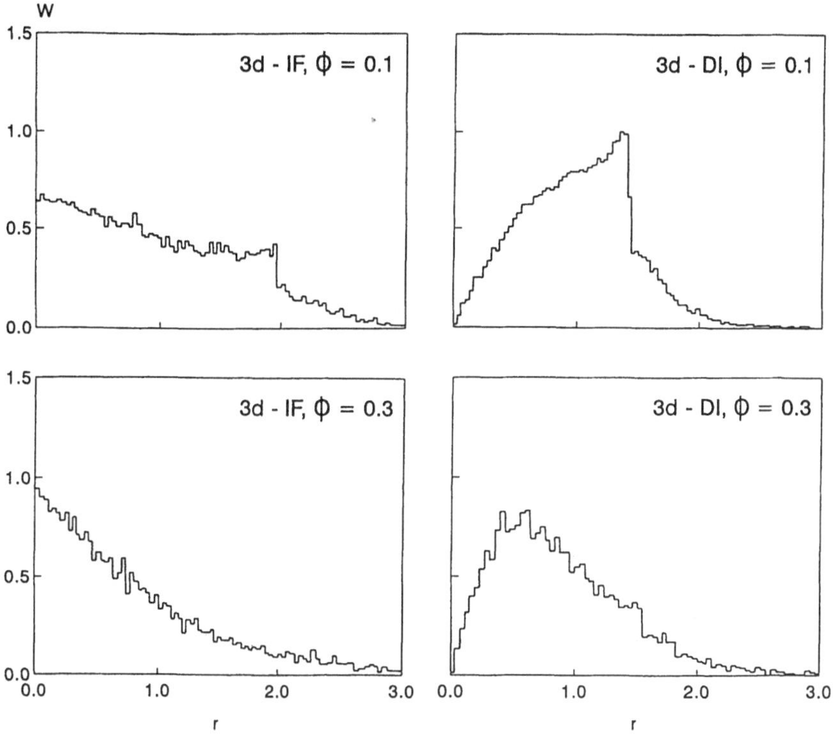

Fig. 5. Cluster size distribution for different growth mechanisms - interface controlled growth **left** and diffusion limited growth **right**.

The assumption of homogeneous nucleation in the case of diffusion limited growth is only valid in a first approximation. Due to the slow diffusion of growth units to the grain surface there appears a concentration gradient around each nucleus. Since the nucleation probability changes drastically with concentration (Christian 1965) there arises a nucleation exclusion zone surrounding the grain. The width of this zone is roughly proportional to the grain size. As a result nonrandom nucleation occurs and the Avrami relation (1) is no longer valid. In a simulation of the diffusion controlled growth process taking into account the effects of nonrandom nucleation, we found that the CSD exhibits a shape as shown in Fig. 5 (Orgzall & Lorenz 1992). However, due to nonrandom nucleation there appear characteristic substructures which are well resolved in our simulation.

It becomes obvious that the CSD's as well as the GSD's show characteristic features of the underlying physical process. Hence, the investigation of the grain structure of igneous rocks yields information on the physical and geological conditions during their formation. Real experimental situations, however, are often more complicated than the models discussed above. Crystallization processes frequently proceed under changing thermodynamic conditions so that the nucleation and growth rates become time dependent due to varying pressure and/or temperature. In order to understand the kinetics under changing external conditions we have developed a theoretical algorithm (Lorenz & Orgzall 1991) that allows for a simulation of the microstructural development in specific experiments using a minimum experimental input (e.g. transformation and nucleation rates). As an example the precipitation kinetics of the B2-phase of potassium chloride under changing pressure successfully could be described by this procedure (Lorenz et al. 1991). From a comparison of CSD's determined experimentally and from numerical simulations it has been shown that the principal growth process is diffusion limited rather than interface controlled.

Percolation Transition

In this section we discuss the critical and fractal properties of the models with constant growth rate and continuous (model M1) as well as simultaneous (M2) nucleation in two and three dimensions. With increasing transformation degree a large cluster occurs, spanning through the whole system. At this percolation point the system undergoes a phase transition and many physical properties change drastically. The exact determination of the percolation threshold is a prerequisite for investigations of universal system properties and for studies of the fractal cluster behaviour.

Φ_c is determined by extrapolating the percolation densities for finite systems of size L to the infinite system according to the scaling law (Stauffer 1985)

$$|\Phi_c(L) - \Phi_c(\infty)| \sim L^{1/\nu} \tag{3}$$

with L - system length and ν - critical exponent at Φ_c.

$\Phi_c(L)$ is derived from the percolation probability $P_L(\Phi)$ for fixed L. For a better extrapolation $\Phi_c(L)$ is determined taking into account periodic as well as nonperiodic boundary conditions where for finite L the threshold value for periodic boundary conditions is the larger one. The difference vanishes for $L \rightarrow \infty$. Details of the applied procedure may be found in Lorenz et al. (1993).

For the model M2 in 2D the results are given in Fig. 6 (lower curve set). The fit to Eq. (3) yields $\Phi_c(\infty) = 0.6764 \pm 0.0009$ and $\nu = 1.37 \pm 0.07$ in excellent agreement with the up to now most accurate results by Rosso (1989) ($\Phi_c = 0.6766 \pm 0.0005$). Although the variance in our analysis is somewhat larger compared with that of Rosso (1989) the accuracy of our procedure is further enhanced by simulation of periodic as well as nonperiodic boundary conditions and the coincidence of both extrapolated threshold values. The exponent ν is comparable with the presumably exact value of 4/3 in two dimensions.

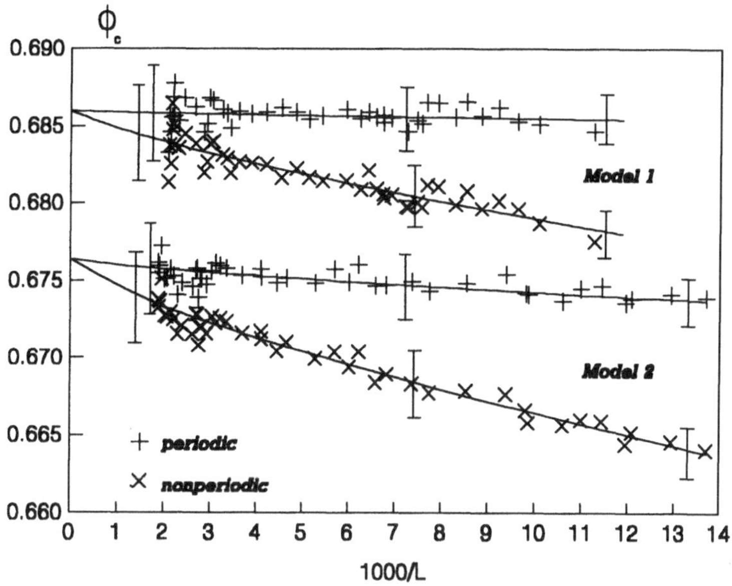

Fig. 6. Finite size scaling plot of percolation thresholds for two dimensional models M1 and M2 and different boundary conditions.

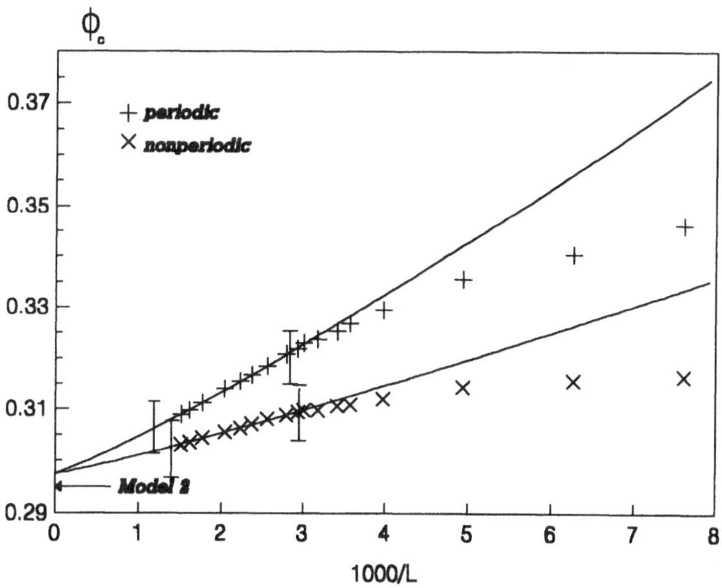

Fig. 7. Finite size scaling plot of percolation thresholds for the three dimensional model M1 and different boundary conditions.

The results for the continuum model M1 with a constant radius distribution are given for two dimensions in Fig. 6 (upper curve set) and for three dimensions in Fig. 7. The length scale L is defined by $L = 1/R_o$ where R_o is the radius of the largest disc.

Table 1. Results for two dimensional models M1 and M2.

	Model M1		Model M2	
Reference	Φ_c	ν	Φ_c	ν
present work	0.6860 ±0.0012	1.28 ±0.08	0.6764 ±0.0009	1.37 ±0.07
Gawlinski &Stanley (1981)			0.676 ±0.003	1.343 ±0.019
Rosso (1989)			0.6766 ±0.0005	
Kertész & Vicsek (1982)	0.70 ±0.02	1.30 ±0.03	0.688 ±0.005	1.35 ±0.07
Phani & Dhar (1984)	0.68 ±0.02			
lattice, exact Stauffer (1985)		$\nu = 4/3$		

In two dimensions we determine the percolation threshold and the critical exponent to be $\Phi_c(\infty) = 0.6860 \pm 0.0012$ and $\nu = 1.28 \pm 0.08$. Within the error limits the exponent ν is equal to the exact value. However, Φ_c is clearly different from that determined for model M2. Thus, for the first time we have shown that the percolation threshold values for M1 and M2 are different and the suggestion that Φ_c might be independent on the radius distribution is ruled out. The results are summarized and compared with some literature data in Table 1.

Fitting the data in the scaling region for the 3D models yields $\Phi_c = 0.297 \pm 0.006$ and $\nu = 0.87 \pm 0.07$. The exponent is in good agreement with 3D lattice and continuum models (Stauffer 1985, $\nu = 0.9$; Balberg & Binenbaum 1985, $\nu = 0.83 \pm 0.09$, M2).

The critical volume fraction for M1 is only slightly larger than that determined for equally sized spheres as evaluated by Margolina & Rosso (1992) $\Phi_c = 0.291 \pm 0.002$. Since the difference of the percolation thresholds for M1 and M2 is smaller than the error limits both values cannot be distinguished in three dimensions. Table 2 summarizes the results.

Using the critical volume fraction Φ_c as previously determined, now the geometric properties of the clusters like volume, surface, hull, anisotropy, etc. as functions of their size R will be determined.

For convenience, dimensionless quantities for volume, hull, and surface are used. For both models M1 and M2 several 10^5 clusters are evaluated and the results are averaged within equally divided intervals of the logarithm of cluster size, ln R, and represented in a double logarithmic plot. The error due to the influence of smaller clusters in the linear fit is systematically excluded.

Due to a large number of papers published concerning fractal properties for model M2 we restrict ourselves here to the discussion of results for M1. Some data for comparison with M2 are given in the tables. At first, results for the 2D model M1 shall be discussed. For large clusters, i.e. in the scaling region, the cluster volume as function of R is given by (Fig. 8) $V \sim R^d$, where the fractal exponent d is estimated as $d = 1.894 \pm 0.003$ in excellent agreement with the best lattice calculations and the presumably exact result $d = 91/48$ (Stauffer 1985).

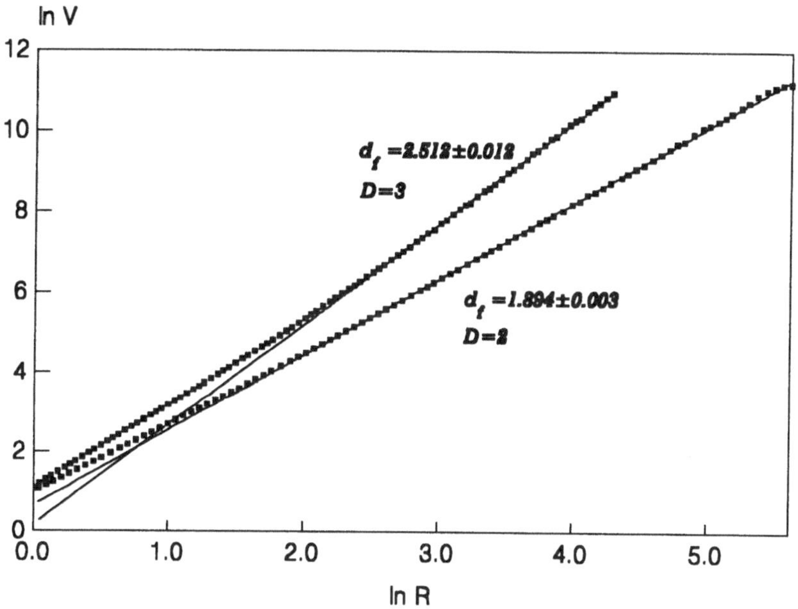

Fig. 8. Dependence of cluster volume V on characteristic length R in two and three dimensions. The fractal exponent d_f is given by the slope for large cluster sizes.

The same exponent is found for the surface, i.e. $S \sim V$ and similarly it is shown that the number of discs N_D within a cluster is proportional to its volume, $N_D \sim V$.

The cluster hull H as function of the linear size R is shown in Fig. 9. The estimated exponent $d_h = 1.751 \pm 0.002$ is close to the two dimensional lattice value 7/4 (Sapoval et al. 1985).

The scaling region is restricted to rather large systems in 3D as already followed from the determination of Φ_c. This is also observed for the fractal properties. The fit in the scaling region yields (Fig. 8) $d = 2.512 \pm 0.012$ in good agreement with corresponding lattice results as for instance determined by Adler et al. (1990), $d = 2.536 \pm 0.047$.

Table 2. Results for three dimensional models M2.

	Model M1		Model M2	
Reference	Φ_c	ν	Φ_c	ν
present work	0.297 ±0.006	0.87 ±0.07		
Lee (1990)				0.83 ±0.09
Margolina & Rosso (1992)			0.291 ±0.002	
Phani & Dhar (1984)	0.303		0.295	
Lattice Stauffer (1985)		$\nu \approx 0.9$		

As in 2D, surface S and number of spheres are both proportional to the cluster volume. The hull was not evaluated due to numerical problems. However, from lattice models it may be concluded that the cluster hull grows with the same exponent as the volume (Bradley et al. 1991).

The models M2 were treated in the same way and the obtained results agree well with recently published literature data and those determined for models M1. Thus, it could be shown that the disorder due to the varying radius distribution is an irrelevant quantity according to the Harris criterion (Harris 1974).

The shape of the clusters is determined measuring its anisotropy. An anisotropy factor f_a is defined dividing the smallest and the largest principal moments of inertia of a cluster. Thus, for an isotropic shape this factor is 1 while for a long and small structure it tends to zero. We average over a large number of equally sized clusters to obtain reliable results.

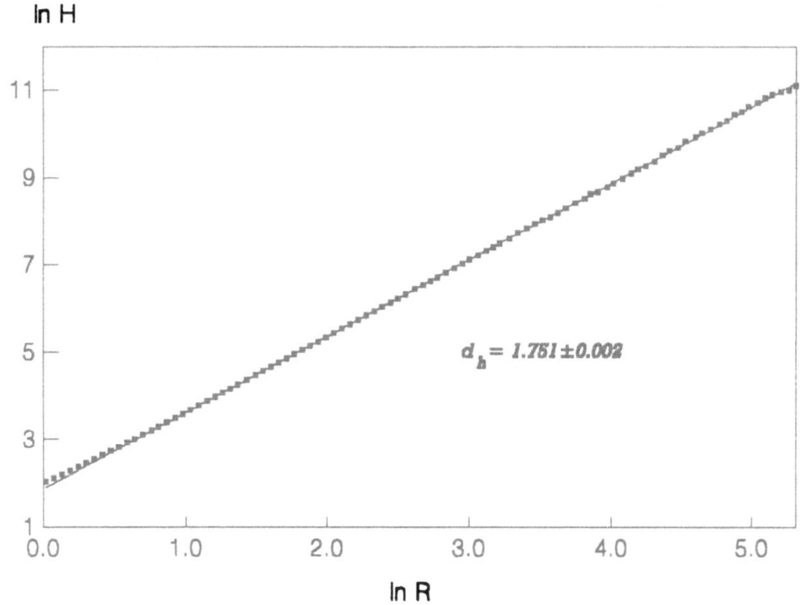

Fig. 9. Fractal dimension d_h for the cluster hull in 2D.

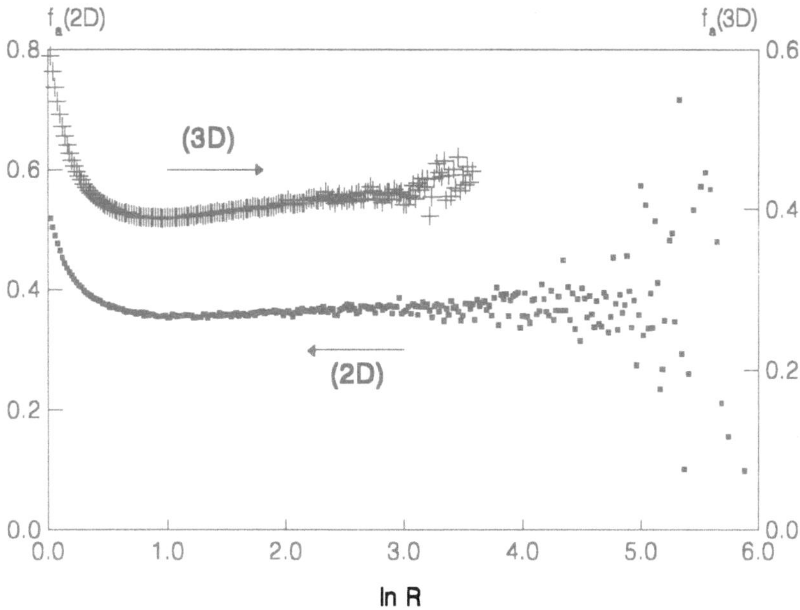

Fig. 10. Cluster anisotropy for two and three dimensional models at Φ_c.

The resulting anisotropy factors are shown in Fig. 10 as function of the cluster size ln R. For large sizes f_a tends to a constant. Averages over the f_a values of the percolating cluster yield $f_a = 0.416$ (2D) and $f_a = 0.503$ (3D). Both values lie well below the isotropic case. Thus, it may be concluded that large clusters up to the percolating cluster are anisotropic near Φ_c.

Conclusions

It is shown that the physical processes of nucleation and growth remarkably influence the properties of the resulting microstructures of the growing phase. Thus, it should be possible to gain reliable information about these basic physical and geologic processes from the investigation of the microstructures. The characteristic differences in grain size distributions should give appropriate knowledge about the formation conditions.

With respect to percolative properties, the comparison of fractal dimensions at the percolation threshold confirmes the conjecture that continuum models with different size distributions of basic units belong to the universality class of corresponding lattice models. The disorder due to the varying size distributions should be a nonrelevant quantity according to the Harris criterion (Harris 1974).

The pore space in sedimentary rocks shows the same fractal properties and basically determines the storage capacity for gas or oil. At the critical density Φ_c the sediments are permeable and the transport properties are mainly influenced by the fractal structure.

References

Adler J, Meir Y, Aharony A, Harris AB (1990) Series study of percolation moments in general dimension. Phys Rev D 41: 9183-9206.

Axe JD, Yamada Y (1986) Scaling relations for grain autocorrelation functions during nucleation and growth. Phys Rev B 34: 1599-1606.

Balberg I, Binenbaum N (1985) Cluster structure and conductivity of three-dimensional continuum systems. Phys Rev A 31: 1222-1225.

Bradley RM, Strenski PN, Debierre J-M (1991) Surfaces of percolation clusters in three dimensions. Phys Rev B 44: 76-84.

Christian JW (1965) The Theory of Transformations in Metals and Alloys. Pergamon Press, Oxford.

Gawlinski ET, Stanley HE (1981) Continuum percolation in two dimensions: Monte Carlo tests of scaling and universality for non-interacting discs. J Phys A: Math Gen 14: L291-299.

Harris AB (1974) Effect of random defects on the critical behaviour of Ising models. J Phys C: Solid State Phys 7: 1671-1692.

Hoshen J, Kopelman R (1976) Percolation and cluster distribution. I. Cluster multiple labeling technique and critical concentration algorithm. Phys Rev B 14: 3438-3445.

Kertész J, Vicsek T (1982) Monte Carlo renormalization group study of the percolation problem of discs with a distribution of radii. Z Phys B: Condensed Matter 45: 345-350.

Lee SB (1990) Universality of continuum percolation. Phys Rev B 42: 4877-4880.

Lorenz B (1987) Grain size distribution in the Kolmogorov model for nucleation and growth: heterogeneous versus homogeneous nucleation. Cryst Res Tech 22: 869-875.

Lorenz B (1989a) Simulation of grain-size distributions in nucleation and growth processes. Acta metall 37: 2689-2692.

Lorenz B (1989b) On the scaling behaviour of correlation functions for nucleation and growth reactions. J Crystal Growth 94: 569-571.

Lorenz B (1990) Influence of nucleation mechanism on the evolution of microstructures during nucleation and growth. Mat Sci Forum 62-64: 737-738.

Lorenz B, Orgzall I (1991) Kinetics of high pressure phase transitions in the diamond anvil cell. In: Hochheimer HD, Etters RD (eds) Frontiers of High Pressure Research. Plenum Press, New York, pp 243-251.

Lorenz B, Orgzall I, Däßler R (1991) Theoretical and experimental investigation of the precipitation kinetics of B2-phase KCl from the solution under high pressure. High Press Res 6: 309-324.

Lorenz B, Orgzall I, Heuer H-O (1993) Universality and cluster structures in continuum models of percolation with two different radius distributions. J Phys A: Math Gen 26: 4711-4722.

Margolina A, Rosso M (1992) Illumination: a new method for studying 3D percolation fronts in a concentration gradient. J Phys A: Math Gen 25 3901-3912.

Orgzall I, Lorenz B (1988) Computer simulation of cluster-size distributions in nucleation and growth processes. Acta metall 36: 627-631.

Orgzall I, Lorenz B (1992) The cluster-size distribution in nucleation and growth processes: Diffusion controlled vs. interface limited growth. Scripta Metall et Mater 26: 889-894.

Phani MK, Dhar D (1984) Continuum percolation with discs having a distribution of radii. J Phys A: Math Gen 17: L645-649.

Rosso M (1989) Concentration gradient approach to continuum percolation in two dimensions. J Phys A: Math Gen 22: L131-136.

Sapoval B, Rosso M, Gouyet JF (1985) The fractal nature of a diffusion front and the relation to percolation. J Physique Lett 46: L149-156.

Spohn T, Hort M, Fischer H (1988) Numerical simulation of the crystallization of multicomponent melts in thin dikes or sills. 1. The liquidus phase. J Geophys Res B 93: 4880-4894.

Stauffer D (1985) Introduction to Percolation Theory. Taylor and Francis, London.

Thompson AH, Katz AJ, Krohn CE (1987) The microgeometry and transport properties of sedimentary rock. Adv Phys 36: 625-694.

Fractal Structure of Quasicrystals

Buming Shen
Institute of Geology, Chinese Academy of Science,
Beijing 100029, PR of China

Abstract. In this paper, the fractal structure model, fractal dimensional calculation and the expressions of quasicrystal lattice with fivefold or eightfold symmetry have been discussed. In a fivefold symmetry quasicrystal, the foundational cell is an icosahedron, and the enlargement coefficient is $1+(\sqrt{5}+1)/2$, and $D=2.6652$. In an eightfold symmetry quasicrystal, the foundational cell is a hexakaidecahedron, and the enlargement coefficient is $1+\sqrt{2}$, and $D=2.7206$. The fractal structure model has many advantages over the Penrose model. These quasicrystal fractal lattices patterns drawn by my mapping program are close to the high-resolution electron microscopic image of real quasicrystal.

Introduction

A quasicrystal with fivefold symmetry in a quenched Al-Mn alloy was first reported (Shechtman et al. 1984), and the quasicrystal with eightfold symmetry was soon discovered also in alloys (Wang et al. 1987, Wang & Kuo 1988). In the past few years, the structure of quasicrystals has been one of the most highly debated topics in physics, because quasicrystals do not show conventional crystallographic translation symmetry (Hiraga et al.1985, Levine et al. 1984, Shechtman 1986, Banced & Heiney 1986, Watanabe et al. 1987, Wittaker et al. 1988, Stephens et al. 1991, Zobetz 1992). The Penrose model and random-tiling model have been considered to be two better models of the quasicrystal structure, where, the random-tiling model is to add some randomness to the Penrose model (Stephens et al. 1991). In my view, the fractal structure model of quasicrystals is also a better model, which has crystallographic significance in addition to the advantages of the above-mentioned two models. This kind of model was first suggested by Peng (1985, 1989). Shen (1989a, 1989b) and Shen et al. (1990, 1991) carried out dimensional calculation for the fractal structure model of quasicrystals with fivefold or eightfold symmetry quasicrystals. Shi et al. (1989, 1991) investigated the configuration of quasicrystal unit cells and deduced their quasilattice. In this chapter, the fractal structure model, fractal dimensional calculation and the expressions of quasicrystal lattice with fivefold or eightfold symmetry are discussed, which is partly identical with a paper published by Shen (1992).

Fractal Structure Model and Expression of Lattice

The fractal structure model of quasicrystals can be described as a spatial pattern that is formed by repetition of a foundational cell according to symmetry operations which follow the enlargement principle of self-similarity in fractal theory. In a fivefold symmetry quasicrystal, the foundational cell is an icosahedron, and the enlargement coefficient is $1+(\sqrt{5}+1)/2$ (Shen 1989a, b). In an eightfold symmetry quasicrystal, the foundational cell is a hexakaidecahedron, and the enlargement coefficient is $1+\sqrt{2}$ (Shen et al. 1990, 1991). The foundational cell and the enlargement coefficient depend on the symmetry of a quasicrystal. The icosahedron, hexakaidecahedron and their projections (decagonal ring and octagonal ring) as foundational cells are shown in Fig. 1 and the responding enlargement coefficients in Fig. 2. As shown in Fig. 3, the lattice of fivefold symmetry quasicrystal can be described as that there is a decagonal ring in the center, and there is a similar decagonal ring, outside of the decagonal ring, all of which constitute a still larger decagonal, and so on.

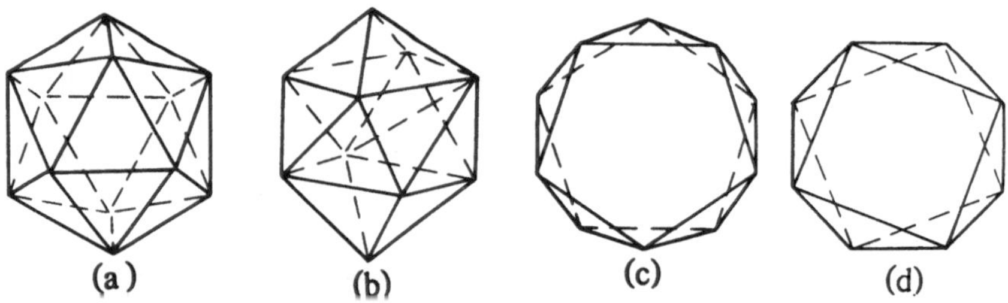

Fig. 1 a-d. The icosahedron, hexakaidecahedron and their projections (decagonal ring and octagonal ring)

The quasicrystal fractal lattice patterns can be drawn in the computer (Proter: SR6602, Japan) by my mapping program. For the fivefold symmetry quasicrystal, the positions of all original decagonal rings can be expressed as G(N,x) and G(N,y):

$$G(N,x) = M\,[\,L_N F^N \cos(36W_N) + L_{N-1} F^{N-1} \cos(36W_{N-1}) + \ldots + L_3 F^3 \cos(36W_3) + L_2 F^2 \cos(36W_2) + L_1 F \cos(36W_1) + \cos(36W_0)] \quad (1)$$

$$G(N,y) = M\,[\,L_N F^N \sin(36W_N) + L_{N-1}F^{N-1}\sin(36W_{N-1}) + \ldots + L_3F^3 \sin(36W_3) + L_2F^2 \sin(36W_2) + L_1F \sin(36W_1) + \sin(36W_0)] \quad (2)$$

$(W_N, W_{N-1}, \ldots, W_3, W_2, W_1, W_0 = 0,1,2,3,\ldots,10)$

where N is the enlarged number of the class and its enlarged time is F; $F = 1 + (1+\sqrt{5})/2$; M is radius of least unit cell; $L_N, L_{N-1}, \ldots, L_1$ can be taken as 0 or 1; W_0, $W_1, W_2, W_3, \ldots, W_N$ correspond to cycle variables of each class for $N=0,1,2,3,\ldots,N$, respectively. The source code of my mapping program is listed in Appendix A.

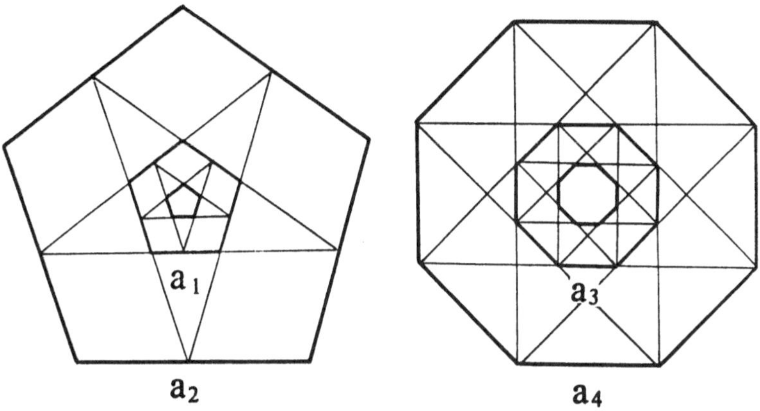

Fig. 2. Enlarge coefficients of icosahedron and hexakaidecahedron
$a2/a1 = 1+(\sqrt{5}+1)/2$; $a4/a3 = 1+\sqrt{2}$

For the eightfold symmetry quasicrystal, the positions of all original octagonal rings can be expressed as G(N,x) and G(N,y):

$$G(N,x) = M\,[\,L_N F^N \cos(45W_N) + L_{N-1}F^{N-1} \cos(45W_{N-1}) + \ldots + L_3F^3 \cos(45W_2) + L_2F^2 \cos(45W_2) + L_1F \cos(45W_1) + \cos(45W_0 - 22.5)] \quad (3)$$

$$G(N,y) = M\,[\,L_N F^N \sin(45W_N) + L_{N-1}F^{N-1} \sin(45W_{N-1}) + \ldots + L_3F^3 \sin(45W_2) + L_2F^2 \sin(45W_2) + L_1F \sin(45W_1) + \sin(45W_0 - 22.5)] \quad (4)$$

$(W_N, W_{N-1}, \ldots, W_3, W_2, W_1, W_0 = 0,1,2,3,\ldots,8)$

where N is the enlarged number of the class and its enlarged time is F^N; $F = 1+\sqrt{2}$; M is radius of least cell; L_N, $L_{N-1}, \ldots, L_1$ can be taken as 0 or 1; W_0, W_1, W_2, $W_3, \ldots, W_N$ correspond to cycle variables of each class for $N = 0,1,2,3,\ldots,N$, respectively. The quasicrystal lattice pattern of fivefold symmetry is shown in Fig. 4, and that of eightfold symmetry in Fig. 6. If every decagonal ring in Fig. 4 is represented by a lattice point, a pattern (Fig. 5) similar

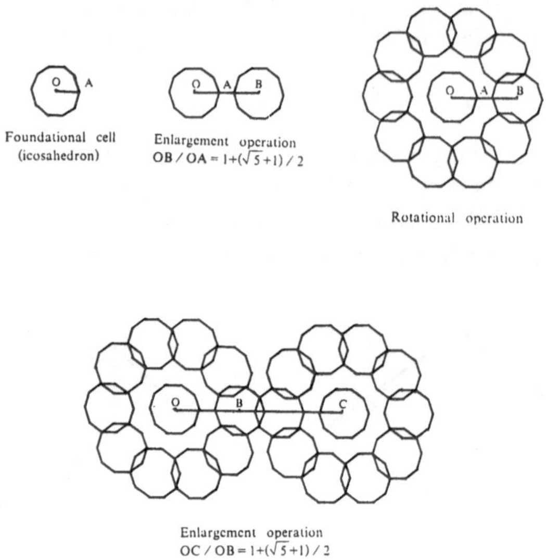

Fig. 3. The deduction of quasicrystal lattice with fivefold symmetry.

to the electron diffraction pattern of fivefold symmetry quasicrystals may be obtained, and if every octagonal ring in Fig. 6 is represented by a lattice point, a pattern (Fig. 7) similar to the electron diffraction pattern of eightfold symmetry quasicrystals may be obtained (Shen 1989a, b, Shen et al. 1990, 1991).

Dimensional Calculation of the Fractal Structure of Quasicrystal

The fractal dimension is an important parameter to express a fractal graph, and it can be calculated by the following formula (Mandelbrot 1983, Takayasu 1986).

$$D = \log N / \mathrm{Log}\ r \tag{5}$$

where r is the self-similarity ratio factor of the graph studied, and N is the number of foundational units included in an original graph. This unit can be obtained by reducing the original graph by 1/r along each independent direction. As above mentioned, in a quasicrystal structure with fivefold symmetry, r is $1+(\sqrt{5}+1)/2$. N is related to space dimension. For example, in a three dimensional space, one icosahedron can be extended by the first order enlargement to 12 equal icosahedrons, which added to that icosahedron in the center makes 13 icosahedrons, thus N = 13.
The fractal dimension is given by

$$D = \log 13 / \mathrm{Log}\ (1+(\sqrt{5}+1)/2) = 2.6652 \tag{6}$$

Fig. 4. Ideal lattice graph of a fivefold symmetry quasicrystal.

For an eightfold symmetry quasicrystal, r is $1+\sqrt{2}$. In a three dimensional space, one hexakaidecahedron can be extended by the first order enlargement to 10 equal hexakaidecahedrons of the same size by the first order enlargement, which added to that hexakaidecahedron in the center makes 11 hexakaidecahedrons, thus, $N = 11$.

The fractal dimension is given by

$$D = \log 11/\log(1+\sqrt{2}) = 2.7206 \tag{7}$$

Obviously, the dimension ($D = 2.7206$) of an eightfold symmetry quasicrystal is larger than that of a fivefold symmetry quasicrystal. This illustrates that the structure of an eightfold symmetry quasicrystal is more close than one of a fivefold symmetry quasicrystal on the microscopic scale equal in size to several unit cells, and that the former is more stable than the latter. It is obvious that the fractal dimension is a synthetic expression of fractal structure of quasicrystals.

Fig. 5. Electron diffraction pattern of a fivefold symmetry quasicrystal.

Discussion

This fractal quasicrystal model has many similarities to that with conventional crystals, because it can also be expressed as a foundational cell that can be repeated according to translational operation. The most important difference between both models, however, is that the former's symmetrical operation is a magnifying or reducing operation of self-similarity, and the latter's is only a translational operation. It is by the self-similar operation that a unit cell can completely fill the space in the quasicrystal fractal structure model. In addition, the fractal structure model has many advantages over the Penrose model.

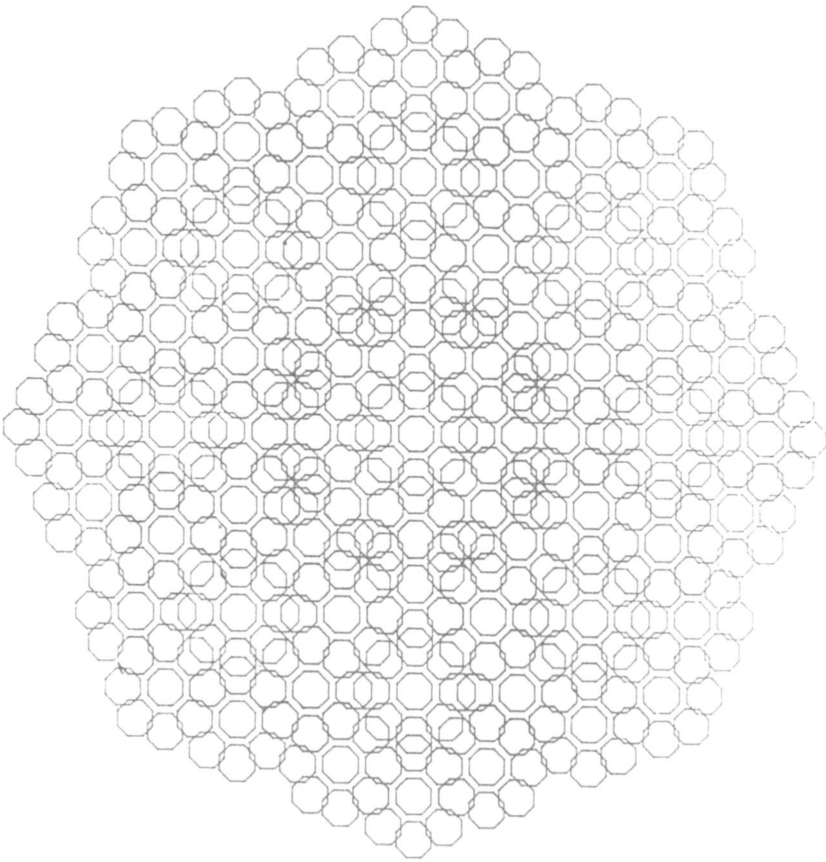

Fig. 6. Ideal lattice graph of an eightfold symmetry quasicrystal.

(1) In the Penrose model a quasicrystal is constructed by repeating and inlaying of two unit cells or more according to some specific rules. For example, for a fivefold symmetry quasicrystal, the unit cell of the Penrose model is a fat rhombus and a thin one, and they cannot reflect the feature of fivefold symmetry, but the unit cell of the fractal structure model Icosahedron, can reflect that. Again, for an eightfold symmetry quasicrystal, the unit cell of the Penrose model is a square and a rhombus, and they cannot reflect the feature of eightfold symmetry, but the unit cell of the fractal model, hexakaidecahedron, can reflect that.

(2) As compared with the ideal lattice of the Penrose model, that of the fractal model is close to the high-resolution electron microscopic image of real quasicrystal. For eightfold symmetry quasicrystals such as $Cr_5Ni_3Si_2$ and $V_{15}Ni_{10}Si$ alloys, the radius of their octagonal rings, 0.27 nm, is changed

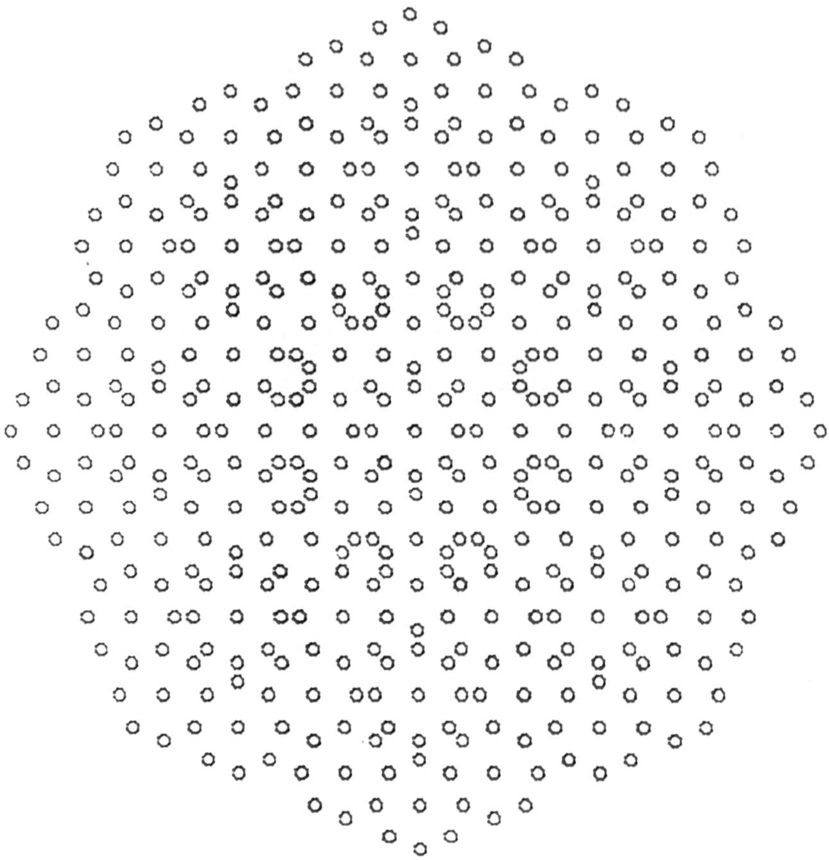

Fig. 7. Electron diffraction pattern of an eightfold symmetry quasicrystal.

by the first-order enlargement of $1+\sqrt{2}$ into 0.65 nm, which is close to the radius (0.63nm) measured from Wang's high-resolution image (Wang 1987). Therefore, Wang's high-resolution image is possibly the first order enlarged pattern of the octagonal rings. We reasonable believe that the least octagonal rings in the Fig. 6 can correspond to black points in the Wang's high-resolution image (from Fig. 3b). Thus, our explanation is better than the explanation which is based on the Penrose graph (from Fig. 3a) made by Wang et al. (1987). If some lines in Wang's Penrose graph are replaced by heavy lines, Fig. 8 will result. Similarly, if some lines in Stephens's Penrose pattern are replaced by heavy lines, Fig. 9 will result. It can be seen from these figures that some intersecting points in Penrose graphs do not necessarily correspond to the lattice point, but all the octagonal or decagonal rings in the Fig. 6 or in Fig. 4 may correspond to the lattice points and so have a crystallographic significance. In a word, for

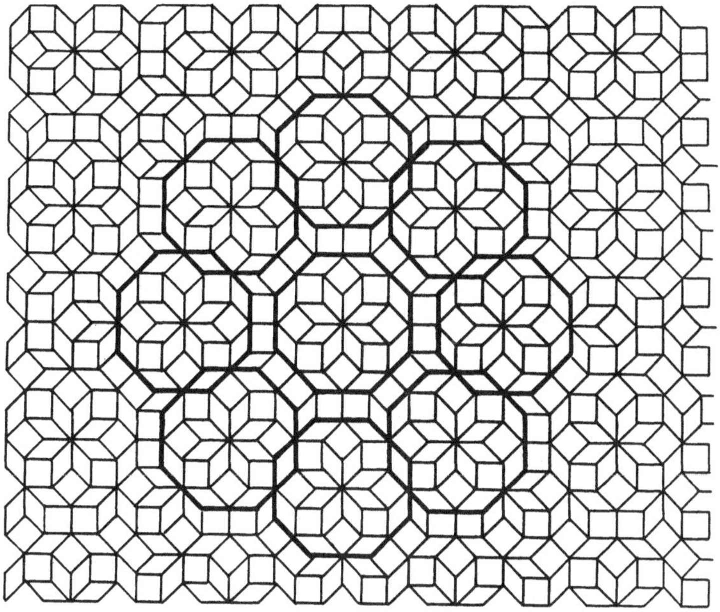

Fig. 8. Lattice graph (heavy line) modified from the Wang's Penrose graph

the fractal structure model, the unit cell and its repeating sufficiently reflect the fivefold or eightfold symmetry element. For the Penrose model, the unit cell (square and rhombi) cannot reflect fivefold or eightfold symmetric elements, but its repetition reflects them and its enlargement coefficient is $1+(\sqrt{5}+1)/2$ or $1+\sqrt{2}$ too. This is the reason why the fractal structure pattern can cover the Penrose pattern and why the fractal structure model is close to the practical crystal structure.

(3) The electron diffraction pattern of quasicrystals is close to the electron diffraction pattern (Fig. 5 and Fig. 7) drawn from the fractal structure model.

(4) The fractal structure model is an ideal model. In fact, some decagonal rings (a plane projection of icosahedrons) or octagonal rings (a plane projection of hexakaidecahedrons) are distorted. Fractal theory is a strong tool to describe a complex pattern, and allow that pattern to have some randomness under the prerequisite of keeping some structureness (self-similarity). Therefore, rigorous matching rules in the Penrose model are not necessary for the fractal model.

It follows that the fractal structure model can explain not only the high-resolution images and electron diffraction patterns, but also has crystallographic significance.

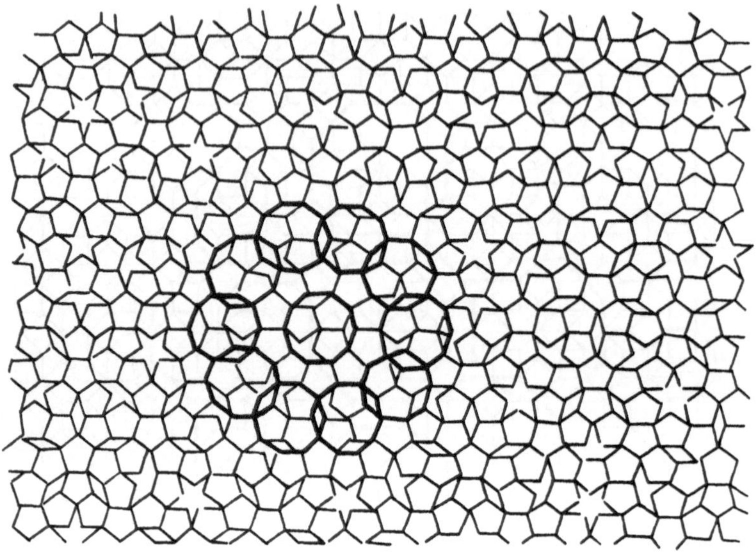

Fig. 9. Lattice graph (heavy line) modified from the Stephens's Penrose graph

Appendix A

Source code of BASIC mapping program of the quasicrystal fractal lattice patterns

```
10 PRINT "IN"
20 DEG
30 F=1+(1+SQRT(5))/2
40 PRINT "AP;1500,1200,-3)"
50 INPUT "Length of minimum for super unit cell = ",M
54 INPUT "1 (Lattice graph), 2 (Electron diffraction pattern) =)",U
56 IF U=2 THEN M=10 : GOTO 60
58 M1=M
60 FOR W3=0 TO 10
70  L1=0 : L2=0
80  IF W3=0 THEN GOTO 100
90  L3=1
100   FOR W2=0 TO 10
110     L1=0
120     IF W2=0 THEN GOTO 140
130     L2=1
140       FOR W1=0 TO 10
160         IF W1=0 THEN GOTO 180
170         L1=1
```

```
180          FOR W0=0 TO 10
190            IF W0=0 THEN D$=",3" : GOTO 220
200            D$=",2"
220            GOSUB 290
230          NEXT W0
240        NEXT W1
250    NEXT W2
260 NEXT W3
270 PRINT "TE"
280 END
290 X2 = M*(L1*F*COS(36*W1) + L2*F**2*COS(36*W2) + L3*F**3*COS
(36*W3)+M1*COS(36*W0)
300 Y2 = M*(L1*F*SIN(36*W1) + L2*F**2*SIN(36*W2) + L3*F**3*SIN(36
*W3)+M1*SIN(36*W0)
310 IF ABS(Y2)<1 THEN Y2=0
320 IF ABS(X2)<1 THEN X2=0
330 PRINT "AP;";
340 PRINT X2;
350 PRINT ",";
360 PRINT Y2;
370 PRINT D$
380 RETURN
```

References

Benced PA, Heiney PA (1986) Structure of rapidly quenched Al-Mn. Phys. (Paris) Colloq 47: C-348.

Hiraga K, Hirabayashi, A (1985) Icosahedral quasicrystals of a melt-quenched Al-Mn alloy. The science reports of the research institutes Tohoku University 32 Series A, 309-314.

Levine D, Steinhardt, PJ (1984) Quasicrystals: a new class of ordered structures. Phys Rev Lett 53: 2477-2480.

Mandelbrot BB. (1983) The Fractal Geometry of Nature. Freeman, New York.

Peng Z (1985) Building principles of quasicrystal and particle fractal structure model. Earth Science 4: 159-174 (in Chinese).

Peng Z (1989) Deduction of quasilattice with fivefold symmetry and particle fractal structure model. Science in China (series B) 32: 215-226.

Shechtman DJ (1986) Quasiperiodic crystal-experimental evidence. Phys. (Paris) Colloq 47: C-2.

Shechtman D, Blech I, Gratias D, Chahn J (1984) A metallic phase with long-range orientational order and no translational symmetry. Phys Rev Lett 53: 1951-1953.

Shen B (1989a) A expression of lattice and calculation of dimension of the fractal structure in quasicrystal. Chinese Science Bulletin 34: 1548-1550.

Shen B (1989b) A expression of lattice and calculation of dimension of the fractal structure in quasicrystal. Chinese Science Bulletin 5: 362-364 (in Chinese).

Shen B (1992) A model of fractal structure of quasicrystals. Scientia Geologica Sinica 1: 63-71.

Shen B, Shi N (1990) The fractal structure for eightfold symmetry quasicrystal. Chinese Science Bulletin 19: 1484-1486 (in Chinese).

Shen B, Shi N (1991) The fractal structure for eightfold symmetry quasicrystal. Chinese Science Bulletin 36: 210-213.

Shi N, Liao L (1989) Point groups and single forms of quasicrystals with eightfold and twelvefold symmetry. Acta Geologica Sinica 2: 39-43.

Shi N, Min L, Shen B (1991) The configuration of quasicrystal unit cell and deduction of quasilattice. Science in China 11: 1216-1223 (in Chinese).

Shi N, Min L, Shen B (1992) The configuration of quasicrystal unit cell and deduction of quasilattice. Science in China 6: 735-744 (in Chinese).

Stephens PW, Goldman AI (1991) The structure of quasicrystal. Scientific American 264: 44-53.

Takayashu H (1986) Fractals (translated by Shen B et al.). Seismological Publishing House, 1989 (in Chinese).

Wang N, Chen H, Kuo K (1987) Two-dimensional quasicrystal with rotational symmetry. Phys Rev Lett 59: 1010-1013.

Wang ZM, Kuo K (1988) The octagonal quasilattice and electron diffraction patterns of the octagonal phase. Acta cryst A44: 857-863.

Watanabe Y, Ito M, Soma T (1987) Noperiodic tessellation with eightfold rotational symmetry. Acta Cryst A43: 133-134.

Whittaker EJW, Whittaker RM (1988) Some generalized Penrose patterns from projections of n-dimensional lattices. Acta Cryst A44: 105-112.

Zobetz E (1992) A pentagonal quasicrystal tiling with fractal acceptance domain. Acta Cryst A48: 328-335.

Protolytic Weathering of Montmorillonite, Described by Its Effective Surface Fractal Dimension

Jürgen Niemeyer[1] & Galina Machulla[2]
[1]Institut für Bodenwissenschaft, Georg-August-Universität Göttingen
D-37075 Göttingen, Germany
[2]Institut für Standortkunde und Agrarraumgestaltung
Martin-Luther-Universität Halle
D-06108 Halle, Germany

Abstract. Octahedral coordinated aluminum is released by the attack of protons from clay minerals like montmorillonite. Apart from other effects, this leads to an alteration of the form of the surface of the clay particles. We used fractal geometry to describe this change in form. The effective surface fractal dimension (D) was determined before and after acid treatment by using the dependency: number of yardsticks necessary to cover the surface versus size of the yardstick area. We used n-alkanols with different cross sectional areas to determine this dependency. This resulted in a D: 2.55 for the untreated and a D: 2.85 for the acid treated clay.

Introduction

Many chemical reactions in soils, like the turnover of nutrients, the catalysis of humification reactions or the adsorption of radionuclides are highly influenced by soil clay minerals. These minerals are subject to chemical reactions like weathering or buffering.

In some areas, due to human activity, the concentration of protons in the soil solution has increased dramatically. This acidification leads to a release of aluminum from the octahedral layers of the clay platelet and consequently to an increase in the content of silanol groups. The free aluminum-aquo-complexes can damage the root system of the plants.

Besides the increase in the surface area, the protolytic decay of the clay results in a change of the surface form, especially at the edges of the clay particle. By finding out the form, it should be possible to obtain a measure of the degree of weathering. We wanted to see if fractal geometry is appropriate to quantify the shape.

Up to now, the fractal dimension of montmorillonite has been measured by gas-adsorption techniques (Sokolowska et al. 1989, Ohoud & van Damme 1990, Srasra et al. 1989).

For experimental simplicity, we used a method developed by Pfeifer et al. (1984) to figure out the effective surface fractal dimension.

Experimental Section

The method is based on the relation $n \propto s^{-D/2}$ [n: monolayer value (moles per unit mass), s: cross-sectional area of the adsorbed molecule, D: effective surface fractal dimension]. The dependency of the accessible surface area from the cross-sectional area of the adsorbed molecule is determined from adsorption measurements in the liquid state. Such an approach has been used by Pfeifer et al. (1984).

The method has been criticised by Drake et al. (1990). In a recent paper Avnir et al. (1992) discuss some arguments used by Drake et al. (1990), that these authors have used some approaches in an erroneous way. For a detailed discussion the reader is referred to the literature cited.

As in Pfeifer et al. (1984) higher alcohols (C_6 - C_{14}) were used to measure the effective surface area. The term 'effective surface area' is used, because due to the mechanism of interaction, mainly H-bonding between the clay surface and the alcohols, the whole surface is not probed.

It is known that in solid smectite preparations up to 50% of the alcohol is located in the micropores between the clay microcrystals (Breen et al. 1993). This effect disturbed the determination of the amount of alcohol truly adsorbed at the surface. To avoid these difficulties associated with the clay coagulation we determined the surface area by adsorption experiments in the liquid state. We used carbon tetrachloride as an unpolar solvent with a high specific density. Suspension of the clay particles was easy to achieve; no aggregates could be observed. The use of CCl_4 allows the determination of the alcohol by using IR-photometry because in the spectral region used (C-H vibration, 2980 cm^{-1}), no IR absorption of the solvent occurs.

Table 1. Percentage of main elements of the clay samples.

Element	Unweathered montmorillonite	Weathered montmorillonite
SiO_2	57 %	70 %
Al_2O_3	19 %	17 %

Clay Weathering

In an Erlenmeyer flask (500 ml), 2 g of the clay and 150 ml 0.001 m hydrochloric acid were mixed and shaken intensively for 5 min. The flask was transferred in a shaken water bath and shaken at 40 rpm for 10 h. After that, multiple centrifugation and redispersion of the clay in distilled water was used to remove the surplus hydrochloric acid. The clay samples were freeze-dried and stored over P_2O_5. Table 1 shows the composition of the samples.

Determination of the Adsorption Isotherm

Five hundred mg of the adsorbent were transferred into a stoppable centrifugation vial; 5 ml of the alcohol-carbon-tetrachloride mixture of the desired concentration was added and shaken overnight. After centrifugation the non-adsorbed alcohol was determined by IR photometry (wavenumber 2980 cm^{-1}, 1 cm pathlength, infrasil cuvettes). The calibration curve was linear over its entire length.

Results

In Figure 1, as an example for an adsorption isotherm obtained, the isotherm of the couple n-octanol untreated montmorillonite is shown. Due to the pronounced plateau, a reliable determination of the monolayer capacity is possible.

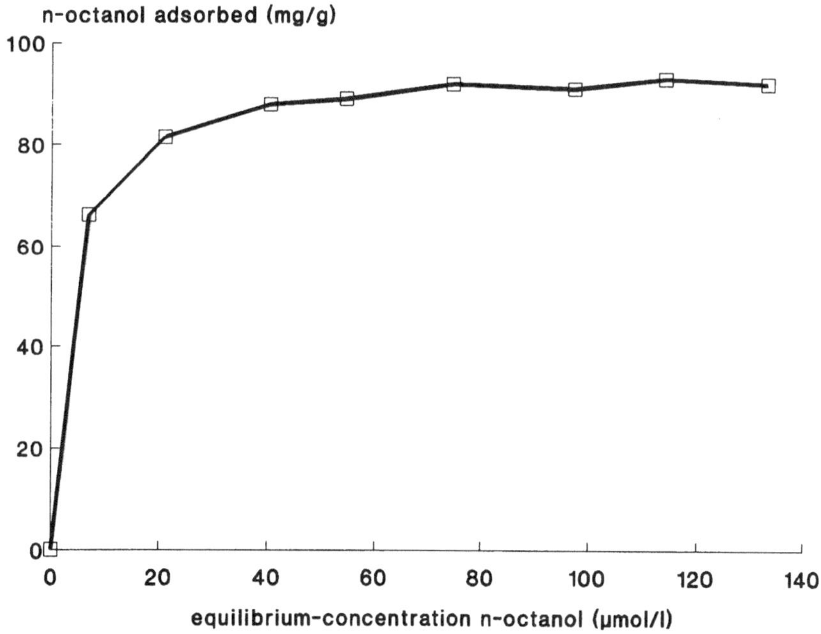

Fig. 1. Adsorption isotherm in the system n-octanol/unweathered montmorillonite.

Tables 2 and 3 give the values of the monolayer capacity of the different alcohols for proton weathered and unweathered clay.

Table 2. Monolayer values of the different alcohols for unweathered montmorillonite.

Alcohol	Monolayer value (μmol/g)
n-Hexanol	575
n-Octanol	537
n-Decanol	436
n-Dodecanol	426
n-Laurylalcohol	380

Table 3. Monolayer values of the different alcohols for weathered montmorillonite.

Alcohol	Monolayer value (μmol/g)
n-Hexanol	776
n-Octanol	707
n-Decanol	588
n-Dodecanol	562
n-Laurylalcohol	489

The dependency of the monolayer capacity from the cross-sectional area of the adsorbates used, in a log-log plot, is shown in Figure 2.

From the slope of the linear regression line D: 2.55 for the unweathered and D: 2.85 for the weathered montmorillonite is obtained.

Discussion

Our results show that the treatment of montmorillonite with protons leads to an increase of the effective surface fractal dimension, i.e. the surface becomes more rough and porous through this treatment.

Srasra et al. (1989) report that acid treatment of montmorillonite results in a decrease of the surface fractal dimension. This is in contradiction to the results presented here. Our findings are supported by electron-microscopical investiga-

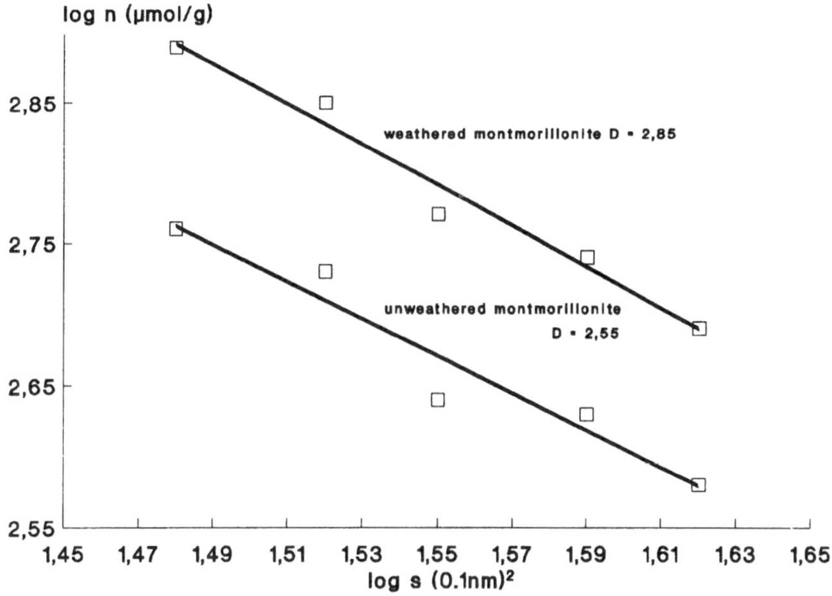

Fig. 2. Monolayer values as a function of the adsorbate cross-sectional area for unweathered and weathered montmorillonite (r^2 of the regression line: 0.98)

tions (Fahn 1973). We assume that these differences can be explained by the different methods used. Basically, Srasra et al. (1989) used the accessibility of the surface for nitrogen molecules which interact with the surface by physisorption. He derives his value from an analysis of the form of the BET isotherm. In contrast to their method we determine the accessibility of sites capable of the formation of H-bonds with the alcohol molecules. These two mechanisms are very different energetically; therefore the two surfaces probed cannot be identical. If the effective surfaces are not the same, then it is not likely that the shape of surfaces is identical.

Conclusions

We have shown that by using fractal geometry one can describe the differences in surface forms between proton-weathered and unweathered montmorillonite particles. Taking these first results as a basis, in might be possible to quantify the degree of weathering of clay particles by further improving the methodology presented here.

References

Avnir D, Farin D, Pfeifer P (1992) A discussion of some aspects of surface fractality and of its determination. New J Chem 16: 439-449.

Breen C, Flynn J, Parkes GMB (1993) Thermogravimetric, infrared and mass-spectroscopic analysis of the desorption of methanol, propan-1-ol, propan-2-ol and 2-methylpropan-2-ol from momtmorillonite. Clay Min 28: 123-138.

Drake JM, Levitz P, Klafter J (1990) A comment on the fractal dilemma in porous silica gels. New J Chem 14: 77-81.

Fahn R (1973): Einfluß der Struktur und der Morphologie von Bleicherden auf die Bleichwirkung bei Ölen und Fetten. Fette Seifen Anstrichmittel 75: 77.

Ohoud MB, Van Damme H (1990) La texture fractale des argiles gonflantes. CR Acad Sci Paris t 311 Série II: 665-670.

Pfeifer P, Avnir D, Farin D (1984) Scaling behaviour of surface irregularity in the molecular domain: from adsorptions studies to fractal categories. J Stat Phys 36: 699-716

Sokolowska J, Stawinski J, Patrykiejew A, Sokolowski S (1989) A note on fractal analysis of adsorption process by soils and soil minerals. Int Agrophysics 5: 3-12.

Srasra E, Bergaya F, Van Damme H, Ariguib NK (1989) Surface properties of an activated bentonite-decolorisation of rape-seed oils. Appl Clay Sci 4: 411-417.

Characteristics and Evolution of Artificial Anisotropic Rocks

Robert Ondrak, Ulf Bayer & Olaf Kahle
GeoForschungsZentrum Potsdam
Projektbereich 3.4 "Modellierung von Geoprozessen"
Telegrafenberg A51, D-14473 Potsdam, Germany

Abstract. Porous media in geologic systems are usually inhomogeneous with respect to porosity and permeability. In addition, these parameters may evolve with time because of the chemical interaction between solids and pore fluid.

We will illustrate how fluxes evolve in strongly inhomogeneous media and how different flow patterns and diagenetic changes depend on the porosity and permeability distribution.

Finally we attempt to characterize anisotropic media by simple scaling rules and power laws as well as to predict the average effective transport coefficients. It is shown that the anisotropic media can be transformed into equivalent isotropic media with respect to the conductivity tensor.

Introduction

Geological media are generally inhomogeneous on all scales from the pore space to bedding and stratigraphic units. Geological media are also mostly anisotropic, being layered or foliated etc. In addition they are not formed randomly or stochastically; instead they usually inherit a dominant deterministic structure related to the process of their formation (Fig. 1). These non-random properties are at the heart of geology, allowing us to classify geological media in a qualitative manner. However, on a sufficiently small scale we always find stochastic fluctuations which modify the properties of rocks. We shall illustrate how the diagenetic evolution of rocks may be affected by inhomogeneities and we will attempt to characterize rock-like media and their conductivity tensor by simple scaling rules and power laws. It should be mentioned that the order of authorship has been determined by a random process.

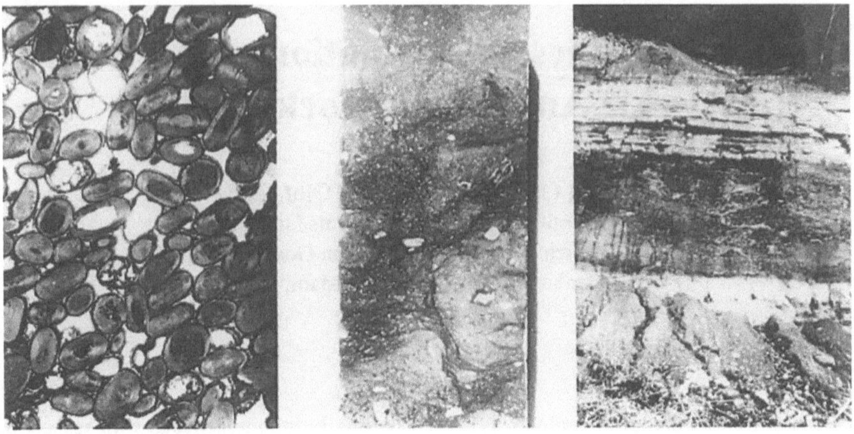

a b c

Fig. 1a-c. Inhomogeneous geological material on different scales: **a** microscopic; **b** macroscopic, **c** outcrop.

Temporal Diagenetic Evolution of Porous Rocks

Rocks are dynamic systems in geological times although they may look very stable. From the time of formation, e.g. by sedimentation, until they are destroyed by weathering and erosion, rocks are affected by chemical alteration. Processes related to fluid flow in the pore space are very important in this regard (Giles & de Boer 1990, Steefel & Lasaga 1990).

The mineralogical composition of a rock is altered due to mineral dissolution and precipitation under the changing thermodynamic conditions during its lifetime. The formation or removal of quartz and calcite cements is one of the features frequently found in sandstones. In this situation, mineral dissolution is controlled by the chemical composition of the pore fluid, and the temperature and pressure conditions under which the reactions take place (Ondrak & Bayer 1993). In order to study and quantify the diagenetic evolution of a sandstone one must consider the chemical evolution of the pore fluid due to fluid flow and mineral reactions (Ondrak 1992, 1993):

$$\frac{d(\phi C_i)}{dt} = -v\frac{dC_i}{dx} + R_i \tag{1}$$

C_i stands for the concentration of the individual species multiplied by the porosity ϕ in order to obtain the actual amount of the dissolved species. The second term describes the changes in concentration due to pore fluid flow, with v denoting the flow velocity. The final term is a reaction term quantifying the reactions which usually result from fluid flow and temperature changes. A common representation of the reaction term is:

$$R_i = A \bullet k_i \bullet \phi \bullet (C_{eq} - C_i) \quad (2)$$

with A representing the reactive surface of the mineral, k_i the kinetic constant of the reaction, and C_{eq} the equilibrium concentration of the species. Equivalent equations without the reactive surface describe the interaction of species dissolved in the pore fluid. Given a set of appropriate equations of type (1) and (2), the evolution of each species in the fluid and the interaction with the solids can be quantified. However, for actual models equation (2) may be simplified, with the reactive surface area A and the reaction constant k being omitted since the flow rates and the associated concentration changes are generally small in the deep subsurface and one can safely assume local equilibrium between the minerals and the pore fluid.

The transport reaction process does not only alter the mineralogy of the rock but also modifies its physical properties especially porosity and permeability. These properties again affect the flow within the rock, resulting in a non-linear system (Chen et al. 1991). The temporal evolution of porosity equals the specific molar volume V_{sp} multiplied by the amount of the dissolved or precipitated mineral:

$$\frac{df}{dt} = \phi \frac{dC_i}{dt} V_{sp} \quad (3)$$

Based on equations (1) to (3) and on an additional relationship between porosity and permeability (Carman-Kozeny-Equation) it is possible to study the principles of diagenesis in sandstones, especially with respect to the effects of temperature and initial inhomogeneities. Fig. 2 illustrates the temporal evolution of porosity in a calcite cemented sandstone. The spatial extent of the sandstone is 30 by 30 km. It gradually dips to the left and temperature increases proportional to depth. Pore fluid enters the layer from the left side, dissolving calcite as the fluid moves along the layer. The initial, homogeneously distributed, porosity of the layer is 25%. With time, dissolution of calcite cement proceeds along a homogeneous front moving along the flow path. The different amounts of calcite removed along the layer result from the varying temperature gradients along the flow path. In a homogeneous medium, the diagenetic evolution is dominated by the temperature distribution as illustrated by this example.

In the second example (Fig. 3) the initial porosity has an inhomogeneous distribution of between 10 and 30%. The resulting inhomogeneous porosity and permeability modify the flow rates along the layer, and finally affect the dissolution rates of calcite cements. Calcite cement is preserved in the shadow zone of areas with reduced permeability. The dissolution fronts here tend to interdigitate along the zones of high flows rates. Flow rate and temperature, however, interact so that the dissolution front becomes disconnected and evolves simultaneously at different space points, particularly where temperature changes

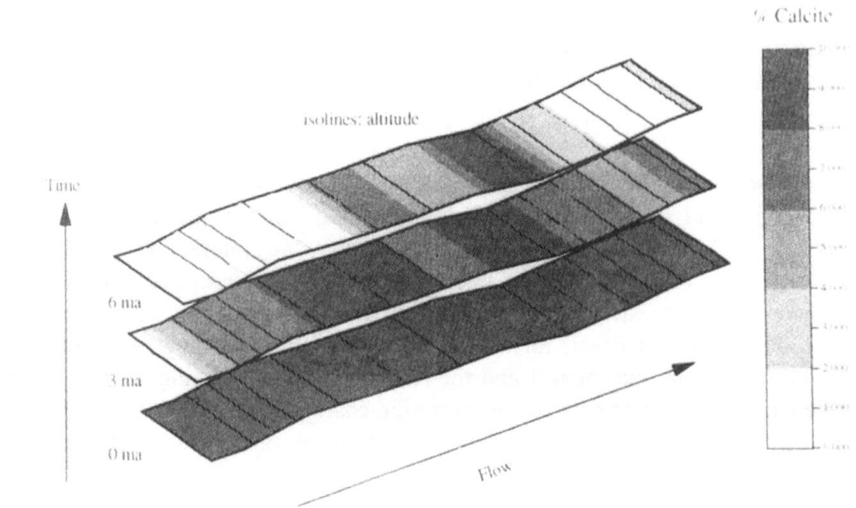

Fig. 2. Temporal evolution of calcite cement in a calcic sandstone with homogeneous porosity distribution over a period of 6 million years. The depth of the layer ranges from 3 - 2 km.

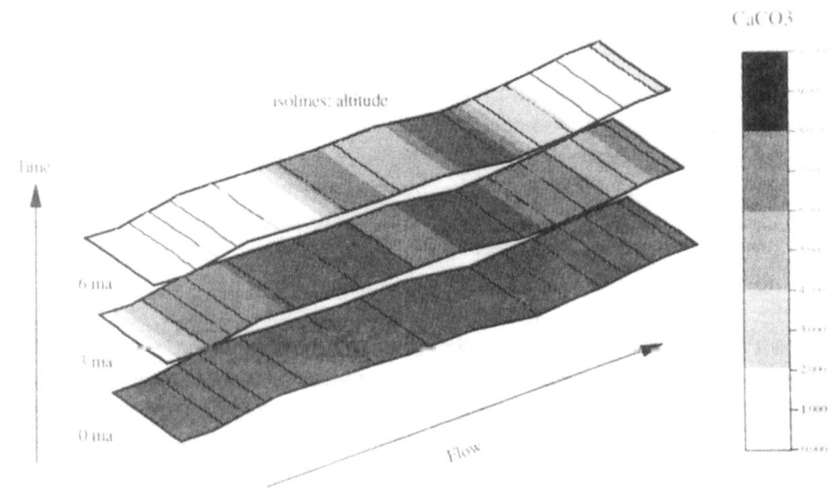

Fig. 3. Temporal evolution of calcite cement in a calcic sandstone with an inhomogeneous porosity distribution over a period of 6 million years. The depth of the layer ranges from 3 - 2 km.

along the flow path are high. In the model, high temperature gradients correlate with the dip of the layer as is the case at the left and the right sides of Fig. 3.

Along zones of low dip the dissolution is mainly controlled by the chemical composition of the pore fluid and thus by fluid flow.

The distribution of porosity and permeability is of considerable interest in many fields of geology. Especially with regard to the exploration of oil and gas it is of interest to predict both the distribution of these parameters and their evolution during geological time. However, information about the distribution of inhomogeneities is usually extremely sparse with regard to practical applications. For practical purposes it would be useful to have trend estimates based on average rock properties, estimates of the temperature history and a scenario for pore fluid flow. One requirement then is to obtain a simple set of measurements characterizing inhomogeneous and anisotropic rocks which can be related to the permeability tensor being an essential parameter as illustrated above. In the next section we will attempt to develop some simple measurements within this context.

Simplified Model for a Two Component System

Rather than approaching the complexity of nature or even of the diagenetic models discussed above, we will restrict ourselves to very simple models for two component systems. These models are closely related to the percolation theory (Stauffer 1985), so that many parameters like percolation threshold, cluster size distribution and fractal properties are well known, at least for certain cases of the model. Pixels are randomly distributed in a 3-dimensional cube (size 64^3) which provides us with a statistically homogeneous medium with well known properties. In order to achieve anisotropies which gradually approach a layered

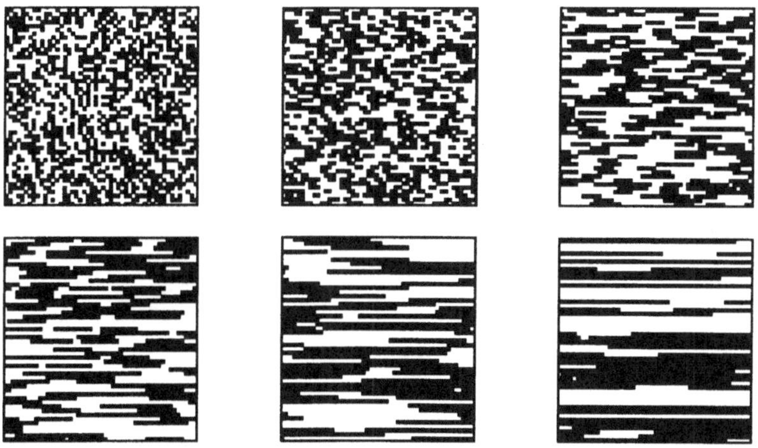

Fig. 4. Increasing anisotropy in generated pattern.

geological medium the pixel sizes are altered [here we use cuboids with one pixel height and a variable generating length l in the (x,y)-plane]. Placing these cuboids randomly into the medium allows overlapping of the blocks and therefore generates rather complex patterns which are similar to geological media (Fig. 4). Because the internal structure of these artificial rocks is well known, we can use them to study the geometrical and transport properties. Similar models have been studied by McCarthy (1991) and layered strata have long been an object of research for various kinds of transport properties (Kunz & Moran 1958, Vozoff 1958, Moran & Finklea 1962).

Characterizing the Medium: Scaling Properties and Power Laws

In order to characterize the medium we can introduce the characteristic mean length $L_{j,i}$ for both materials $j = 1,2$ and every direction $i = x,y,z$; which, of course, is the average length of continuous lines consisting only of material 1 or 2 (p_j: portion of material $j = 1,2$). Then there exists a definite correlation between the generating length l and the mean length $L_{j,i}$. (Fig. 5):

$$L_{2,i} = l_i(1-p_2)^{-[\alpha+(1-\alpha)\frac{l_0}{l_i}]} \quad , \text{for } l_i > l_0 \tag{4}$$

$$L_{2,i} = l_i(1-p_2)^{-1} \quad , \text{for } l_i \le l_0 \tag{5}$$

$$\text{and} \quad \frac{p_1}{L_{1,i}} = \frac{p_2}{L_{2,i}} \quad , (p_1+p_2=1) \tag{6}$$

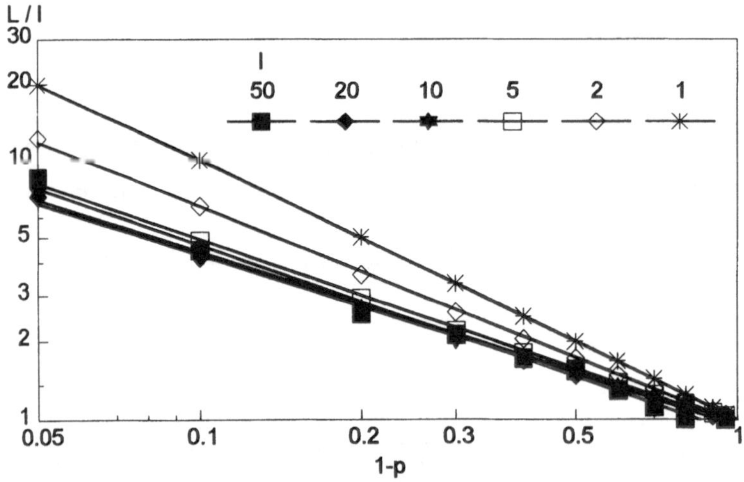

Fig. 5. Power law dependence for characteristic length L of portion p for different generating lengths l

The parameter α is a constant characteristic (α = 0.58..) for this kind of model. The parameter l_0 is the basic unit, e.g. one pixel. While the equation is valid for all possible sizes of generating cuboids within an infinite medium, we only use cuboids with size $l_x = l_y = l$ and height l_z=1 to reduce variable interdependency and we work with a finite medium which makes the determination of L difficult for high values of l and p_2. It should be noted, that the mean length L is symmetrical if material 1 is exchanged with material 2, while the distribution of geometric sizes is asymmetrical (for material 2 all sizes are greater or equal to l, but not for material 1). The characteristic mean length L can be considered the average free way length for a particle under Brownian or diffusive motion.

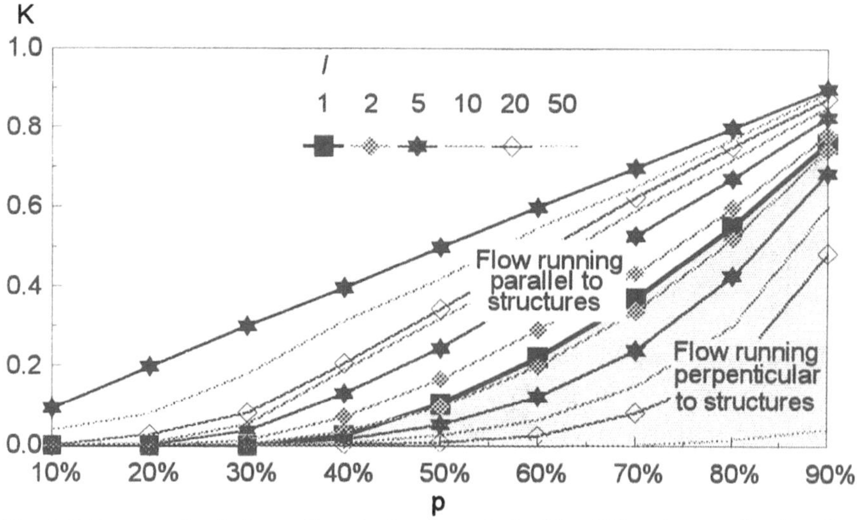

Fig. 6. Influence of anisotropy (layering) – characterised by different generating lengths l – on average conductivity K, using different material ratios p with one material being impermeable.

Figure 6 elucidates the influence of the anisotropic effect on the conductivities (permeabilities) parallel and perpendicular to the "layering" with regard to the generating length l and the relative frequency p of the randomly distributed material. The calculated conductivities are based on the solution of the diffusion equation for the different model substances. These calculations are performed for a range of contrasting conductivities, thus providing a continuum formulation on the macroscopic scale. The algorithms for computing the fluxes are similar to the method of Farrell & Larson (1972).

The percolation example, however, cannot be directly related to the pore space model. Natural rocks are permeable even for very low porosities, porosities far less than the percolation threshold of our models. However, the models may be used for microscopic considerations if the site oriented approach is replaced by a bond oriented model (e.g. Dias & Payatakes 1986, Rotheburg et al. 1987). The bonds then are the necks connecting two pores, while every site may represent a certain pore space (cf. the bond shrinking model). Alternatively, analytical (Berryman 1983, 1985) or percolation methods may be used, whereby aspects of diagenesis may be incorporated (Roberts & Schwartz 1985, Schwartz & Kimminau 1987).

The conductivity tensors defined so far are related to the relative frequency p and to the generation length l which causes anisotropy. However, while p may be measured on the model medium, l is disturbed and cannot be reconstructed from the pattern arising from the generation process.

We must, therefore, consider some additional geometrical properties of the models.

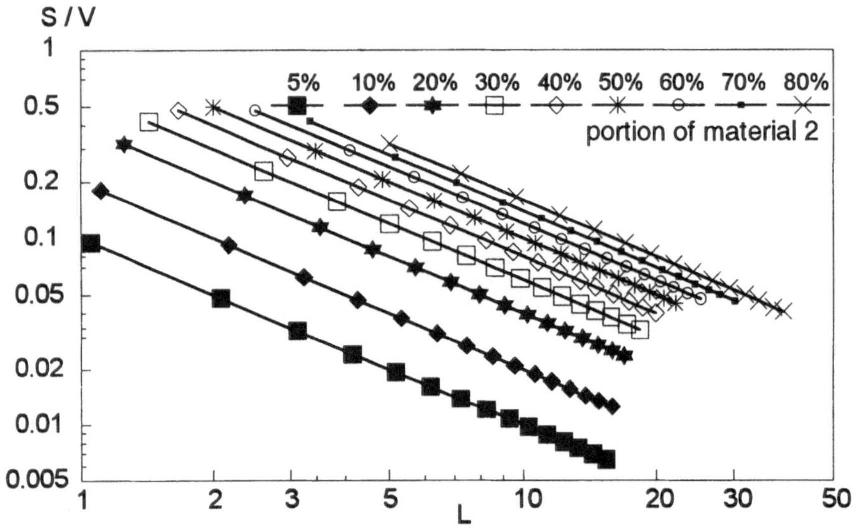

Fig. 7. Dependence of the inner surface S on the characteristic free length l of material 2.

In addition to the mean free way length L, we can introduce an inner surface S (surface between material 1 and 2) to characterize the medium. The three projections S_i of S into the principle directions (i=x,y,z) provide another measurement for anisotropy . The relationship between the characteristic lengths (Fig. 7) and the surface projections may be expressed as:

$$\frac{S_i}{V}=\frac{2p_2}{L_{2,i}}=\frac{2p_1}{L_{1,i}}=\frac{p_1}{L_{1,i}}+\frac{p_2}{L_{2,i}} \qquad (7)$$

with V the volume of the medium. This is a simple scaling relationship with L^{-1}. There is, however, no inter-dependence between the different directions. The total inner surface is the sum of these three directional parts.

The parameters L or S are useful for describing the geometric anisotropy tensor of the medium, where the form factors are taken from an analogy in magnetism (cf. Damm 1988).

$$P = \frac{L_{\max}}{L_{\min}} = \frac{S_{\max}}{S_{\min}} \qquad \text{(degree of anisotropy)} \qquad (8)$$

$$L = \frac{L_{\max}}{L_{\text{int}}} = \frac{S_{\max}}{S_{\text{int}}} \qquad \text{(degree of lineation)} \qquad (9)$$

$$F = \frac{L_{\text{int}}}{L_{\min}} = \frac{S_{\text{int}}}{S_{\min}} \qquad \text{(degree of foliation)} \qquad (10)$$

$$T = 2\frac{\ln\frac{L_{\text{int}}}{L_{\min}}}{\ln\frac{L_{\max}}{L_{\min}}} - 1 = 2\frac{\ln\frac{S_{\text{int}}}{S_{\min}}}{\ln\frac{S_{\max}}{S_{\min}}} - \qquad \text{(factor of shape)} \qquad (11)$$

$$E = \frac{L_{\text{int}}^2}{L_{\max}L_{\min}} = \frac{S_{\text{int}}^2}{S_{\max}S_{\min}} \qquad \text{(another shape factor)} \qquad (12)$$

Two of the form factors are sufficient to describe the whole tensor. In our case the coordinate system is by definition the principle coordinate system and we have just two different principle values (mutatis mutandis for S):

$$L_{max} = L_{int} = L_x = L_y \qquad \text{and} \quad L_{min} = L_z. \qquad (13)$$

It follows that:

$$P = L = E \qquad \text{and} \quad F = T = 1 \qquad (14)$$

and thus one parameter will completely describe the geometry of the tensor.

Transport Coefficients

The next step is to consider the effective transport coefficient of the model media with regard to the anisotropy effects. Fig. 6 illustrates the relationship of the average conductivity K from the generating line and the frequency p of the permeable material as discussed above. In the isotropic percolation case (l = 1,

$k_1 = k,\ k_2 = 0$) the material will only be permeable above the percolation threshold. For this special case, the average conductivity was found to be:

$$K = k\left(\frac{p - p_c}{1 - p_c}\right)^D \qquad p > p_c, \tag{15}$$

$$K = 0 \qquad p \le p_c \tag{16}$$

with $p_c \approx 0.317$ and $D \approx 1.72$ for our models. The equation is closely related to the probability P_∞ that a lattice site is a member of the maximal cluster which is given by:

$$P_\infty(p) \approx (p - p_c)^\beta \tag{17}$$

(Peitgen et al. 1992). The term $(1 - p_c)$ on the other hand is related to the mean length $L_{2,i}$ at the percolation threshold by:

$$L_c = (1 - p_c)^{-1}\ ;\ l_i = l_0 = 1 \tag{18}$$

If we allow both conductivities to be non-zero, the relationships become more complicated because there exists no simple power law allowing us to calculate the average conductivity K. However, in the case of a perfectly stratified medium ($l \to \infty$), the mean conductivities are sufficiently approximated by the weighted arithmetic and harmonic means:

$$K = p_1 k_1 + p_2 k_2 \equiv A \qquad \text{(for flow } \parallel \text{ structure)}, \tag{19}$$

$$K = \frac{k_1 k_2}{p_2 k_1 + p_1 k_2} \equiv H \qquad \text{(for flow } \perp \text{ structure)}. \tag{20}$$

If we consider the medium to be a mixture of serial and parallel resistors, the mean conductivity could be expected to be the limit of an iteration between these means. With:

$$k_1^n = f(k_1^{n-1}, k_2^{n-1}, p_2) = H \quad \text{and} \quad k_2^{n'} = g(k_1^{n-1}, k_2^{n-1}, p_2) = A, \tag{21}$$

we find

$$k_1^{n-1} < k_1^{m} < k_2^{m} < k_2^{n-1}, \tag{22}$$

with the limit

$$\lim k_1^\infty = \lim k_2^\infty = M(k_1, k_2, p_2). \tag{23}$$

For the probabilities $p_1 = p_2 = 0.5$ we find $M = G = k_1^{p1} k_2^{p2}$, i.e. the geometric mean (Schönberg 1982).

For isotropic media and conductivity contrasts of less than 20:1 there is a good correlation between the mean M and the average conductivity calculated from a flow simulation. The errors are less than 2% as long as a portion of the higher permeable material is below 50%. In the case of one material being impermeable the approximation fails totally and equations (15) and (16) must be applied. This observation gives rise to the assumption that the geometric anisotropy modifies the isotropic case by some scaling function so that the mean conductivity is given by the limits:

$H \leq K \leq I$ for flow ^ to structure,

$I \leq K \leq A$ for flow || to structure,

with
- H: harmonic mean,
- A: arithmetic mean,
- I: mean conductivity for the isotropic case, which currently can only be estimated from flow simulations.

A first approach to describe K as a function of (H, I, A) is given by the power law:

$$K_{\parallel} = I^{\alpha} * A^{1-\alpha} \qquad \parallel \text{ to structure,}$$

$$K_{\perp} = I^{\beta} * H^{1\beta} \qquad \perp \text{ to structure}$$

which indeed provides an excellent approximation for the data as illustrated by Fig. 8. The parameters α and β can be expected to depend on the degree of anisotropy, although a simple functional relationship with the form functions defined above has not yet been determined.

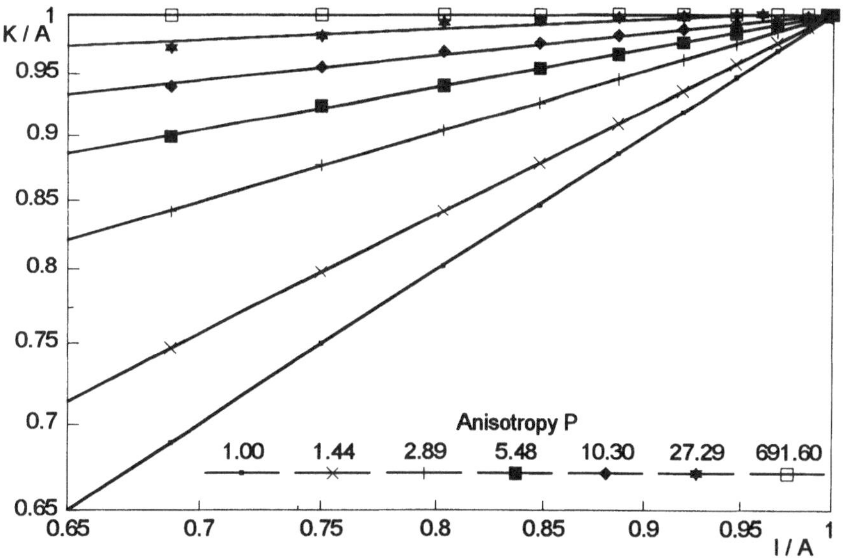

Fig. 8. Dependence of K/A over I/A to fit equation $K = I^{\alpha} A^{(1-\alpha)}$ in the case of p_2=80%.

Conclusions

The results indicate that anisotropies as defined in this chapter are related to the isotropic percolation case by a power law which can be understood as a scaling property. It seems, therefore, reasonable to reduce the infinite number of anisotropy models to a smaller set of isotropic models by a scaling law. Our results for randomly generated media amplify the results of Kunz & Moran (1958), which showed "that a wide class of potential problems involving anisotropic media can be transformed into equivalent problems involving only isotropic media". Further work, however, is needed to more fully understand the influence of geometrical properties on the effective transport coefficients, and it may be a major task to incorporate the changes due to diagenetic processes.

References

Berryman JG (1983) Computing Variational Bounds for Flow through Random Aggregates of Spheres. J Comp Phys 52: 142-162.

Berryman JG (1985) Bounds on fluid permeability for viscous flow through porous media. J Chem Phys 82: 1459-1467.

Bomberg M (1972) Similitude requirements for moisture flow through the porous materials. Proceedings of the second symposium on fundamentals of transport phenomena in porous media 1: 88.

Chen W, Ghaith A, Park A, Ortoleva P (1991) Diagenesis Through Coupled Processes: Modeling Approach, Self-Organization, and Implications for Exploration. AAPG Mem 49: 103-130.

Damm V (1988) Suszeptibilitätsanisotropien in Sedimentiten und Magmatiten aus dem Gebiet der DDR und Metamorphiten der Schirmacher-Oase (Antarktika). Veröffentlichung des Zentralinstituts für Physik der Erde 95.

Dias MM, Payatakes AC (1986) Network models for two-phase flow in porous media, Part 1. Immiscible microdisplacement of non-wetting fluids. J Fluid Mech 164: 305-336.

Farrell DA, Larson WE (1972) Computer Analysis of the Pore Structure of Isotropic Porous Media. In: Proceedings of the second symposium on fundamentals of transport phenomena in porous media 1: 74-87.

Giles MR, de Boer RB (1990) Origin and significance of redistributional secondary porosity. Marine Petrol Geol 7: 378-397.

Kunz KS, Moran JH (1958) Some effects of formation anisotropy on resistivity measurements in boreholes. Geophysics 23: 770-794.

McCarthy JF (1991) Analytical models of the effective permeability of sand-shale reservoirs. Geophys J Int 105: 513-527.

Moran JH, Finklea EE (1962) Theoretical Analysis of Pressure Phenomena Associated with the Wireline Formation Tester. Journal of Petroleum Technology 8: 899-908.

Ondrak R (1992) Mathematical simulation of water-mineral interaction and fluid flow with respect to the diagenetic evolution of sandstones. In: Kharaka & Maest (eds), Water-Rock Interaction. Balkema, Rotterdam, pp 1187-1191.

Ondrak R (1993) Untersuchungen zur mathematischen Simulation der Sandstein-diagenese und der damit verbundenen Veränderungen der Porosität und Permeabilität. Jül-Bericht 2270, KFA, Jülich.

Ondrak R, Bayer U (1993) Dissolution and Cementation in Basin Simulation. In: Merriam D, Harff J (eds): Computerized Basin Analysis: the Prognosis of Energy and Mineral Resources. Plenum Publ Corp, New York, pp 59-82.

Peitgen HO, Jürgens H, Saupe D (1992) Chaos and Fractals - New Frontiers of Science. Springer, New York Berlin Heidelberg.

Prieur du Plessis J, Masliyah JH (1988) Mathematical Modelling of Flow Through Consolidated Isotropic Porous Media. Transport in Porous Media 3: 145-161.

Roberts JN, Schwartz LM (1985) Grain consolidation and electrical conductivity in porous media. Phys Rev B 31: 5990-5996.

Rothenburg L, Matyas E, Ambrus SZ (1987) Statistical aspects of flow in a random network of channels. Stochastic Hydrol Hydraul 1: 217-240.

Schoenberg IJ (1982) Mathematical Time Exposures. The Mathematical Association of America, Washington.

Schwartz LM, Kimminau St (1987) Analysis of electrical conduction in the grain consolidation model. Geophysics 52: 1402-1411.

Stauffer D (1985) Introduction to percolation theory. Taylor & Francis, London Philadelphia.

Steefel CI, Lasaga AC (1990) Evolution of dissolution patterns: permeability change due to coupled flow and reaction. In: Melchior DC & Bassett RL (eds) Chemical Modeling in Aqueous Systems II, ACS Symposium Series No. 416, American Chemical Society, Washington, 212-225.

Vozoff K (1958) Numerical resistivity analysis: Horizontal Layers. Geophysics 23: 536-556.

Characteristic Multifractal Element Distributions in Recent Bioactive Marine Sediments

Jürgen Kropp[1], Arthur Block[1], Werner von Bloh[1], Thomas Klenke[2] & Hans-Joachim Schellnhuber[1]
[1]Potsdam-Institut für Klimafolgenforschung e.V.(PIK)
Postfach 601203, D-14412 Potsdam, Germany
[2]Institut für Chemie und Biologie des Meeres (ICBM)
Carl von Ossietzky Universität Oldenburg
Postfach 2503, D-26111 Oldenburg, Germany

Abstract. It is shown that SEM/EDX imaging is a powerful tool for detecting the spatial distribution of different elements and minerals in recent bioactive sediments. The evaluation of these distribution patterns by multifractal analysis provides a quantitative assessment of the characteristic inhomogeneities.

The combination of optical and multifractal techniques leads to the discrimination of intricate structures, the description of element - element correlations, and the identification of specific processes: e.g. the formation of (i) microbially induced calcite or (ii) authigenic clay-minerals indicated by an evident local correlation between aluminium and silicium.

Introduction

The geometry of complex geosystems like recent sediments or sedimentary rocks is a spatial result of various interacting processes. These include genetic and early diagenetic events and processes, which express themselves in characteristic structural features of a sediment, *observed on many scales*.

One powerful tool to describe the geometry of complex geosystems is the concept of fractality (Mandelbrot 1983): Geometries with statistical scale invariance over several orders of magnitude can be characterized by a fractal dimension (Fig. 1).
The combination of this method with modern image-analysis techniques has been the basis for recent progress in quantifying the arrangement of minerals as well as their surface and pore space properties in sedimentary geosystems and rocks (Krohn 1988, Hansen & Skjeltorp 1988).

Physical, chemical or biological processes, or probabilities and weight distributions, however, are controlled by the geometry of the pore space itself and the interaction between solid matrix components and the compounds transported by percolating fluids. As a consequence, one has to regard a complex spatial distribution as defining a measure on the sedimentary microstructure, which is it-

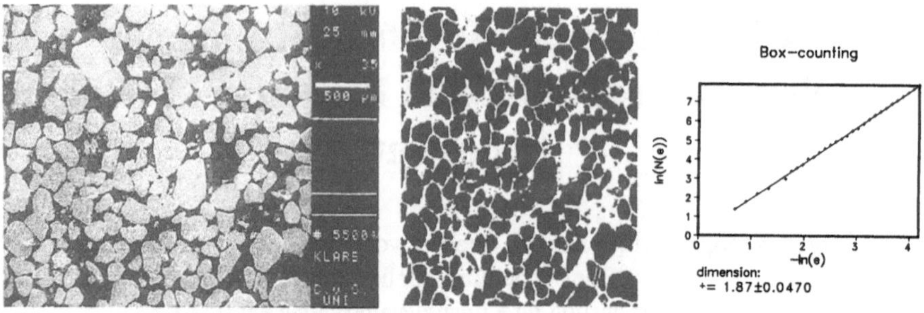

Fig. 1. Scanning electron photomicrograph of a recent bioactive marine sediment (sandlayer). In a second step this image is digitized (512 x 512 pixel) by grayscale filtering in order to prepare it for box - counting analysis.

self a self-similar object. One single number, namely the "ordinary" fractal dimension is not sufficient for the scaling characterization of such complex distributions: they have to be described by the more general concept of multifractality (e.g. Tel 1988). This formalism, is based on the so-called *generalized* (box-counting) *fractal dimensions* D_B (q), $q \in \mathbb{R}$ (Hentschel & Procaccia 1983), defined by

$$D_B(q) = \lim_{\varepsilon \to 0} \frac{1}{q-1} \frac{\ln \sum_{\nu=1}^{N(\varepsilon)} \left[p_\nu(\varepsilon)\right]^q}{\ln \varepsilon} \tag{1}$$

Here the index ν labels the individual boxes of an ε cover of the object and $p_\nu(\varepsilon)$ denotes the relative weight of the i - th box.

In practice, $p_\nu(\varepsilon)$ is encoded by a finite ensemble of points embedded in the Euclidean space (in our case, a 2 - dimensional coordinate list of elemental dots). $p_\nu(\varepsilon)$ can be calculated according to

$$P_\nu(\varepsilon) = \frac{N_\nu(\varepsilon)}{N} \tag{2}$$

where $N_\nu(\varepsilon)$ is the number of points falling into the ν - th box and N is the total number of elemental dots.

Methodology

The highly porous and unconsolidated sedimentary specimens are embedded in epoxy resin. By using a SEM/EDX equipment (ZEISS DSM 940/LINK QX 2000) and an image processing unit (SERIES 151 image processor by IMAGING

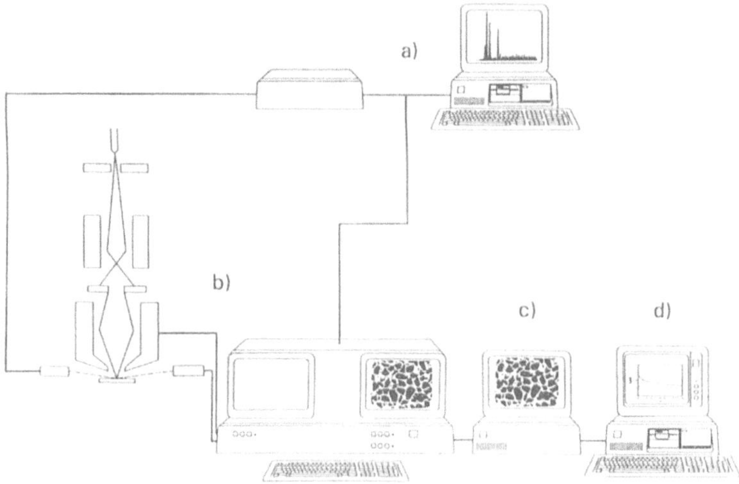

Fig. 2a-d. Arrangement of experimental equipment: **a** energy dispersive x-ray spectrometer, **b** scanning electron microscope, **c** image processor, **d** transputer system installed in a personal computer.

TECHNOLOGY INC.) elemental dot maps with a resolution of 512 x 512 pixels in 256 gray scales can be obtained. A box - counting algorithm (Block et al. 1990) is implemented on 4 transputers from INMOS (T800 + 1 MB RAM), which are installed in an AT personal computer controlling the image processing unit and allowing fast processing of the SEM/EDX images (Fig. 2).
The elemental dots in the image represent the local occurrence of an element in the sedimentary sample. Applying multifractal methods these occurrence maps are quantified by the associated multifractal spectra. For further details and tests see Block et al. (1991).

Sedimentological Implications

Discrimination of Substructures

We have investigated lower supratidal versicolored tidal flat sediments of Mellum Island situated in the southern part of the German Bight/North Sea. Our results demonstrate the capability of combined optical-multifractal techniques to resolve the fine structure of such specimens.

Within a profile of a recent sedimentary column (sequence of sandlayers and microbial mats) different sedimentary substructures exhibit a characteristic multifractal behavior: the sandlayers can be characterized by a monofractal

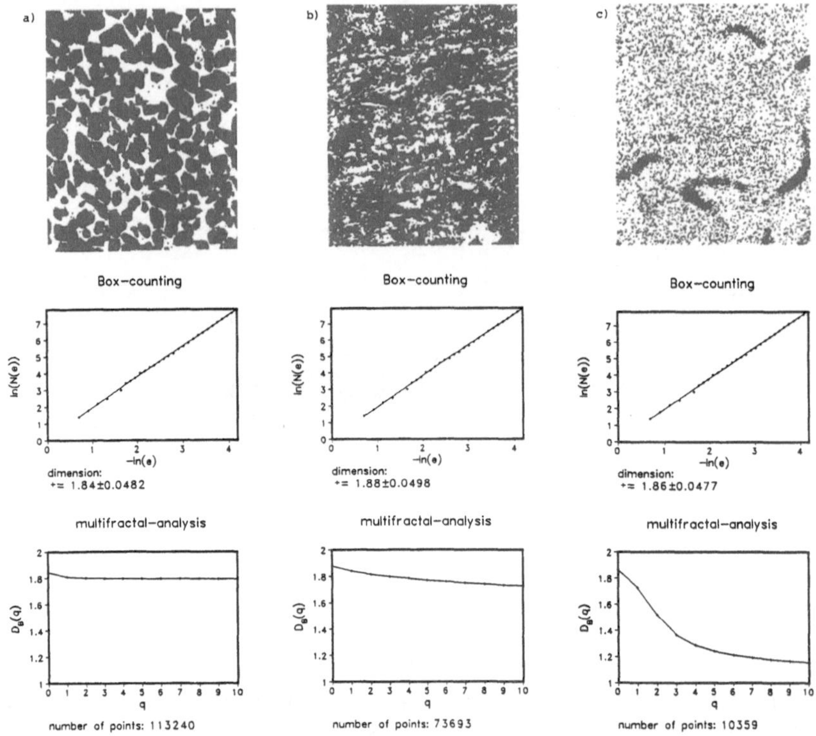

Fig. 3a-c. Silicium distribution maps of three different sedimentary substructures and associated multifractal spectra: **a** sandlayer; **b** subrecent microbial mat; **c** recent microbial mat.

silicium distribution, whereas recent and subrecent microbial mats are distinguishable by a clear-cut and specific multifractal behavior of silicium (Fig. 3).

The spectra of elemental maps obtained from other elements like aluminium, iron, sulphur, potassium and calcium also exhibit specific multifractal behavior, depending on the substructure under investigation.

Quantification of Processes

Two early diagenetic processes in bioactive marine sediments can be analyzed with our method:

(i) In subrecent microbial mats the comparison of multifractal spectra received from primary and processed secondary maps provides new insights into sediment dynamic processes. Assuming that the inert siliciclastic detritus is

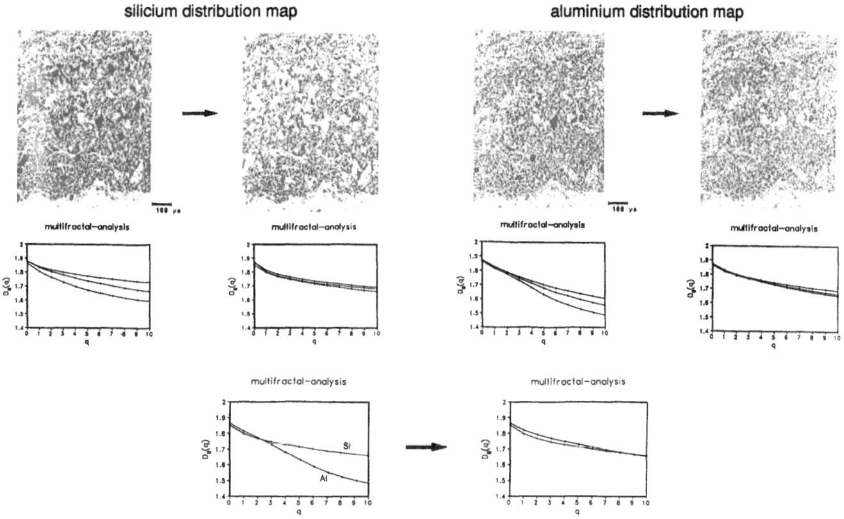

Fig. 4a-d. Multifractal analysis of silicium and aluminium distribution maps obtained from subrecent microbial mats before and after extraction of the siliciclastic sediment components: **a** primary silicium map, 47898 points; **b** secondary silicium map, 32294 points; **c** primary aluminium map, 26821 points; **d** secondary aluminium map, 24006 points).

not involved in early diagenetic processes, the extraction of these components clearly reveals a correlation between silicium and aluminium (Fig. 4).

Such effects are only found in subrecent microbial mats. This indicates that the early diagenetic formation of alumosilicate minerals in such substructures is connected to the onset of the decomposition of organic matter!

(ii) In recent microbial mats microscopic variations of the calcium distribution are detected within a layer of a few millimeter thickness (Fig. 5). Ca - rich horizons contain syndepositional and early diagenetic magnesium - calcite $[(Ca_{1-x}Mg_x)CO_3, x \simeq 0.13]$.

Its distribution and formation is coupled to the break down and replacement of organic matter due to the activities of heterotrophic microbial communities. Assuming that calcium dots trace authigenic calcite the obtained multifractal spectra of these distribution patterns can be conceived as a measure of the local intensity of this specific and dominating mineralization process.

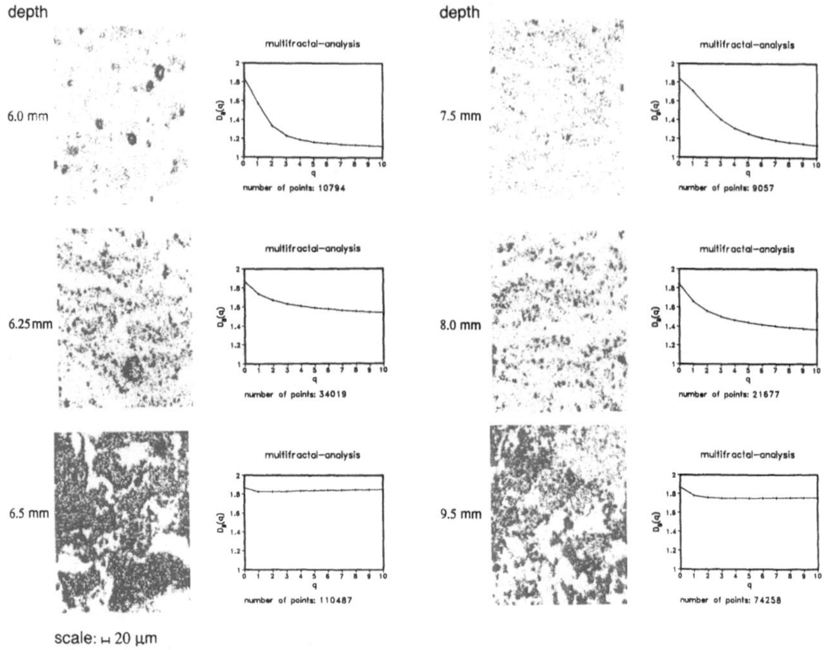

Fig. 5a-f. Calcium distribution maps obtained from different horizons of a recent microbial mat and associated multifractal spectra: **a** 6.0 mm, 10794 points; **b** 6.25 mm, 34019 points; **c** 6.5 mm, 110487 points; **d** 8.0 mm, 9057 points; **e** 8.5 mm, 21677 points; **f** 9.25 mm, 74258 points.

Conclusions

The multifractal analysis of elemental density distributions in recent marine sediments provides new insights into both the geometrical organisation of sedimentary substructures and intrinsic early diagnetic processes.

We have demonstrated that the $D_B(q)$ - spectra calculated from the data, enable a quantitative measurement of order and disorder in elemental distributions. They are also valuable information regarding the geochemical and biological support of sediment intrinsic processes (e.g., microbial induced calcite formation).

Future work will focus on the investigation of other mineralization processes like the formation of authigenic sulphides, the determination of discrimination criteria for substructures and the modelling of structure forming processes by cellular automata methods (Wolfram 1983).

References

Block A, von Bloh W, Schellnhuber HJ (1990) Efficient box-counting determination of generalized fractal dimensions. Phys Rev A 42: 1869-1874.

Block A, von Bloh W, Klenke T, Schellnhuber HJ (1991) Multifractal analysis of the microdistribution of elements in sedimentary structures using images from scanning electron microscopy and energy dispersive x-ray spectrometry. J Geophys Res 96: 16223-16230.

Hansen JP, Skjeltorp AT (1988) Fractal pore space and rock permeability implications. Phys Rev B 38: 2635-2638.

Hentschel HGE, Procaccia I (1983) The infinite number of generalized dimensions of fractals and strange attractors. Physica D 8: 435-444.

Krohn CE (1988) Sandstone fractal and Euclidean pore volume distributions. J Geophys Res 93: 3286-3296.

Mandelbrot BB (1983) The fractal geometry of nature. Freeman, New York.

Tel T (1988) Fractals, multifractals and thermodynamics. An introductory review. Z Naturforsch 43A: 1154-1174.

Wolfram S (1983) Statistical mechanics of cellular automata. Rev Mod Phys 55: 601-644.

Fractal Analyses of Pleistocene Marine Oxygen Isotope Records

Michael Schulz[1], Manfred Mudelsee[2] & Thomas C.W. Wolf-Welling[1]
[1]GEOMAR, Forschungszentrum für marine Geowissenschaften, Wischhofstr. 1-3, D-24148 Kiel, Germany
[2]Geologisch-Paläontologisches Institut, Universität Kiel, Olshausenstr. 40, D-24118 Kiel, Germany

Abstract. Fractal dimensions of oxygen isotope ($\delta^{18}O$) data obtained from deep-sea records were estimated. The globally distributed data reflect climate variability during the late Pleistocene.

All fractal dimensions fall into the range of $1 < D < 2$. However, the records do not show self-affinity over the entire range of investigated time scales, instead two significantly different fractal dimensions can be observed. This result contrasts with previous suggestions that time series obtained from climate proxies might possess self-affinity with a single fractal dimension over a range of 10 to 10^5 years.

The estimated fractal dimensions of the $\delta^{18}O$ records for time scales between 3 and ~20 ka fall into a narrow range, with an average value of $D = 1.51 \pm 0.06$.

Estimated fractal dimensions for longer time scales may be biased by the presence of periodic components in the time series. The limited range of time scales which can be described by a *single* fractal dimension raises questions about the applicability of simple fractal theory to this kind of data.

Introduction

A very common characteristic of many natural phenomena is their scale invariance, that is, an object looks the same under a change of resolution. Typical examples include the shape of clouds and rock surfaces. Mandelbrot (1967) introduced the term fractal to describe this scale invariance. A fractal can be described as an object made of parts similar to the whole in some way, i.e. in a statistical sense (Feder 1988). A statistically self-similar fractal in a two-dimensional (x,y) space can be defined as follows (Turcotte 1992, p 74): $f(x,y)$ is statistically similar to $f(rx,ry)$, where r is a scaling factor. In this case both variables are scaled by the same scaling-factor.

When dealing with time series similarity transformations usually require different scaling factors for time axis and the variable being measured as a function of time. A time series can be described as a self-affine fractal if $f(t,y)$ is statistically similar to $f(rt,r^{H}y)$, where r is again a scaling factor and H is the Hausdorff dimension. The latter is related to the fractal dimension D by:

$D = 2 - H$ (Turcotte 1992, p 75). If D falls into the range of $1 < D < 2$ the time series is called a 'fractional Brownian function' (Mandelbrot 1977).

Here we investigate whether late Pleistocene time series of oxygen isotopes ($\delta^{18}O$) can be described as fractals. The $\delta^{18}O$ records reflect changes in continental ice volume and to a lesser degree variations in ocean water temperature, hence, they can be regarded as climate proxy data. The fractal approach is purely descriptive, therefore, no *direct* information about the underlying physical processes documented by the $\delta^{18}O$ data is obtained.

In addition, we address the question as to whether or not a global fractal dimension exists describing climatic fluctuations over the entire Pleistocene as previously suggested by Fluegeman and Snow (1989).

Method

To determine the fractal dimension of the various time series we follow a method developed by Higuchi (1988). Given an evenly spaced time series $Y_1, Y_2, \ldots, Y_N$, first new time series are constructed:

$$Y_k^m = Y_m, Y_{m+k}, Y_{m+2k}, \ldots, Y_{m+\left[\frac{N-m}{k}\right]\cdot k} \quad (m = 1, 2, \ldots, k),$$

where [...] denotes integer notation, m is the initial time and k the interval time, both being integers. For any time interval k, k new time series can be constructed. Now the *length* of each time series Y_k^m is given by:

$$L_m(k) = \left\{ \left(\sum_{i=1}^{\left[\frac{N-m}{k}\right]} \left| Y_{m+ik} - Y_{m+(i-1)\cdot k} \right| \right) \cdot \frac{N-1}{\left[\frac{N-m}{k}\right] \cdot k} \right\} \cdot \frac{1}{k} \tag{1}$$

The length of the curve for a time interval k, $\langle L(k) \rangle$ is defined as the average over k sets of $L_m(k)$:

$$\langle L(k) \rangle = \frac{1}{k} \cdot \sum_{m=1}^{k} L_m(k) \tag{2}$$

For a statistically self-affine curve in a (x,y)-plane

$$\langle L(k) \rangle \propto k^{-D}. \tag{3}$$

where D is the fractal dimension. The interval time k is related to the time scale τ by: $\tau = \Delta t \cdot k$, where Δt denotes the time interval between the equidistant data points (e.g. 3 ka).

Compared to the spectral method (e.g. Turcotte 1992; p 74), the advantage of the Higuchi method is that relatively short time series will result in stable estimations of the fractal dimension (Higuchi 1988). In addition, a change of the fractal dimension over the investigated time range can be identified more easily.

Before determining the fractal dimensions of the $\delta^{18}O$ records, all time series were linearly interpolated to obtain equidistant time series. The interpolation was performed in such a manner that the number of points in the interpolated time series is equal to the number of original data points. Subsequently, a linear trend was subtracted from the data. Instead of fitting a linear function in log-log space to the calculated $[k,\langle L(k)\rangle]$ data we fitted the nonlinear model (Eq. 3) to the points, since any linearization of the data may lead to wrong results when using a least-square method for fitting a function to the transformed points (Appendix). Fitting the nonlinear model was carried out by numerical approximation using the Levenberg-Marquardt method (e.g. Press et al. 1989).

Fractal dimensions were calculated for all possible sets of $[k,\langle L(k)\rangle]$ with ≥ 3 data pairs. To evaluate the goodness-of-fit of the resulting models, the χ^2 probability for each fit was calculated. To do so, first the uncertainties associated with the determined $\langle L(k)\rangle$ must be evaluated. With the known errors of the measured parameters (section "Data Material") this was done by applying the laws of error propagation to Equations 1 and 2. The probability of a significant fit decreases with increasing number of $[k,\langle L(k)\rangle]$ points used for a certain fitting procedure. Hence, one would obtain the highest probabilities if the entire set of $[k,\langle L(k)\rangle]$ were divided into relatively short segments (e.g. 3 points per segment). This would result in a large number of fractal dimensions, each valid for a short time scale only. Since we are interested in describing the time series by as few fractal dimensions as possible, one has to find a compromise between the goodness-of-fit of the fitting function and the length of the segments for which the fractal dimension are estimated. Evaluation of the fitting results was guided by the following criteria:

- χ^2 'Goodness-of-fit' probability ≥ 0.1,
- interval-length $[\tau_{min}, \tau_{max}]$ comparable between the records.

Data Material

The $\delta^{18}O$ time series used for the fractal analyses were obtained from published Pleistocene deep-sea records and comprise the last 380-1000 ka with resolutions in the order of 2-5 ka (a typical $\delta^{18}O$ record is shown in Figure 1). Core locations and characteristics are summarized in Table 1. Depending on the depth of the core sites and the depth habitat (benthic vs. planktonic) of the foraminifera

Table 1. Summary of the core locations and characteristics. * denotes a modified stratigraphy based on (Shackleton et al. 1990); T_{cont}: continuous part of the record used for the present study; Δt: time interval of the linearly interpolated time series; N: number of data points; Foram: indicates whether $\delta^{18}O$ data were measured on *P*lanktonic or *B*enthic foraminifera. Last column indicates the source of the data: 1. Nelson et al. 1986; 2. Ruddiman et al. 1989; 3. Wolf 1991; 4. Tiedemann 1991; 5. deMenocal et al. 1993; 6. Shackleton and Hall 1989; 7. Shackleton et al. 1990; 8. Mix et al. 1991; 9. Clemens et al. 1991.

Core	Latitude	Longitude	Depth [m]	T_{cont} [ka]	Δt [ka]	N	Foram	'Sampled' Ambient Water Mass	Ref.
DSDP 594*	45°31'S	174°75'E	1204	9-972	3.0	322	P	Sub-Antarctic Surface Water	1
DSDP 594*	45°31'S	174°75'E	1204	8-968	4.0	241	B	Antartic Intermediate Water	1
ODP 607*	41°00'N	32°58'W	3427	245-998	3.0	251	B	North-Atlantic Deep Water	2
ODP 643*	67°43'N	01°02'E	2780	4-501	3.5	143	P	Norwegian-Sea Water	3
ODP 658	20°45'N	18°35'W	2263	76-661	1.8	326	B	North-Atlantic Deep Water	4
ODP 663	01°12'N	11°53'W	3708	0-915	3.0	306	P	South-Atlantic Central Water	5
ODP 677	01°12'N	83°44'W	3461	5-993	2.6	381	B	Pacific Deep Water	6,7
RC 13-110	00°06'N	95°39'W	3231	3-735	3.0	245	B	Pacific Deep Water	8
RC 27-61	17°00'N	60°00'E	1893	6-432	3.0	143	P	Arabian-Sea Water	9
V 19-27	00°28'S	82°04'W	1373	2-376	2.0	188	B	Antarctic Intermediate Water	8

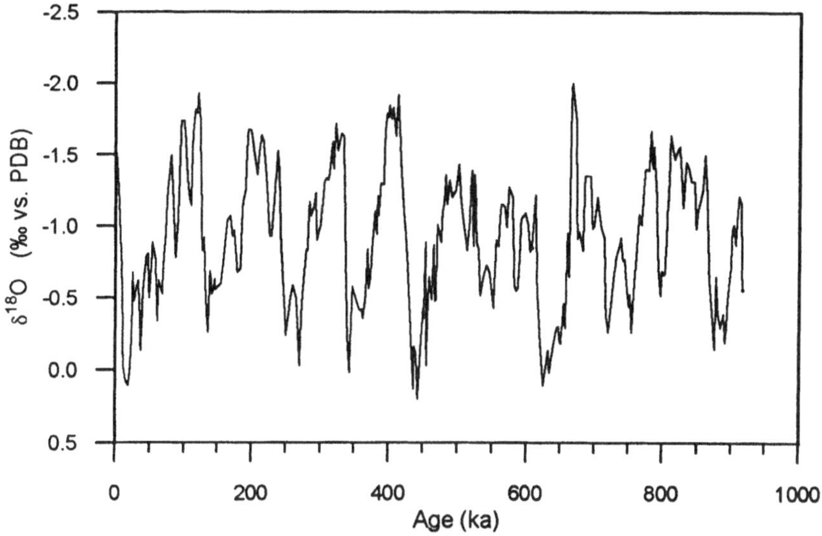

Fig. 1. Oxygen isotope record spanning the last 900 ka obtained from deep-sea core ODP 663 (deMenocal et al. 1993). The $\delta^{18}O$ time series mainly reflects changes in continental ice volume, with more negative values indicating less ice (Note that the ordinate scale is inverted).

used for the $\delta^{18}O$ determination, the data document the $\delta^{18}O$ variations of the ambient water masses of the core sites, i.e. upper, intermediate and deep water masses.

All $\delta^{18}O$ data are reported relative to the PDB standard. Random errors associated with the $\delta^{18}O$ determination arise from analytical uncertainties (0.05 to 0.07‰) and random variations in the sample population (e.g. habitat effects). Mix (1992) estimated the combined random error to be in the order of 0.1‰. This value was used for all data sets to evaluate the goodness-of-fit (above). The age models of all $\delta^{18}O$ records are based on oxygen isotope stratigraphy. Whenever a $\delta^{18}O$ record contains a hiatus only the continuous parts of the time series was used (Table 1). Although some of the time series document climatic variations throughout the entire Pleistocene and upper Pliocene, only those parts younger than 1 Ma were used for the present study, because there is evidence for a fundamental change in the mode of the climatic variations around this time (e.g. Ruddiman et al. 1986).

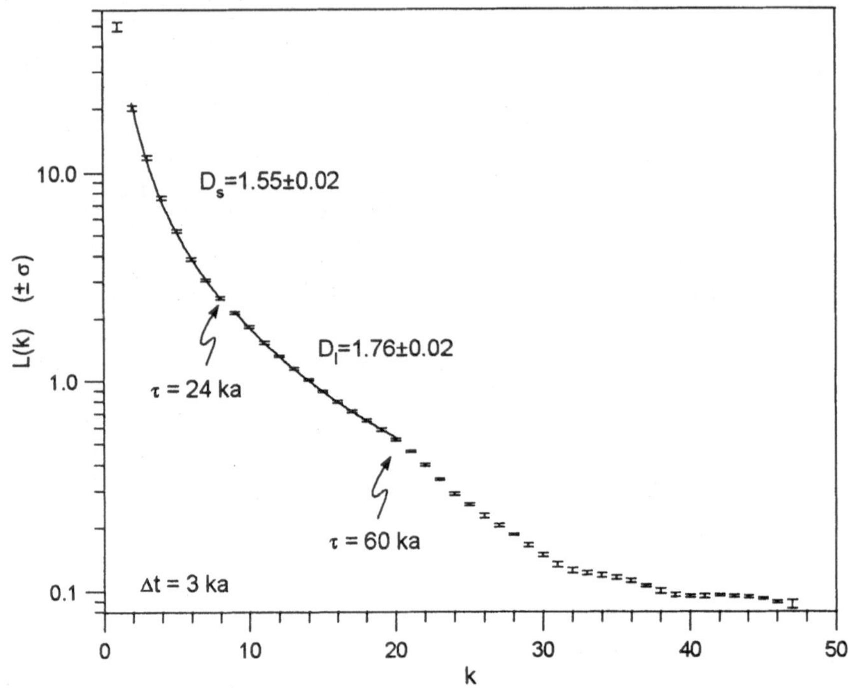

Fig. 2. Determination of the fractal dimensions for the $\delta^{18}O$ record of ODP 663. $\langle L(k)\rangle$ is the 'length' of the time series as a function of a 'time yardstick' k. The dimensions were obtained by fitting a nonlinear model to the $[k,\langle L(k)\rangle]$ points. The parameter k is related to the actual time scale, τ by $\tau = k \cdot \Delta t$, where Δt is the linearly interpolated sampling interval. Note that the fractal dimension describing the time series changes at $\tau = 24$ ka and that for $\tau > 60$ ka no significant fit is possible.

Results of the Fractal Analyses

Figure 2 shows a typical example of a $[k,\langle L(k)\rangle]$ plot and the fitted nonlinear function used to determine the fractal dimension. Figure 3 summarizes all fractal dimensions obtained from the time series which pass the above mentioned requirements. The estimated fractal dimensions cover an interval between $1.42 < D < 1.93$, hence all time series may be described as fractional Brownian functions.

However, it is also obvious from Figure 3 that the $\delta^{18}O$ records do not show a unique fractal dimension over the entire time scale ($k_{max} \cdot \Delta t$) being investigated. Instead, two significantly different fractal dimensions can be observed at each core (except data from core DSDP 594): a lower fractal dimension for short time scales ranging from ~3-20 ka (τ_s group) and a significantly higher dimension for longer time scales between ~20-60 ka (τ_l group). For time scales longer than approximately 60 ka it becomes impossible to yield a significant fit of a model according to Equation 3.Considering the τ_s group the lowest fractal dimensions can be observed for two of the deep-water $\delta^{18}O$ records (ODP 658 and RC 13-110). An exceptionally high fractal dimension was determined for core ODP 643, falling into the range of the τ_l group. The average fractal dimension of this group (excluding the outlier from the poor resolution record ODP 643) is $D_s = 1.51 \pm 0.06$. (Mean values are given as weighted means and errors as weighted sample standard deviations throughout the text.)

The τ_l group is characterized by slightly more scattered fractal dimensions as compared to the τ_s group ($D_l = 1.79 \pm 0.08$). There seems to be no obvious relation between fractal dimensions and water masses. A comparison of the difference between the fractal dimensions, i.e. $D_{t_s} - D_{t_l}$ for each core, shows that the largest differences occur for deep-water $\delta^{18}O$ data.

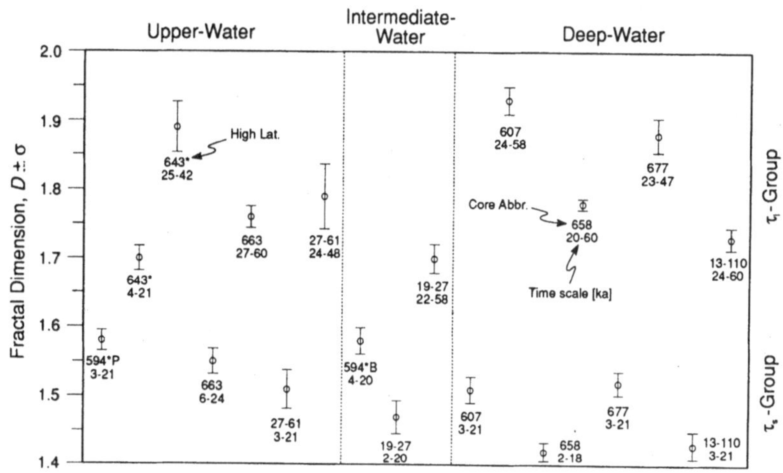

Fig. 3. Synopsis showing the fractal dimensions determined for the various $\delta^{18}O$ records. Cores have been sorted according to the type of ambient water mass (Table 1). Note that only the core numbers from Table 1 are used to identify the cores. High latitude cores are marked by '*'. Also shown is the time scale interval for which the fractal dimension was determined. Errors are reported as sample standard deviations. The horizontal line (dotted) separates the τ_s group from the τ_l group (see text for further explanations).

Discussion and Conclusions

The concept of fractals was introduced in order to describe scale *invariance*, i.e. the notion that $f(t, ry)$ is statistically similar to $f(rt, r^H y)$ even if the scaling factor r changes over several orders of magnitude. Considering the nature of the $\delta^{18}O$ data, one is confronted with the fact that the range of possible time scales comprises less than two orders of magnitude (3 ka < τ < 150 ka). Moreover, since it is not possible to find fractal dimensions for the $\delta^{18}O$ records for time scales larger than 60 ka, the scale variation decreases to about one order of magnitude. This limitation raises questions concerning the applicability of simple fractal theory to this kind of data and should be kept in mind during the following discussion.

The finding that the $\delta^{18}O$ records cannot be adequately described by a single fractal dimension over the entire range of investigated time scales contrasts with previous suggestions that time series obtained from climate proxies might possess self-affinity with a single fractal dimension over a range of 10 to 10^5 years (Fluegeman & Snow 1989).

Considering the global distribution of the core locations, the variety of water masses and the different foraminifera species used for measuring the isotopic values, it is remarkable that the fractal dimensions estimated for the $\delta^{18}O$ records for time scales < 20 ka fall into a fairly narrow range. We interpret this as an expression of the dominance of the global ice-volume signal over the regional signal (water temperature, habitat). This is in accordance with our observation that the fractal dimension (for the same time scale using, the method described above) obtained from the SPECMAP stack (Imbrie et al. 1984), which can be assumed as ice-volume proxy, is $D = 1.48 \pm 0.01$ and, thus, coincides with our average fractal dimension for the τ_S group.

From the observation that the lowest fractal dimensions in the τ_S group stem from deep-water records, one may hypothesize that this is caused by the slower response of the deep ocean compared to the surface-water masses. Similarly one may argue that the high fractal dimension obtained from the high latitude surface record (ODP 643) is due to the proximity of this site to the location of ice formation. However, the average present-day deep-ocean mixing time is on the order of 1 ka, and surface water mixes in a few decades (e.g. Murray 1992). Since these time scales are shorter than the minimum time scale in our investigation, even the deep ocean can be assumed to be well mixed, thus refuting the above implications. Since the southern hemisphere high latitude surface-water record (DSDP 594) does not show a similar high fractal dimension, we regard the high fractal dimension estimated for core ODP 643 as an outlier. The inference about deep-water masses is also invalidated by the fact that different fractal dimensions arise from cores that document the same water mass (ODP 607/658: North Atlantic deep water; ODP 677/RC 13-110: Pacific deep water). We attribute these differences either to sampling errors, which are higher than those being assumed, or to errors in the core stratigraphy.

Looking at the fractal dimensions for time scales longer than ~20 ka, the finding that $D_l > D_S$ is somewhat curious. In terms of a fractional Brownian function this could be interpreted as a trend towards higher antipersistence (Feder 1988) of the underlying physical system or, in terms of a spectral representation, towards a white-noise spectrum. However, this interpretation is unreasonable if one considers the underlying climate system. Especially the ocean should act as a low-pass filter and thus lead towards higher persistence, resulting in a red-noise spectrum. The additional observation that no evidence for self-affinity was found for time scales exceeding ~60 ka points to the possible effect that periodic components in the time series bias the estimated fractal dimension. A likely source of such periodic elements are the Milanković frequencies [i.e. periodic variations of the Earth's orbit, which are well documented in the deep-sea sedimentary record (e.g. Imbrie et al. 1984)], particularly the two strongest components with main periods of 41 and 100 ka . So far we have not been able to discriminate whether the above observations are artifacts or whether they imply that over certain time scales no self-affinity exists.

Finally, the estimated fractal dimensions are, due to the goodness-of-fit test, a function of measurement uncertainties. Assuming larger errors would result in a significant fit of the model (Eq. 3) over a *larger range* of time scales (and at the same time the number of fractal dimensions necessary to describe the time series would decrease). Due to the shape of the model function this would result in higher fractal dimensions than those given.

Appendix

Fractal sets can be scaled according to power laws as in Equation 3. It is very common to determine the fractal dimension by the linearization of such power laws. Instead of fitting a nonlinear model to the data (which requires extensive numerical calculations), the data are transformed (i.e. by taking logarithms) in order to yield a linear regression problem (which is computationally easy to handle). However, estimating fractal dimensions from linearized data is only an approximate method, since linearizing transformations change the statistical distribution of the errors associated with data (Bard 1974, p. 79).

Assume the following power-law function:

$$\mu_i = f(\mathrm{x}) = \beta_0 x_i^{\beta_1}$$

This nonlinear model can be linearized by means of a logarithmic transformation:

$$\ln \mu_i = \ln \beta_0 + \beta_1 \ln x_i \qquad \text{(A1)}$$

Let y_i be realizations of this function and ε an error term accounting for the noise in the data. The least-squares regression method assumes ε to be additive and Gaussian distributed, N(0,σ). Hence, Equation A1 becomes:

$$\ln y_i = \ln \beta_0 + \beta_1 \ln x_i + \varepsilon$$

This means, however, that in terms of the original data the error becomes *multiplicative*:

$$y_i = \beta_0 x_i^{\beta_1} \cdot e^{\varepsilon} \tag{A2}$$

Since the assumption concerning the error term in Equation A2 does not hold true for the data used in this study, the linearization transformation may lead to significantly biased estimations of the parameters β_0 and β_1.

Comparison of fractal dimensions obtained by a linear regression in log-log space with those obtained by fitting a nonlinear model to the data points reveal differences on the order of 10%. Therefore, determinations of fractal dimensions should be done by the use of nonlinear models instead of simple, but inaccurate, linear models.

Acknowledgments. We thank J.H. Kruhl and an anonymous reviewer for their comments, which helped to improve the final manuscript. We are also grateful to M. Raymo and R. Tiedemann for providing us their $\delta^{18}O$ data. This work was supported by the *Deutsche Forschungsgemeinschaft*: grant TH 200/10-2 and *Graduiertenkolleg 'Dynamik globaler Kreisläufe im System Erde'*.

References

Bard Y (1974) Nonlinear parameter estimation. Academic Press, New York,

Clemens S, Prell W, Murray D, Shimmield G, Weedon G (1991) Forcing mechanisms of the Indian Ocean monsoon. Nature 353: 720-725.

deMenocal PB, Ruddiman WF, Pokras EM (1993) Influences of high- and low-latitude processes on African terrestrial climate: Pleistocene eolian records from equatorial Atlantic Ocean Drilling Program Site 663. Paleoceanogr 8: 209-242.

Feder J (1988) Fractals. Plenum Press, New York.

Fluegeman RH Jr, Snow RS (1989) Fractal analysis of long-range paleoclimatic data: Oxygen isotope record of Pacific core V28-239. PAGEOPH 131: 307-313.

Higuchi T (1988) Approach to an irregular time series on the basis of the fractal theory. Physica D 31: 277-283.

Imbrie J, Hays JD, Martinson DG, McIntyre A, Mix AC, Morley JJ, Pisias NG, Prell WL, Shackleton NJ (1984) The orbital theory of Pleistocene climate: support from a revised chronology of the marine $\delta^{18}O$ record. In: Berger A, Imbrie J, Hays J, Kukla G, Saltzman B (eds) Milankovitch and climate, vol Part I. Reidel, Dordrecht, pp 269-305.

Mandelbrot BB (1967) How long is the coast of Britain? Science 155: 636-638.

Mandelbrot BB (1977) Form, chance, and dimension: fractals. Freeman, San Francisco.

Mix AC (1992) The marine oxygen isotope record: Constraints on timing and extent of ice-growth events (120-65 ka). In: Clark PU, Lea PD (eds) The last interglacial-glacial transition in North America. Geol Soc Amer Spec Paper 270, pp 19-30.

Mix AC, Pisias NG, Zahn R, Rugh W, Lopez C, Nelson K (1991) Carbon 13 in Pacific deep and intermediate waters, 0-370 ka: implications for ocean circulation and Pleistocene CO_2. Paleoceanogr 6: 205-226.

Murray JW (1992) The oceans. In: Butcher SS, Charlson RJ, Orians GH, Wolfe GV (eds) Global biogeochemical cycles, International Geophysics Series 50. Academic Press, London, pp 175-211.

Nelson CS, Hendy CH, Cuthbertson AM, Jarrett GR (1986) Late Quaternary carbonate and isotope stratigraphy, subantarctic Site 594, southwest Pacific. In: Kennett JP, von der Borch CC, et al. (eds) Init. Repts. DSDP Leg 90. Washington, U.S. Govt. Printing Office, pp 1425-1436.

Press WH, Flannery BP, Teukolsky SA, Vetterling WT (1989) Numerical recipes in Pascal. Cambridge Univ Press, Cambridge.

Ruddiman WF, Raymo M, McIntyre A (1986) Matuyama 41,000-year cycles: North Atlantic Ocean and northern hemisphere ice-sheets. Earth Planet Sci Lett 80: 117-129.

Ruddiman WF, Raymo M, Martinson, DG, Clement, BM, Backman, J (1989) Pleistocene evolution: northern hemisphere ice sheets and north Atlantic circulation. Paleoceanogr 4: 353-412.

Shackleton NJ, Hall MA (1989) Stable isotope history of the Pleistocene at ODP Site 677. In: Becker K, Sakai H, et al. (eds) Proc. ODP, Sci Results 111. College Station, TX (Ocean Drilling Program), pp 295-316.

Shackleton NJ, Berger A, Peltier WR (1990) An alternative astronomical calibration of the lower Pleistocene timescale based on ODP Site 677. Trans R Soc Edinburgh: Earth Sci 81: 251-261.

Tiedemann R (1991) Acht Millionen Jahre Klimageschichte von Nordwest Afrika und Paläo-Ozeanographie des angrenzenden Atlantiks: Hochauflösende Zeitreihen von ODP-Sites 658-661. Ber Geol-Paläont Inst Univ Kiel 46: 1-190

Turcotte DL (1992) Fractals and chaos in geology and geophysics. Cambridge Univ Press, Cambridge.

Wolf TCW (1991) Paläo-ozeanographisch-klimatische Entwicklung des nördlichen Nordatlantiks seit dem späten Neogen (ODP Legs 105 und 104, DSDP Leg 81). Geomar Rep 5: 1-92.

IV
Methods

"Counter-Scaling" Method for Estimation of Fractal Properties of Self-Affine Objects

Sergey S. Ivanov
P.P. Shirshov Institute of Oceanology,
23 Krasikova Str., 117218 Moscow, Russia

Abstract. A new method of estimation of fractal properties of self-affine sets is suggested which allows the analysis of scale invariance of many natural processes, objects, and phenomena. The method is based on scale tracing of the values of two different kinds of dispersions of the set - relative to the smallest and to the largest possible scales. They are called internal and external dispersions respectively. Some case histories of its use are discussed showing that the "counter-scaling" method provides reliable estimates of fractal dimensions for various geophysical sets.

Introduction

Mandelbrot (1986) showed that standard methods of fractal geometry are not applicable to many natural objects and processes and introduced the notion of "self-affine" sets. He also showed from a theoretical point of view that such self-affine sets must be described with two different values of fractal dimension - local and global ones, which coincide in the case of pure self-similarity. Since then many efforts have been made to define adequate and convenient methods for estimation of fractal properties of such sets. The principal challenge which faces the fractal analysis of self-affine sets is non-commensurability of scale units along different axes. Therefore in such situations one cannot define such geometric fundamentals as length, area, square, sphere, and so on. Thus, generally accepted methods of studying of self-similarity (perimetrical, box-covering and others) lose their geometrical background and become inadequate.

Different methods were developed in order to overcome this difficulty. The first elaborated by Mandelbrot & Wallis (1969) is known as the *R/S* method (of renormalized range). It was shown that the renormalized amplitude of self-affine sets R/S calculated for different scales obeys self-similarity power law $R/S \sim \tau^H$, where H is the so called Hurst parameter. In this way one can obtain a value of the fractal dimension $D = E - H$, where E is the Euclidean dimension of the problem. With the aid of this method many important results were obtained, especially in the field of statistical analyses of various hydrological time series.

There are also some cases of its successful application in solving geological problems (for instance Fluegeman & Snow 1989).

A less successful approach to overcoming the difficulty of non-commensurability was by means of separate analysis of dispersion of coordinates of points of a set (Matsushita & Ouchi 1989). The authors, in my opinion, did not succeed in avoiding the distortions caused by scale changes, as they themselves state. Regradless of these distortions the approach is a rather common in the practice of fractal analysis. For example, it leads to a systematic error in estimations of the fractal dimension of sea floor relief (Barenblatt et al. 1984).

The most reliable method of fractal analysis of self-affine sets offered to date is the method of variations. It is based on a well-known theoretical concept of scaling of dispersion of the so called fractional Brownian motion (random walk). It is stated that $<(z_i-z_j)^2> \sim \tau^{2H}$, where z_i, z_j are coordinates of a Brownian particle separated by interval τ, and triangular brackets mean averaging. Having expanded this concept with respect to other self-affine sets Mark & Aronson (1984) presented a convenient method of estimation of their fractal properties. Additionally, it serves as a good tool to reveal latent periodicities. The method of variations allows one to obtain simple and readable results, as in my analysis of the self-similar properties of global topography (Ivanov 1994).

Baseline Considerations

Here I would like to suggest a new method for the estimation of scale invariance of self-affine sets based on scaling studies of dispersions and average values. Its principal outlines are as follows.

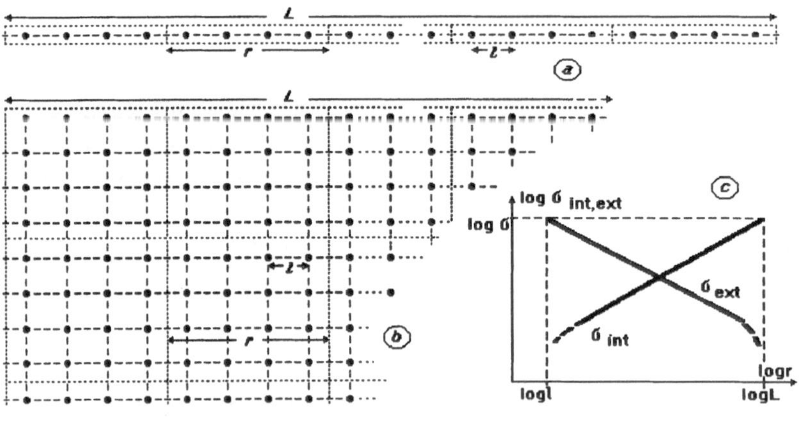

Fig. 1a-c. Scale units used in estimation of self-similar properties of **a**) one-dimensional and two-dimensional sets and predicted scaling behaviour of internal and external dispersions in the case of self affinity **c**).

Let us regard a set of N values of a certain function F on a one-dimensional net with total length L and resolution l (Fig. 1a). As far as we can take into account only a limited number of values N, this gives us the total sample scale range $[l \ldots L]$, where $L = N/l$. We may describe this set by its mean value $<F>$ and dispersion σ^2. Imagine that the whole set is divided into m equal parts of size r, each containing a subset of n values so that $m/n = N$, $r = L/m = n/l$. Each subset F_i is characterized by its mean value $<F_i>$, and dispersion of F-values within the subset (internal dispersion σ^2_{int}). The subset of m values of $<F_i>$ has its own average, equal to $<F>$ and these values have a dispersion that may be called external (σ^2_{ext}). So we state that each scale within the range of the whole set is characterized by two different values of dispersion - the external one (dispersion of averages) and the internal one, that is an average of m values of internal dispersion. For the smallest scale l each subset contains only one value and there is only external dispersion equal to total dispersion σ^2. For the largest scale L only internal dispersion may be defined and it is also equal to the total set dispersion. Therefore we can state that $\sigma_{int}(L) = \sigma_{ext}(l) = \sigma$.

It is obvious that the same considerations may be easily applied to two-dimensional (Fig. 1b) or multidimensional nets.

Qualitative analysis leads to the conclusion that while r increases from l to L the external dispersion decreases from σ^2 and the internal one rises to σ^2 (Fig. 1c). One may suppose that for self-affine sets both of these dispersions must be scaled so that $log(\sigma_{int}) \sim a \cdot log(r)$ and $log(\sigma_{ext}) \sim -b \cdot log(r)$ and the values of a and b are closely connected with the fractal dimension of the whole set d. More than that, we expect that $a = b$ and $d = E - a$, where E = Euclidean dimension.

Thus, the suggested method must be useful for the analysis of self-affine properties of "fields" - sets of numerical values of an arbitrary parameter defined on a net of an arbitrary dimension. It is necessary to emphasize that the "field" here means nothing more than a set of values and does not demand either continuity or differentiability. On the contrary, the self-affinity theoretically implies the roughness of this "field" at infinitely small scales.

Case Histories

I tested these considerations on a variety of geophysical sets defined on one-dimensional and two-dimensional nets. Here I shall contribute some examples to illustrate the method. The results are summarized in Table 1, presenting the values of fractal dimensions obtained by direct and inverse scaling of dispersions.

1. Specific power of seismic energy release. I used available catalogs of earthquakes for the Appenine peninsular (with an observation period of 90 years) and Kuril-Kamchatka seismic zone (25 years). For cells approximately 0.1°x 0.1° in size the release of seismic energy was estimated through recalculation of magnitudes of earthquakes registered within each cell. Then the annual value

Table 1. *Fractal Dimensions Values* estimated by the dispersion counter-scaling method.

Region	Scale range	Fractal dimension	
		from internal scaling	from external scaling
		Specific seismic power	
Kurils	10-2000 km	1.36 ± 0.02	1.31 ± 0.02
Kamchatka	10-1000 km	1.32 ± 0.02	1.26 ± 0.02
Appenines	10-1000 km	1.30 ± 0.04	1.24 ± 0.03
		Regional topography	
Kurils	100-2000 km	1.42 ± 0.02	1.90 ± 0.01
Kamchatka	100-1000 km	1.47 ± 0.02	1.75 ± 0.03
	Equatorial belt (East of Greenwich)		
0° - 40°	100-4000 km	1.45 ± 0.02	1.92 ± 0.01
40° - 80°	100-4000 km	1.51 ± 0.03	1.84 ± 0.03
80° - 120°	100-4000 km	1.57 ± 0.03	1.84 ± 0.03
120°- 160°	100-4000 km	1.63 ± 0.02	1.90 ± 0.01
160°- 200°	100-4000 km	1.68 ± 0.03	1.80 ± 0.07
200°- 240°	100-4000 km	1.57 ± 0.03	1.85 ± 0.02
240°- 280°	100-4000 km	1.58 ± 0.03	1.86 ± 0.01
280°- 320°	100-4000 km	1.34 ± 0.03	1.93 ± 0.01
320°- 0°	100-4000 km	1.31 ± 0.03	1.84 ± 0.03
		Geomagnetic field reversals	
-	20-500 ka	0.50 ± 0.01	0.51 ± 0.00
-	0.5-40 Ma	0.95 ± 0.01	0.88 ± 0.01

was determined representing the flux of seismic energy per area unit, i.e. specific power of seismic process. Thus two-dimensional sets of numerical power values were obtained. These data were processed by the above described method of "dispersion counter-scaling". The results are presented in Fig. 2a. One can see that the fractal dimensions determined from internal and external dispersions practically coincide within the estimated accuracy.

2. Earth's surface topography. For two of the three regions listed above the "counter-scaling" method was applied to analyze the Earth's surface elevations. The computations were performed on the base of two-dimensional maps of heights and depths, averaged in one-degree cells. The results show that the scaling behaviour of internal dispersion is not similar to that of the external one (Fig. 2b). The differences between fractal dimension values determined from direct and inverse scaling significantly exceed the estimated error range. I regard

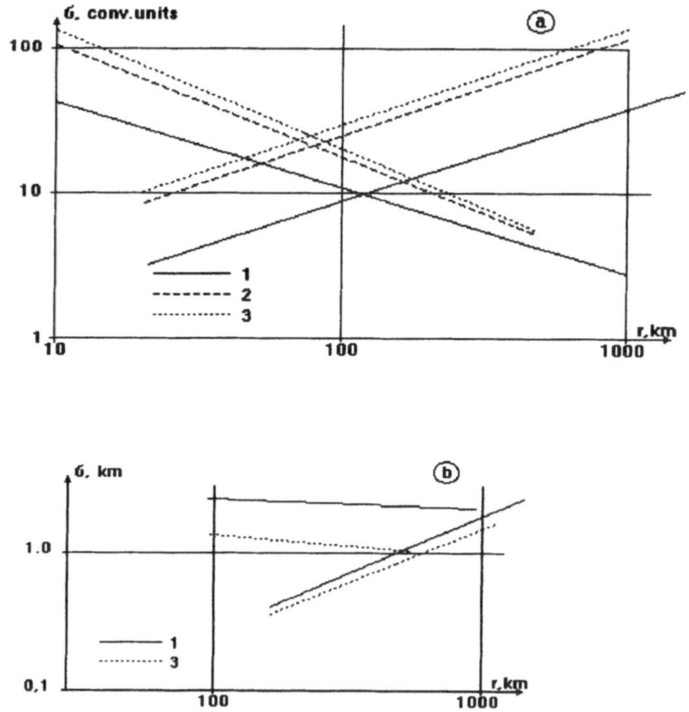

Fig. 2a, b. Graphs of scale dependence of internal and external dimension values of various sets: **a** the field of specific power of seismic process; **b** surface topography; *1* - Kuril zone; *2* - Kamchatka zone; *3* - Appenine peninsular.

this as evidence that this difference really exists and is characteristic of regional topography.

In order to confirm this conclusion I applied this method of fractal study to one-degree averages of surface elevations within the equatorial belt of the Earth from 20°N to 20°S. The whole belt was subdivided into 9 non-overlapping squares with a side of 40°. The results of the computations (Table 1) show that the above mentioned difference is observed in all cases and the values of fractal dimensions obtained from direct scaling (i.e. from internal dispersion) are closer to the estimates of fractal dimension of global relief obtained by perimetric (Ivanov 1994) and spectral (Turcotte 1989) methods.

3. Sequence of geomagnetic field reversals during the last 160 Ma In this case I analyzed the number of inversions of geomagnetic field during a certain time interval divided by the length of the interval. This value may be referred to as density of inversions and directly characterize the non-linear processes in the Earth's core. It was found that the fractal dimensions estimated by the scaling of internal and external dispersions are approximately equal (Fig. 3). But in this case another feature is of interest - an abrupt change in scaling properties at

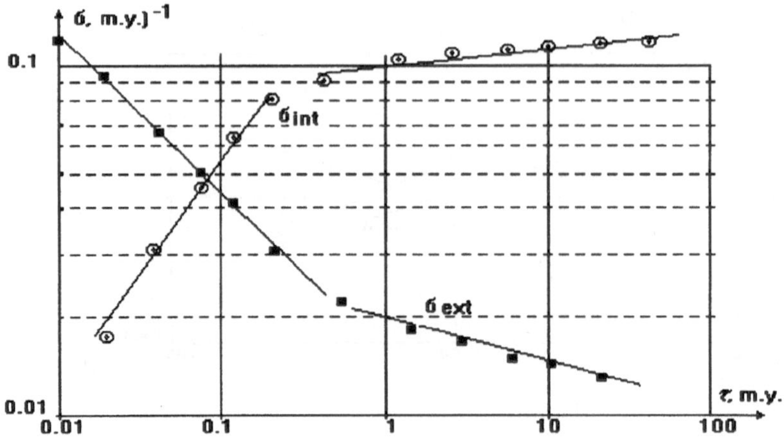

Fig. 3. Graphs of time scale dependence of internal and external dimension values of geomagnetic reversals density.

intervals of about 0.5 Ma. For shorter intervals fractal dimension of the process is 1/2 (with an error less than 0.01) that is an evidence of a random regime (for more details see Ivanov 1993), and in the scale range greater than 0.5 Ma the process reveals certain signs of self-organization. This change is strongly illustrated in the scaling of both dispersions.

Conclusions

These examples from different branches of the geosciences prove that self-similarity (or self-affinity) seems to be an intrinsic property of various natural processes and objects and the method suggested in this chapter provides its adequate and reliable estimate. In the cases when counter-scaling of the two kinds of dispersion reveals significant asymmetry (as for regional topography) one can suppose a more complex structure of the set under study. Fractal dimensions of such sets are different when studied from inside the set or from outside it. In this situation it seems reasonable to identify them as Mandelbrot's local and global fractal dimensions of a self-affine set correspondingly.

Acknowledgements. These researches are supported by Russian Foundation of Fundamental Studies, grant 93-05-8943. I would like to thank Prof. Andrey S. Monin and Prof. Vladilen F. Pisarenko for helpful discussions.

References

Barenblatt GI, Zhivago AV, Neprochnov YuP, Ostrovsky AA (1984) Fractal dimension: a quantitative characteristic of ocean bottom relief. Oceanology 24: 695-697.

Fluegeman R, Snow S (1989) Analysis of oxigen isotopic record. PAGEOPH 131: 307-313.

Ivanov SS (1993) On the self-similarity of the sequence of inversions of geomagnetic field. Geomagnetism i Aeronomia 5: 201-206 (in Russian).

Ivanov SS (1994) Global Relief: Evidence of Fractal Geometry. In: Kruhl JH (ed) Fractals and Dynamic Systems in Geoscience. Springer, Berlin Heidelberg New York, pp 221-230 (this volume).

Mandelbrot BB (1986) Self-affine fractal sets. In: Pietronero L, Tozatti L (ed) Fractals in Physics. North-Holland, Amsterdam, pp. 3-28.

Mandelbrot BB, Wallis J (1969) Some long-run properties of geophysical records. Water Resources Res 5, 2: 321-340.

Mark D, Aronson P (1984) Scale-dependent fractal dimensions of topographic surfaces: an empirical investigation, with applications in geomorphology and computer mapping. Math Geol 16, 7: 671-683.

Matsushita M, Ouchi S (1989) On the self-affinity of various curves. Physica D 38: 246-251.

Turcotte D (1989) Fractals in Geology and Geophysics. PAGEOPH 131: 171-195.

Application of the Grassberger-Procaccia Algorithm to the $\delta^{18}O$ Record from ODP Site 659: Selected Methodical Aspects

Manfred Mudelsee & Karl Stattegger
Graduiertenkolleg "Dynamik globaler Kreisläufe im System Erde",
Geologisch-Paläontologisches Institut und Museum, Universität Kiel,
Olshausenstr. 40, D-24118 Kiel, Germany

Abstract. The Grassberger-Procaccia algorithm, which allows one to estimate the correlation dimension of a chaotic system, was investigated on oxygen isotope data from the deep-sea sediment core ODP Site 659 from the eastern North Atlantic. These data provide uniform coverage of the last 5.2 Ma and show an average spacing of 4.5 ka (1170 points). They are assumed to reflect global ice volume as a proxy for global climate.

This paper focusses on methodical aspects. One important point consists in the use of nonlinear fits instead of linear fits after a logarithm transformation. The latter overestimates the correlation dimension. An advantage is that the comparable large amount of original data allowed the number of (equidistant) working-data points (generated by linear interpolation) to be held equal to the number of original data points in order to strive for small statistical dependences between the working-points.

The Pliocene and Pleistocene climatic signatures can clearly be distinguished in the reconstructed space of climatic states.

The behavior of the estimated correlation dimension on the embedding dimension does not indicate the existence of a low-dimensional attractor. The behavior on a variable time lag does not indicate a definite lower boundary for the time lag above 4.5 ka. The behavior on the fit region shows a strong dependence on the upper boundary for the region, whereas the dependence on the lower boundary is weak.

Introduction

Correlation Dimension and the Grassberger-Procaccia Algorithm

The correlation dimension (D) has become an important descriptive parameter of chaotic systems; it is sensitive to the dynamical processes of coverage of the attractor shown by this system, in contrast to the fractal dimension which describes the geometry of the attractor. (The latter has an upper boundary in D.) The correlation dimension can be of importance for constructing models of such systems, e.g., a low value of D would indicate the possibility of describing the system by a low number of deterministic equations. Advantageously, the correlation dimension can be calculated from a single time series of a system by reconstructing the system's trajectory in an "embedding space" (Packard et al. 1980, Takens 1981, Ruelle 1981, Grassberger & Procaccia 1983). To construct the embedding space, a time lag is introduced by which the time series is successively shifted. The dimension of the trajectory, and thus the correlation dimension, can be calculated using the algorithm of Grassberger & Procaccia (1983) (in addition to different algorithms). Hereafter, the whole procedure is called the GP-method. It is outlined briefly below.

Given an equidistant time series,

$$X(1), X(2), \ldots, X(N),$$

with a certain spacing and a certain number N of data points, the embedding space is constructed by means of Y vectors:

$$\begin{aligned} Y(1) &= [X(1), X(1+L), X(1+2L), \ldots, X(1+(M-1)L)], \\ Y(2) &= [X(2), X(2+L), X(2+2L), \ldots, X(2+(M-1)L)], \\ &\vdots \\ Y(p) &= [X(p), X(p+L), X(p+2L), \ldots, X(p+(M-1)L)]; \end{aligned} \tag{1}$$

where L denotes a time lag (measured in units of the data spacing) and M is the embedding dimension. Notice that the number p of vectors is given by N-(M-1)L. Each vector Y corresponds to a point in the embedding space and defines a certain state of the system which is described by the time series (ODP 659 data, Fig. 4).

The Grassberger-Procaccia algorithm introduces the correlation integral, also called correlation function, C(r),

$$C(r) = \text{factor} \times \text{sum of distances } |Y(i)-Y(j)| \text{ smaller than } r.$$

(The symbol r denotes the distance between points in the embedding space and should not be confused with the correlation coefficient.) For our purposes, the (normation) factor is set to 1/2 accounting for the symmetry in i, j of distances.

Grassberger & Procaccia (1983) showed that for "large" M and "small" r the following relation holds true provided an attractor exists:

$$C(r) \propto r^{D} \tag{2}$$

In practice, the correlation dimension D is determined by setting M=1,2,3, ... and estimating a power-law relation between r and C(r). This is usually carried out by means of linear least-squares regression on the logarithmically transformed data. If the value of the estimated exponent of the power-law relation saturates with growing *M*, this saturation value is taken as the best estimate of D. On the contrary, a white noise process, for which no attractor exists, would not produce a saturation behavior of the exponent (Fig. 1). Strictly speaking, one should use the term "correlation dimension" only when saturation occurs. In the text, however, we have adopted it synonymously to the exponent of the power-law relation.

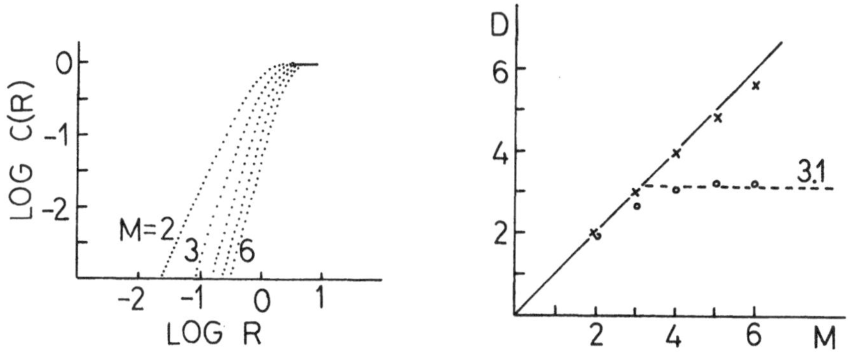

Fig. 1. Redrawn after Nicolis & Nicolis (1984), with modified notation. Left panel: Correlation functions in double-log plot for different choices of M (note that the normation factor (section "Correlation dimension...") is different to that used in this chapter); the linear range gives the slope, D. Right panel: D versus M; the saturation-like behavior for the climate data (circles) led to the conclusion of Nicolis & Nicolis (1984) that an attractor exists with $D \approx 3.1$ for the long-term climatic system. The white noise process (x) does not show a saturation behavior.

The Grassberger-Procaccia (GP) Method Applied to Experimental Data

Describing complex natural systems is often done by starting with basic equations given by the natural laws considered as necessary for the description of the system. From these a set of, usually coupled, differential equations is derived which describe the behavior of the system. One example is the Lorenz model in meteorology (Lorenz 1963). From the differential equations a, theoretically, unlimited number of noise-free data points can be derived. Those "theoretical" systems allow efficiently converging calculations of the correlation dimension with the GP-method (Ben-Mizrachi et al. 1984).

However, when the governing equations of a natural system are generally unknown, calculation of the correlation dimension with the GP-method must be done using experimental data which are generally limited in number and influenced by noise. Therefore, some points of critique have been raised concerning the GP-method applied to experimental data:

(a) *Number of data points*, N. Figure 5a, from ODP 659 data, shows the saturation behavior of C(r) for large r, where the power-law (Eq. 2) does not hold true anymore. Regression comprising this region of r would yield a dimension, D, that is too low. The fit region, therefore, has an upper boundary, r_U. It has been shown that for too high choices of M (with N fixed) the value of r_U diminishes and, hence, the possibility to determine the exponent also diminishes. The maximum meaningful value of M grows with N; however, a uniformly accepted formula N(M) does not exist yet. For example, if $N \approx 1000$, a maximum value of $M \approx 4$ can be derived following

Theiler (1986), a maximum value of D ≈ 5.5 can be derived from the slightly corrected formula of Eckmann & Ruelle (1992).

(b) *Time lag*, L. In order to cover an attractor without any preference for a certain region of it, the vectors Y(i) should be statistically independent of each other. This can be achieved by using a high value of L. Alternatively, as suggested by Theiler (1986), the summation in the correlation integral about the distances $|Y(i)-Y(j)|$ could involve the condition that $|i-j|$ should be high enough to exclude statistically dependent pairs Y(i), Y(j). On the other hand very high values of L lead to the loss of information on finer time scales. A generally accepted relation is (e.g. Maasch 1989):

$$L \approx \text{correlation time of the time series}$$

(for the determination of the correlation time, see below). In addition, we stress that another source of statistical dependence between the vectors Y(i) is introduced by the initial interpolation of the data points in order to get equidistant points. This dependence increases if more working-points are generated than data points existed originally.

(c) *Experimental noise*. It is comprehensible that a fit region below a certain value, r_L, does not make sense because a finite distance between two vectors, Y(i) and Y(j), is already given by the error of the experimental data. It has been shown (Ben-Mizrachi et al. 1984) that r_L should be in the order of this experimental noise amplitude.

More details about the role of N, L and the "scaling region" (that is the interval $[r_L, r_U]$) are given in: Lochner et al. (1989), Nerenberg & Essex (1990).

(d) *Colored noise*. A basic point of critique has been raised by Osborne and Provenzale (1989) who showed that some stochastic processes (colored noise) also exhibit a finite correlation dimension. Onto that, we emphasize studying the following topics more intensively: the multiplicative interaction between variables describing complex natural systems, the impact of this interaction on the shape of the distribution function, i.e. the color of the noise, of these variables and the significance of the lognormal distribution (Aitchison & Brown 1957) for the description of such systems.

Since 1983, the GP-method has been applied to experimental data from natural systems. The data cover different temporal and spatial scales. Nicolis & Nicolis (1984) claimed the existence of a definite, low-dimensional attractor (D = 3.1) for the long-term global climate system (Fig. 1). They used the $\delta^{18}O$ record of the deep-sea sediment core V28-238 which covers the last 800 ka and consists of ≈200 data points. This result was subsequently criticized by Grassberger (1986) who warned that "spuriously small dimension estimates can be obtained from using too few, too finely sampled and too highly smoothed data". Maasch (1989) investigated 14 Pleistocene $\delta^{18}O$ records and concluded that an attractor exists with D in the range 4-6.

In this work we have concentrated on selected methodical aspects of the application of the GP-method to experimental data: From the above critique we have investigated point (b) and, less extensively, points (a) and (c). As data source we have chosen a $\delta^{18}O$ record covering a longer time interval and containing more data points than the records used in the publications mentioned above. This experimental time series contains the maximum of data currently available in paleoclimatology, but still is relatively „short“ in comparison to "theortical" systems.

Additionally, of significant importance, we show that the commonly practised way of fitting the power-law behavior of C(r) (Eq. 2) in order to estimate the correlation dimension systematically overestimates the latter. Instead of using double-log plots and linear fits we use nonlinear fits.

Global Long-Term Climatic Variations Recorded from Deep-Sea Core ODP Site 659

Time Series

Data source is the deep-sea sediment core ODP Site 659, recovered during the Ocean Drilling Program, from the eastern North Atlantic (18°04.63'N, 21°01.57'W) at a water depth of 3070 m (Sarnthein & Tiedemann 1989). Specimens of the benthic foraminifer *Cibicidoides wuellerstorfi* ($CaCO_3$ shell), size fraction 250-315µm, were taken to measure by means of a mass spectrometer the oxygen isotope ratios relative to a standard (PDB):

$$\delta^{18}O\ [‰] = [(^{18}O/^{16}O)_{sample} - (^{18}O/^{16}O)_{PDB}]/(^{18}O/^{16}O)_{PDB} \times 1000$$

The average, absolute uncertainty of the $\delta^{18}O$ values is 0.07 ‰.

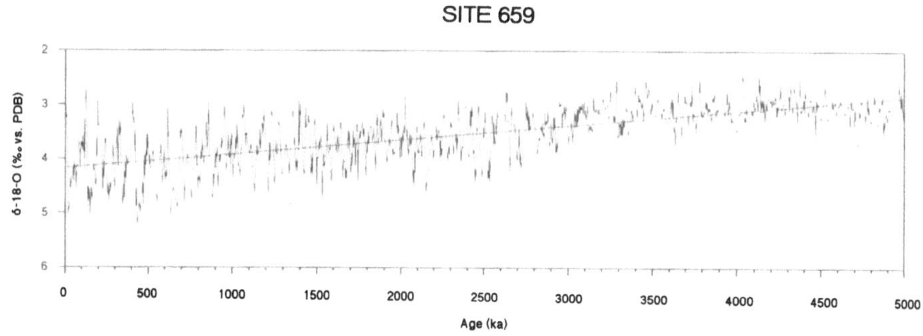

Fig. 2. The time series: oxygen isotope record for the ODP Site 659, 0-5000 ka (Tiedemann et al., in press), linear trend; for further explanation see the text.

Autocorrelation Function

These data are assumed to reflect the global ice volume as a proxy for global climate (low values: "warm"/"less land-ice", high values: "cold"/"much land-ice"). The time scale (Figure 2, Tiedemann et al., in press) was achieved by determining magnetic and biostratigraphic events in the core and then tuning the time scale between these events by means of the calculated temporal behavior of the parameters of the Earth's orbit (which act as "pacemakers" for the global long-term climate, see, e.g., Hays et al. 1976, Imbrie et al. 1984). The core shows no hiatuses, the number of original data points, N, is 1170, the maximum age is 5268 ka, the average spacing is 4.5 ka. The linear trend (≈- 0.266 ‰ /ka) is also shown in Fig. 2.

Subtracting the linear trend permitted calculation of the autocorrelation function of the time series (Figure 3); its first zero crossing, which is generally recommended as an estimate of the correlation time (e.g., Maasch 1989), is 23.4 ka. This value corresponds closely to the Milankovitch precession period of 23 ka (Berger & Loutre 1991).

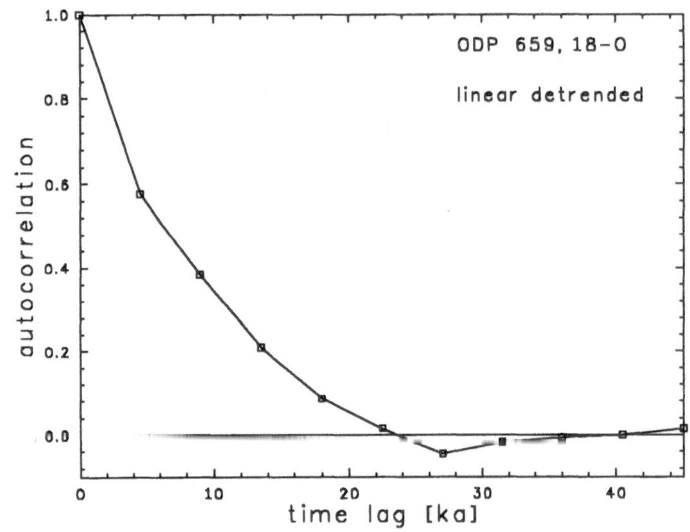

Fig. 3. Autocorrelation function of the linear detrended time series (Fig. 2); the first zero crossing occurs at the time lag 23.4 ka.

Embedding Space

In order to obtain equidistant working-points, the original data (not detrended) were interpolated by a linear interpolation. This seemed to be acceptable since it reduced the variance, and, hence, the stored information, of the original data by

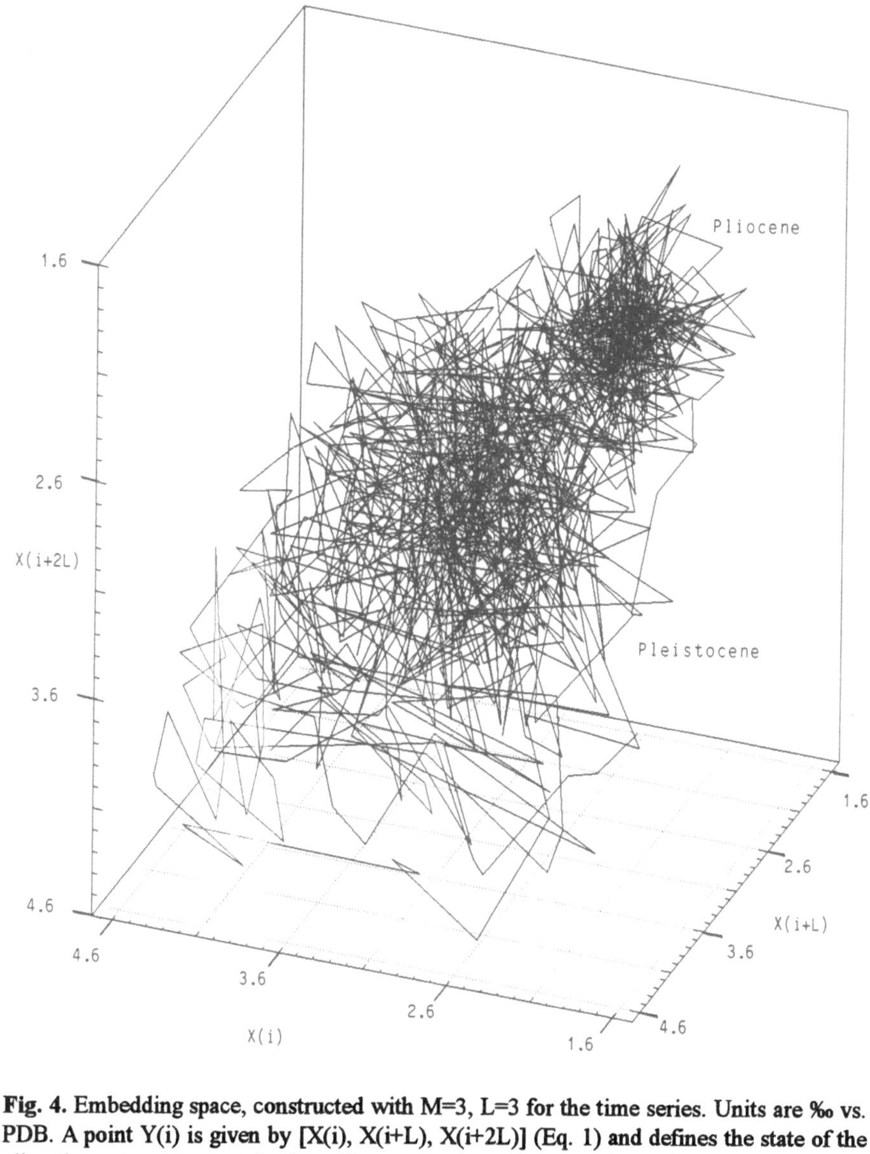

Fig. 4. Embedding space, constructed with M=3, L=3 for the time series. Units are ‰ vs. PDB. A point Y(i) is given by [X(i), X(i+L), X(i+2L)] (Eq. 1) and defines the state of the climatic system at a certain time. The connection of consecutive points ("trajectory") in the embedding space reflects the temporal behavior of the system. The correlation dimension describes such sometimes strange-looking objects ("strange attractors") not only geometrically but as well notices the "density" of points in the embedding space.

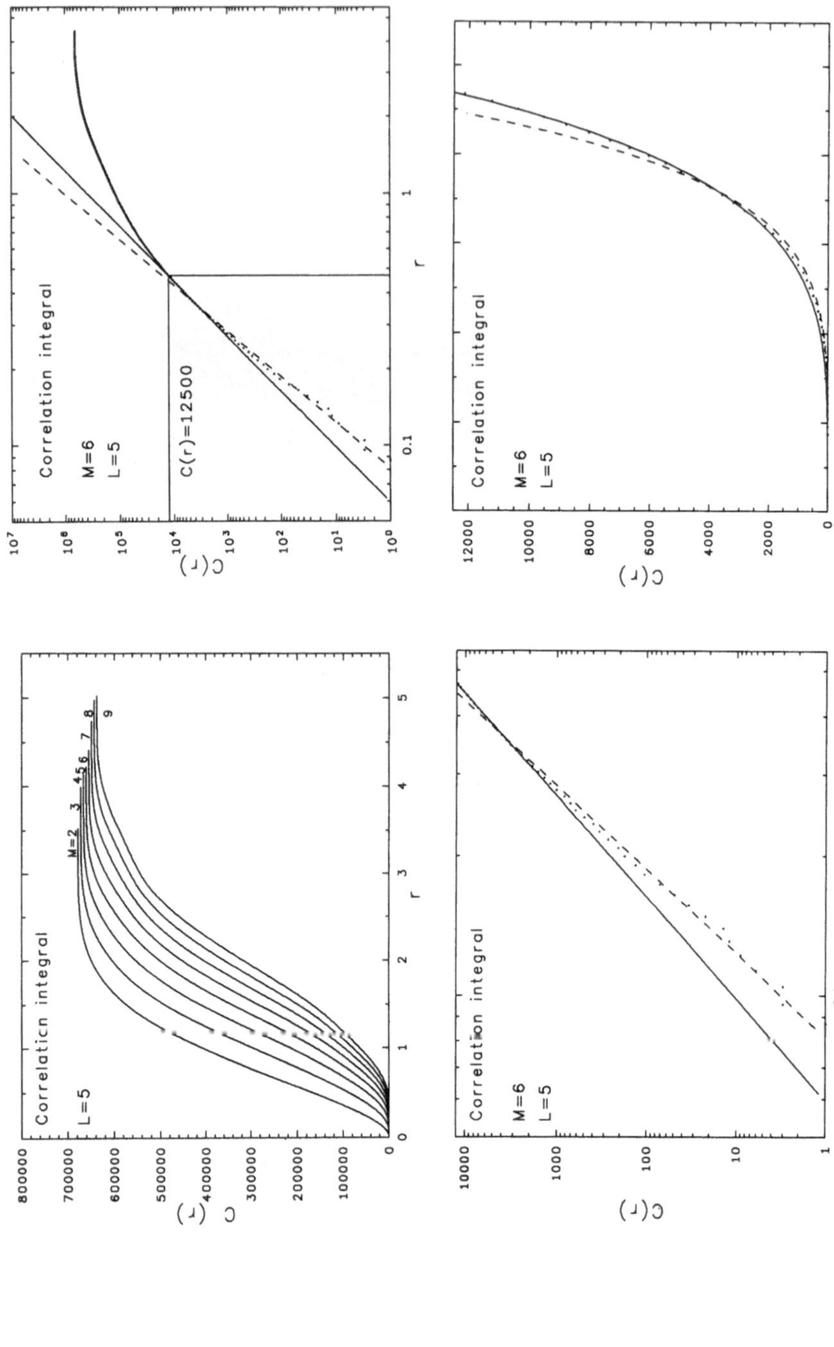
Correlation integral
M=6
L=5
C(r)=12500
r
C(r)
Correlation integral
M=6
L=5
r
C(r)
Correlaticn integral
L=5
M=2 3 4 5 6 7 8 9
r
C (r)
Correlation integral
M=6
L=5
r
C(r)

Fig. 5 b d
a c

only 4%. The number of working-points was held equal to the number of original data points in order to avoid the enhancement of the statistical dependence between the vectors Y(i) (above, section "The GP-Method...", point b). Figure 4 shows the system's trajectory in the embedding space for a certain choice of M and L. There appear to be *two* clouds of points, representing two climatic states ["Pliocene": older part of the time series (Fig. 2), showing lower oxygen isotope values and smaller oscillations; "Pleistocene": younger part of the time series, showing higher oxygen isotope values and larger oscillations]. One possibility would be to substract the linear trend from the time series, the other would be to investigate these distinct time intervals separately. First of all, however, we show the results using the entire range of the original data.

Correlation Integral

The correlation integral, C(r), was determined for M = 2,...,9 and L = 1,...,9 (Figure 5a). L is measured in units of the spacing (4.5 ka) of the equidistant points.

Double-log plots and linear fits: overestimating the correlation dimension

Figure 5b shows the correlation function for a certain choice of M and *L* in a double-log plot. The latter is the basis for the commonly practised way of regressing the original model

$$C(r) \propto r^D$$

Fig. 5a-d. a Correlation integral, calculated with L=5, M=2,...,9 in linear plot for the entire scale of distances, r. Note the power-law-like growth of C(r) for small r, the saturation behavior for large r (which corresponds to the limited extension of objects like those shown in Fig. 4), and the saturation value of p(p-1)/2 (*p* is the number of vectors in the embedding space, see section "Correlation Dimension..."). Increasing values of M generally yield larger distances and a lower saturation value. **b** Correlation integral (dots), calculated with L=5, M=6 in double-log plot. [Because of the unequally spaced values of log (r) the dots representing higher values of r are not resolved and appear as a thick line. In Fig. 5d, using a linear plot, the r-values are equally spaced.] The scaling region corresponds to C(r) < 12500 and is marked. The best linear fit (dashed) using this scaling region yields the slope, i.e. D = 5.53±0.05; the best power-law fit (solid; see Eq. 2), using the same scaling region and the linear plot, yields D = 4.57±0.03. **c** Same as in Figure 5b, focussing on the scaling region. **d** Same as in Figure 5c, but using linear scales for r and C(r). The superiority of the nonlinear power-law fit over the linear fit after the logarithm transformation is demonstrated. The sum of squared residuals of 3.70×10^5 from the nonlinear fit in comparison with 3.83×10^7 from the linear fit after the logarithm transformation underlines this observation. The superiority was noticed for all other values of L and M.

and consists of taking the logarithms and then fitting a straight line to the transformed data:

$$\log[C(r)] = \text{const.} + D \times \log[r].$$

Thus, the slope estimates the correlation dimension. The alternative regression of the model consists in nonlinear (power-law) fits to the original model. This method demands numerical calculations and is generally neglected in favor of the double-log plots.

Also plotted in Fig. 5b is the best linear fit (dashed) using the following scaling region:
r_L = minimum value of r, r_U corresponds to $C(r) < 12500$. This fit yields a slope, i.e. $D = 5.53 \pm 0.05$ (1 sigma) and a sum of squared residuals of 3.83×10^7. The sum of squared residuals measures the deviation of the fit function from the data points. The nonlinear regression (solid), using the same scaling region yields $D = 4.57 \pm 0.03$ and a sum of squared residuals of 3.70×10^5. Thus, *the same data points are more precisely described by the nonlinear regression than by the linear regression after logarithm transformation*. Figures 5c and 5d focus on thescaling region, whereby Fig. 5d demonstrates visually the superiority of the nonlinear regression for the "true" relationship, $C(r) \propto r^D$. The statement that the loss of the equal spacing between the r-values (when they are logarithmically transformed) would produce this difference is refuted: The same difference results if in the double-log plot the spacing between the r-values is equidistant (compare Fig. 5b). The fact that the linear fit after a log-log transformation systematically overestimates the correlation dimension is explained by the following two points: (1) When assuming the original power-law relationship for true, the log-log transformation produces a violation of the Gauss-Markov conditions for least-squares regression, such that a biased estimation results (Sen & Srivastava 1990, page 35; see also Schulz et al. 1994). (2) The *over*estimation results from the monotonic growth of the power-law function for positive exponents. Additionally, the systematic deviation from the power-law, i.e. the saturation behavior, growing with r, may also contribute to this effect. We, therefore, choose the nonlinear regression method in order to estimate D. The problem of the inadequacy of linear transformations for regression purposes (e.g. for a power-law function) has been demonstrated, e.g. by Rützel (1976).

Scaling Region

Concerning the scaling region: as the experimental noise is considerably weak, the results from taking r_L in the magnitude of the noise or instead taking r_L = minimum value of r differ only very little from each other; the important value is r_U. Here we decided to determine r_U corresponding to several permitted maximum values of C(r), i.e.:

C(r) < 50000 (or 25000, 12500, 6250, 3125.) We know that this is only a first attempt. An indepth investigation of the problem of the proper choice of the scaling region must choose whether to define the scaling region via boundaries in C(r) (as we have done) or boundaries in r, or to measure the quality of the regression and using that interval which yields the best fit. Also the spacing for the distances, r, is still arbitrary and should be investigated.

Results

The following dependences of the estimated correlation dimension on the parameters time lag, scaling region, and embedding dimension resulted:

(i) *The dependence of the correlation dimension*, D, *on the time lag*, L, (Fig. 6) shows that the value of D is not strongly influenced by L above 1 (4.5 ka). Therefore, the above stated recommendation, to choose L to be in the order of the correlation time (23.4 ka), does not seem to be a necessary condition for the determination of D.

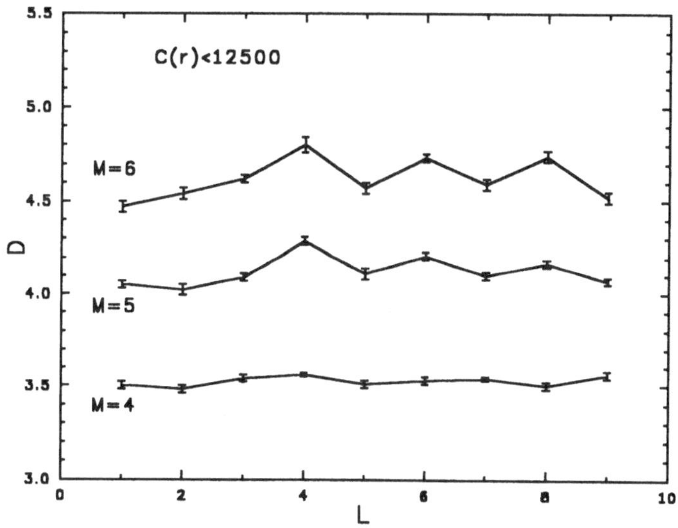

Fig. 6. Dependence of the estimated correlation dimension, D, on the time lag, L, for three different values of M. The values of D result from nonlinear regression and r_U corresponds to C(r) < 12500. L is measured in units of the spacing (4.5 ka) of the time series. The error bars represent the average standard deviation of the estimated correlation dimension.

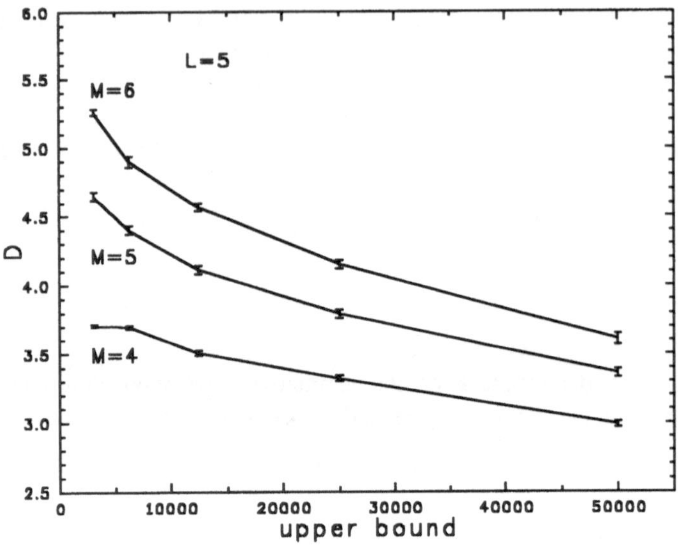

Fig. 7. Dependence of the estimated correlation dimension, D, on the fit range (scaling region) for three different values of M. The values of D result from nonlinear regression and the time lag L=5. The fit range [r_L, r_U] was: r_L = minimum value of r; the upper boundary, r_U, corresponds to C(r) < 50000, resp., 25000, 12500, 6250, 3125 (x-axis).

(ii) *The dependence on the scaling region* (Fig. 7) underlines the strong influence of the upper boundary, r_U. However, our method of handling the problem of the proper choice of the scaling region is not sufficient to give a detailed recommendation.

(iii) *The dependence of the correlation dimension*, D, *on the embedding dimension*, M, (Fig. 8), displays no saturation behavior. Hence, no low-dimensional attractor exists for the investigated climate system. This statement holds true also when the uncertainties concerning the proper choice of the time lag, *L*, and the upper boundary of the scaling region, r_U, are considered. As Figures 6 and 7 show, for a variety of values of *L* and r_U the correlation dimension, *D*, increases when *M* is increased. When following the restriction M < 5, respectively D < 5.5, from above (section "The GP-method...", a) due to the limited number of data points, the statement changes to: there is no attractor with D < 4 (fig. 8), respectively D < 5.5. This high dimension of the attractor compounds further the problem of „short" time series.

(iv) In addition, all calculations were performed on the *linear detrended* data points (compare Fig. 3) yielding the *same* results.

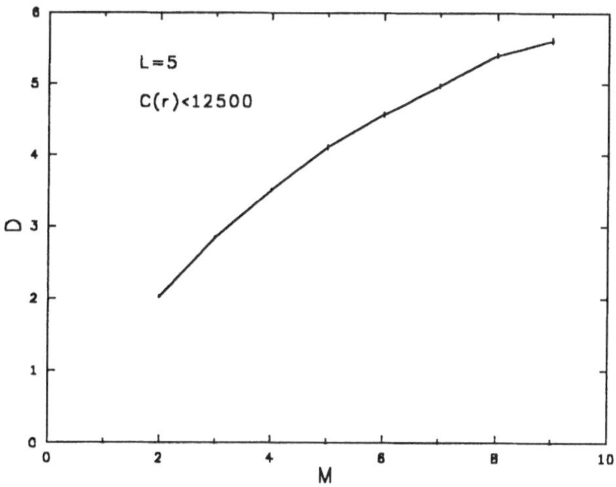

Fig. 8. Dependence of the estimated correlation dimension, D, on the embedding dimension, M. The values of D result from nonlinear regression. The time lag, *L*, is 5, and r_U corresponds to C(r) < 12500.

Conclusions

The Grassberger-Procaccia method to estimate the correlation dimension from a time series was applied to paleoceanographic data which reflect the global climatic variations over the last five million years.

It can be shown that the conventionally used double-log plots and linear fits systematically overestimate the correlation dimension. Therefore, nonlinear fits need to be calculated.

The value of the time lag, L, while greater than 4.5 ka, does not have a strong influence on the estimated dimension.

The choice of the upper boundary of the scaling region, r_U, however, does have a strong influence on the estimated dimension. This effect will be studied in detail in the future.

Based on the present state of the GP-method we conclude: the long-term climatic system described by our deep-sea sediment core does not show a low-dimensional attractor.

Finally, considering the points of critique from above, we suspect that by the means of the GP-method, at the present state of knowledge, only the rejection of low-dimensional attractors in natural systems may be possible.

Our future research will include the investigation of (i) distinct time intervals, e.g. the Pliocene ("warm") and the Pleistocene ("cold" / "oscillating"); (ii) additional high-resolution paleoclimatic data records; and (iii) synthetic time series for comparison purposes.

Acknowledgements. We gratefully acknowledge R. Tiedemann for providing the data material, a diagram and information about ODP Site 659. K. Herterich and K.A. Maasch made valuable review comments. P. Dingle refined the English of the manuscript. The Graduiertenkolleg "Dynamik globaler Kreisläufe im System Erde" is sponsored by the Deutsche Forschungsgemeinschaft.

References

Aitchison J, Brown JAC (1957) The lognormal distribution. Cambridge University Press, Cambridge.

Ben-Mizrachi A, Procaccia I, Grassberger P (1984) Characterization of experimental (noisy) strange attractors. Phys Rev A 29: 975-977.

Berger A, Loutre MF (1991) Insolation values for the climate of the last 10 million years. Quat Sci Rev 10: 297-317.

Eckmann, J-P, Ruelle D (1992) Fundamental limitations for estimating dimensions and Lyapunov exponents in dynamical systems. Physica D 56: 185-187.

Grassberger P (1986) Do climatic attractors exist? Nature 323: 609-612.

Grassberger P, Procaccia I (1983) Characterization of strange attractors. Phys Rev Lett 50: 346-349.

Hays JD, Imbrie J, Shackleton NJ (1976) Variations in the Earth's orbit: pacemaker of the ice ages. Science 194: 1121-1132.

Imbrie J, Hays JD, Martinson DG, McIntyre A, Mix AC, Morley JJ, Pisias NG, Prell WL, Shackleton NJ (1984) The orbital theory of Pleistocene climate: support from a revised chronology of the marine $\delta^{18}O$ record. In: Berger A, Imbrie J, Hays JD, Kukla G, Saltzman B (eds) Milankovitch and Climate, Part I. Reidel, Dordrecht, pp 269-305.

Lochner JC, Swank JH, Szymkowiak AE (1989) A search for a dynamical attractor in Cygnus X-1. Astrophys J 337: 823-831.

Lorenz EN (1963) Deterministic nonperiodic flow. J Atmos Sci 20: 130-141.

Maasch KA (1989) Calculating climate attractor dimension from $\delta^{18}O$ records by the Grassberger-Procaccia algorithm. Clim Dynam 4: 45-55.

Nerenberg MAH, Essex C (1990) Correlation dimension and systematic geometric effects. Phys Rev A 42: 7065-7074.

Nicolis C, Nicolis G (1984) Is there a climatic attractor? Nature 311: 529-532

Osborne AR, Provenzale A (1989) Finite correlation dimension for stochastic systems with power-law spectra. Physica D 35: 357-381.

Packard NH, Crutchfield JP, Farmer JD, Shaw RS (1980) Geometry from a time series. Phys Rev Lett 45: 712-716.

Ruelle D (1981) Chemical kinetics and differentiable systems. In: Pacault A, Vidal C (eds) Nonlinear phenomena in chemical dynamics. Springer, Berlin Heidelberg New York, pp 30-37.

Rützel E (1976) Zur Ausgleichsrechnung: Die Unbrauchbarkeit von Linearisierungsmethoden beim Anpassen von Potenz- und Exponentialfunktionen. Archiv für Psychologie 128: 316-322.

Sarnthein M, Tiedemann R (1989) Toward a high-resolution stable isotope stratigraphy of the last 3.4 million years: sites 658 and 659 off Northwest Africa. In: Ruddiman W, Sarnthein M, et al. (eds). Ocean Drilling Program, College Station, pp 167-185 (Proceedings of the Ocean Drilling Program, scientific results, vol 108).

Schulz M, Mudelsee M, Wolf-Welling TCW (1994) Fractal Analyses of Pleistocene Marine Oxygen Isotope Records. in: Kruhl JH (ed) Fractals and Dynamic Systems in Geoscience. Springer, Berlin Heidelberg New York, pp 377-387 (this volume).

Sen A, Srivastava M (1990) Regression analysis. Springer, Berlin Heidelberg New York.

Takens F (1981) Detecting strange attractors in turbulence. In: Rand DA, Young L-S (eds) Dynamical systems and turbulence. Springer, Berlin Heidelberg New York, pp 366-381 (Lecture notes in mathematics, vol 898).

Theiler J (1986) Spurious dimension from correlation algorithms applied to limited time-series data. Phys Rev A 34: 2427-2432.

Tiedemann R, Sarnthein M, Shackleton NJ (in press) Astronomic time scale for the Pliocene Atlantic $\delta^{18}O$ and dust flux records of ODP Site 659. Paleoceanography.

Appendix: List of Reviewers

D. Avnir	(Jerusalem, Israel)
P. Bak	(Uptown, U.S.A)
D. N. Baker	(Greenbelt, U.S.A.)
U. Bayer	(Potsdam, Germany)
H. Berckhemer	(Frankfurt/M., Germany)
S. Berndt	(Frankfurt/M./, Germany)
T. Blenkinsop	(Harare, Zimbabwe)
A. Bunde	(Hamburg, Germany)
C. A. Carlson	(Tucson, U.S.A.)
J. Dubois	(Paris, France)
C.J.G. Evertsz	(Bremen, Germany)
R. H. Fluegeman	(Muncie, U.S.A.)
T. Fowler	(Ottawa, Canada)
J. M. García-Ruiz	(Granada, Spain)
V. Haak	(Potsdam, Germany)
H. Hara	(Sendai, Japan)
K. Herterich	(Bremen, Germany)
C. Hooge	(Montréal, Canada)
K. Ito	(Nada, Japan)
B. Jamtveit	(Oslo, Norway)
Y. Kagan	(Los Angales, U.S.A.)
O. Kahle	(Potsdam, Germany)
B. H. Kaye	(Sudbury, Canada)
L. Knopoff	(Los Angeles, U.S.A.)
G. Korvin	(Adelaide, Australia)
E. Kozak	(Lublin, Poland)
J. H. Kruhl	(Frankfurt/M., Germany)
H.-J. Kümpel	(Bonn, Germany)
M. Kupkovà	(Košice, Slovakia)
K. Maasch	(Onono, U.S.A.)
G. Mandl	(Graz, Austria)
W. Martienssen	(Frankfurt/M., Germany)

G. Mayer-Kress	(Urbana, U.S.A.)
R. Meissner	(Kiel, Germany)
H. Nagahama	(Shizuoka, Japan)
D. Norton	(Tucson, U.S.A.)
A. Ord	(Perth, Australia)
P. Ortoleva	(Bloomington, U.S.A.)
E. Papadimitriou	(Thessaloniki, Greece)
H. Pape	(Hannover, Germany)
W. L. Power	(Perth, Australia)
G. Purcaru	(Frankfurt/M., Germany)
G. Rossi	(Trieste, Italy)
H. Schaeben	(Metz, France)
D. Schertzer	(Paris, France)
C. H. Scholz	(Palisades, U.S.A.)
V. V. Silberschmidt	(Perm, Russia)
S. Sokolowski	(Lublin, Poland)
A. Sornette	(Nice, France)
Steurer	(Hannover, Germany)
J. Strehlau	(Kiel, Germany)
E.F. Stumpfl	(Leoben, Austria)
R. Sultan	(Beirut, Lebanon)
D.L. Turcotte	(Ithaca, U.S.A.)
T. Vicsek	(Budapest, Hungary)
M. Vignes-Adler	(Meudon, France)
R. Wang	(Potsdam, Germany)

Subject Index

The page numbers indicate only the first appearance of a subject in each chapter.